Springer Series in Statistics

Advisors:
P. Bickel, P. Diggle, S. Fienberg, U. Gather,
I. Olkin, S. Zeger

Springer Series in Statistics

Alho/Spencer: Statistical Demography and Forecasting.
Andersen/Borgan/Gill/Keiding: Statistical Models Based on Counting Processes.
Atkinson/Riani: Robust Diagnostic Regression Analysis.
Atkinson/Riani/Cerioli: Exploring Multivariate Data with the Forward Search.
Berger: Statistical Decision Theory and Bayesian Analysis, 2nd edition.
Borg/Groenen: Modern Multidimensional Scaling: Theory and Applications,
   2nd edition.
Brockwell/Davis: Time Series: Theory and Methods, 2nd edition.
Bucklew: Introduction to Rare Event Simulation.
Cappé/Moulines/Rydén: Inference in Hidden Markov Models.
Chan/Tong: Chaos: A Statistical Perspective.
Chen/Shao/Ibrahim: Monte Carlo Methods in Bayesian Computation.
Coles: An Introduction to Statistical Modeling of Extreme Values.
David/Edwards: Annotated Readings in the History of Statistics.
Devroye/Lugosi: Combinatorial Methods in Density Estimation.
Efromovich: Nonparametric Curve Estimation: Methods, Theory, and Applications.
Eggermont/LaRiccia: Maximum Penalized Likelihood Estimation, Volume I: Density
   Estimation.
Fahrmeir/Tutz: Multivariate Statistical Modelling Based on Generalized Linear
   Models, 2nd edition.
Fan/Yao: Nonlinear Time Series: Nonparametric and Parametric Methods.
Farebrother: Fitting Linear Relationships: A History of the Calculus of Observations
   1750-1900.
Federer: Statistical Design and Analysis for Intercropping Experiments, Volume I:
   Two Crops.
Federer: Statistical Design and Analysis for Intercropping Experiments, Volume II:
   Three or More Crops.
Ghosh/Ramamoorthi: Bayesian Nonparametrics.
Glaz/Naus/Wallenstein: Scan Statistics.
Good: Permutation Tests: A Practical Guide to Resampling Methods for Testing
   Hypotheses, 2nd edition.
Good: Permutation Tests: Parametric and Bootstrap Tests of Hypotheses, 3rd edition.
Gouriéroux: ARCH Models and Financial Applications.
Gu: Smoothing Spline ANOVA Models.
Györfi/Kohler/Krzyżak/Walk: A Distribution-Free Theory of Nonparametric
   Regression.
Haberman: Advanced Statistics, Volume I: Description of Populations.
Hall: The Bootstrap and Edgeworth Expansion.
Härdle: Smoothing Techniques: With Implementation in S.
Harrell: Regression Modeling Strategies: With Applications to Linear Models, Logistic
   Regression, and Survival Analysis.
Hart: Nonparametric Smoothing and Lack-of-Fit Tests.
Hastie/Tibshirani/Friedman: The Elements of Statistical Learning: Data Mining,
   Inference, and Prediction.
Hedayat/Sloane/Stufken: Orthogonal Arrays: Theory and Applications.
Heyde: Quasi-Likelihood and its Application: A General Approach to Optimal
   Parameter Estimation.

(continued after index)

Roger B. Nelsen

An Introduction
to Copulas

Second Edition

 Springer

Roger B. Nelsen
Department of Mathematical Sciences
Lewis & Clark College, MSC 110
Portland, OR 97219-7899
USA
nelsen@lclark.edu

ISBN   978-1-4419-2109-3   e-ISBN   978-0-387-28678-5

Printed in the United States of America.   (SBA)

9 8 7 6 5 4 3 2 1

springeronline.com

To the memory of my parents
Ann Bain Nelsen
and
Howard Ernest Nelsen

Preface to the First Edition

In November of 1995, I was at the University of Massachusetts in Amherst for a few days to attend a symposium held, in part, to celebrate Professor Berthold Schweizer's retirement from classroom teaching. During one afternoon break, a small group of us were having coffee following several talks in which copulas were mentioned. Someone asked what one should read to learn the basics about copulas. We mentioned several references, mostly research papers and conference proceedings. I then suggested that perhaps the time was ripe for "someone" to write an introductory-level monograph on the subject. A colleague, I forget who, responded somewhat mischievously, "Good idea, Roger—why don't *you* write it?"

Although flattered by the suggestion, I let it lie until the following September, when I was in Prague to attend an international conference on distributions with fixed marginals and moment problems. In Prague, I asked Giorgio Dall'Aglio, Ingram Olkin, and Abe Sklar if they thought that there might indeed be interest in the statistical community for such a book. Encouraged by their responses and knowing that I would soon be eligible for a sabbatical, I began to give serious thought to writing an introduction to copulas.

This book is intended for students and practitioners in statistics and probability—at almost any level. The only prerequisite is a good upper-level undergraduate course in probability and mathematical statistics, although some background in nonparametric statistics would be beneficial. Knowledge of measure-theoretic probability is not required.

The book begins with the basic properties of copulas and then proceeds to present methods for constructing copulas and to discuss the role played by copulas in modeling and in the study of dependence. The focus is on bivariate copulas, although most chapters conclude with a discussion of the multivariate case. As an introduction to copulas, it is not an encyclopedic reference, and thus it is necessarily incomplete—many topics that could have been included are omitted. The reader seeking additional material on families of continuous bivariate distributions and their applications should see (Hutchinson and Lai 1990); and the reader interested in learning more about multivariate copulas and dependence should consult (Joe 1997).

There are about 150 exercises in the book. Although it is certainly not necessary to do all (or indeed any) of them, the reader is encouraged to read through the statements of the exercises before proceeding to the next section or chapter. Although some exercises do not add anything to the exposition (e.g., "Prove Theorem 1.1.1"), many present examples, counterexamples, and supplementary topics that are often referenced in subsequent sections.

I would like to thank Lewis & Clark College for granting me a sabbatical leave in order to write this book; and my colleagues in the Department of Mathematics, Statistics, and Computer Science at Mount Holyoke College for graciously inviting me to spend the sabbatical year with them. Thanks, too, to Ingram Olkin for suggesting and encouraging that I consider publication with Springer's *Lecture Notes in Statistics*; and to John Kimmel, the executive editor for statistics at Springer, for his valuable assistance in the publication of this book.

Finally, I would like to express my gratitude and appreciation to all those with whom I have had the pleasure of working on problems related to copulas and their applications: Claudi Alsina, Jerry Frank, Greg Fredricks, Juan Quesada Molina, José Antonio Rodríguez Lallena, Carlo Sempi, Abe Sklar, and Manuel Úbeda Flores. But most of all I want to thank my good friend and mentor Berthold Schweizer, who not only introduced me to the subject but also has consistently and unselfishly aided me in the years since and who inspired me to write this book. I also want to thank Bert for his careful and critical reading of earlier drafts of the manuscript and his invaluable advice on matters mathematical and stylistic. However, it goes without saying that any and all remaining errors in the book are mine alone.

Roger B. Nelsen
Portland, Oregon
July 1998

Preface to the Second Edition

In preparing a new edition of *An Introduction to Copulas*, my goals included adding some topics omitted from the first edition while keeping the book at a level appropriate for self-study or for a graduate-level seminar. The major additions in the second edition are sections on:
- a copula transformation method;
- extreme value copulas;
- copulas with certain analytic or functional properties;
- tail dependence; and
- quasi-copulas.

There are also a number of new examples and exercises and new figures, including scatterplots of simulations from many of the families of copulas presented in the text. Typographical errors in the first edition have been corrected, and the references have been updated.

Thanks again to Lewis & Clark College for granting me a sabbatical leave in order to prepare this second edition; and to the Department of Mathematics and Statistics at Mount Holyoke College for again inviting me to spend the sabbatical year with them. Finally, I would like to thank readers of the first edition who found numerous typographical errors in the first edition and sent me suggestions for this edition.

Roger B. Nelsen
Portland, Oregon
October 2005

Contents

Preface to the First Edition vii

Preface to the Second Edition ix

1 Introduction 1

2 Definitions and Basic Properties 7
  2.1 Preliminaries 7
  2.2 Copulas 10
  Exercises 2.1-2.11 14
  2.3 Sklar's Theorem 17
  2.4 Copulas and Random Variables 24
  Exercises 2.12-2.17 28
  2.5 The Fréchet-Hoeffding Bounds for Joint Distribution
       Functions of Random Variables 30
  2.6 Survival Copulas 32
  Exercises 2.18-2.26 34
  2.7 Symmetry 36
  2.8 Order 38
  Exercises 2.27-2.33 39
  2.9 Random Variate Generation 40
  2.10 Multivariate Copulas 42
  Exercises 2.34-2.37 48

3 Methods of Constructing Copulas 51
  3.1 The Inversion Method 51
    3.1.1 The Marshall-Olkin Bivariate Exponential Distribution 52
    3.1.2 The Circular Uniform Distribution 55
  Exercises 3.1-3.6 57
  3.2 Geometric Methods 59
    3.2.1 Singular Copulas with Prescribed Support 59
    3.2.2 Ordinal Sums 63
  Exercises 3.7-3.13 64
    3.2.3 Shuffles of M 67
    3.2.4 Convex Sums 72
  Exercises 3.14-3.20 74
    3.2.5 Copulas with Prescribed Horizontal or Vertical Sections 76
    3.2.6 Copulas with Prescribed Diagonal Sections 84

Exercises 3.21-3.34 86
3.3 Algebraic Methods 89
    3.3.1 Plackett Distributions 89
    3.3.2 Ali-Mikhail-Haq Distributions 92
    3.3.3 A Copula Transformation Method 94
    3.3.4 Extreme Value Copulas 97
Exercises 3.35-3.42 99
3.4 Copulas with Specified Properties 101
    3.4.1 Harmonic Copulas 101
    3.4.2 Homogeneous Copulas 101
    3.4.3 Concave and Convex Copulas 102
3.5 Constructing Multivariate Copulas 105

4 Archimedean Copulas 109
4.1 Definitions 109
4.2 One-parameter Families 114
4.3 Fundamental Properties 115
Exercises 4.1-4.17 132
4.4 Order and Limiting Cases 135
4.5 Two-parameter Families 141
    4.5.1 Families of Generators 141
    4.5.2 Rational Archimedean Copulas 146
Exercises 4.18-4.23 150
4.6 Multivariate Archimedean Copulas 151
Exercises 4.24-4.25 155

5 Dependence 157
5.1 Concordance 157
    5.1.1 Kendall's tau 158
Exercises 5.1-5.5 165
    5.1.2 Spearman's rho 167
Exercises 5.6-5.15 171
    5.1.3 The Relationship between Kendall's tau
          and Spearman's rho 174
    5.1.4 Other Concordance Measures 180
Exercises 5.16-5.21 185
5.2 Dependence Properties 186
    5.2.1 Quadrant Dependence 187
Exercises 5.22-5.29 189
    5.2.2 Tail Monotonicity 191
    5.2.3 Stochastic Monotonicity, Corner Set Monotonicity,
          and Likelihood Ratio Dependence 195
Exercises 5.30-5.39 204
5.3 Other Measures of Association 207
    5.3.1 Measures of Dependence 207
    5.3.2 Measures Based on Gini's Coefficient 211
Exercises 5.40-5.46 213
5.4 Tail Dependence 214

Exercises 5.47-5.50 216
5.5 Median Regression 217
5.6 Empirical Copulas 219
5.7 Multivariate Dependence 222

6 Additional Topics 227
6.1 Distributions with Fixed Margins 227
Exercises 6.1-6.5 233
6.2 Quasi-copulas 236
Exercises 6.6-6.8 240
6.3 Operations on Distribution Functions 241
6.4 Markov Processes 244
Exercises 6.9-6.13 248

References 251

List of Symbols 263

Index 265

1 Introduction

The study of copulas and their applications in statistics is a rather modern phenomenon. Until quite recently, it was difficult to even locate the word "copula" in the statistical literature. There is no entry for "copula" in the nine volume *Encyclopedia of Statistical Sciences*, nor in the supplement volume. However, the first update volume, published in 1997, does have such an entry (Fisher 1997). The first reference in the *Current Index to Statistics* to a paper using "copula" in the title or as a keyword is in Volume 7 (1981) [the paper is (Schweizer and Wolff 1981)]—indeed, in the first eighteen volumes (1975-1992) of the *Current Index to Statistics* there are only eleven references to papers mentioning copulas. There are, however, 71 references in the next ten volumes (1993-2002).

Further evidence of the growing interest in copulas and their applications in statistics and probability in the past fifteen years is afforded by five international conferences devoted to these ideas: the "Symposium on Distributions with Given Marginals (Fréchet Classes)" in Rome in 1990; the conference on "Distributions with Fixed Marginals, Doubly Stochastic Measures, and Markov Operators" in Seattle in 1993; the conference on "Distributions with Given Marginals and Moment Problems" in Prague in 1996; the conference on "Distributions with Given Marginals and Statistical Modelling" in Barcelona in 2000; and the conference on "Dependence Modelling: Statistical Theory and Applications in Finance and Insurance" in Québec in 2004. As the titles of these conferences indicate, copulas are intimately related to study of distributions with "fixed" or "given" marginal distributions. The published proceedings of the first four conferences (Dall'Aglio et al. 1991; Rüschendorf et al. 1996; Beneš and Štěpán 1997; Cuadras et al. 2002) are among the most accessible resources for the study of copulas and their applications.

What are copulas? From one point a view, copulas are functions that join or "couple" multivariate distribution functions to their one-dimensional marginal distribution functions. Alternatively, copulas are multivariate distribution functions whose one-dimensional margins are uniform on the interval (0,1). Chapter 2 will be devoted to presenting a complete answer to this question.

Why are copulas of interest to students of probability and statistics? As Fisher (1997) answers in his article in the first update volume of the *Encyclopedia of Statistical Sciences*, "Copulas [are] of interest to statisticians for two main reasons: Firstly, as a way of studying scale-free measures of dependence; and secondly, as a starting point for constructing families of bivariate distributions, sometimes with a view to simulation." These topics are explored and developed in Chapters 3, 4, and 5.

The remainder of this chapter will be devoted to a brief history of the development and study of copulas. Readers interested in first-hand accounts by some of those who participated in the evolution of the subject should see the papers by Dall'Aglio (1991) and Schweizer (1991) in the proceedings of the Rome conference and the paper by Sklar (1996) in the proceedings of the Seattle conference.

The word *copula* is a Latin noun that means "a link, tie, bond" (*Cassell's Latin Dictionary*) and is used in grammar and logic to describe "that part of a proposition which connects the subject and predicate" (*Oxford English Dictionary*). The word copula was first employed in a mathematical or statistical sense by Abe Sklar (1959) in the theorem (which now bears his name) describing the functions that "join together" one-dimensional distribution functions to form multivariate distribution functions (see Theorems 2.3.3 and 2.10.9). In (Sklar 1996) we have the following account of the events leading to this use of the term copula:

> Féron (1956), in studying three-dimensional distributions had introduced auxiliary functions, defined on the unit cube, that connected such distributions with their one-dimensional margins. I saw that similar functions could be defined on the unit n-cube for all $n \geq 2$ and would similarly serve to link n-dimensional distributions to their one-dimensional margins. Having worked out the basic properties of these functions, I wrote about them to Fréchet, in English. He asked me to write a note about them in French. While writing this, I decided I needed a name for these functions. Knowing the word "copula" as a grammatical term for a word or expression that links a subject and predicate, I felt that this would make an appropriate name for a function that links a multidimensional distribution to its one-dimensional margins, and used it as such. Fréchet received my note, corrected one mathematical statement, made some minor corrections to my French, and had the note published by the Statistical Institute of the University of Paris as Sklar (1959).

But as Sklar notes, the functions themselves predate the use of the term copula. They appear in the work of Fréchet, Dall'Aglio, Féron, and many others in the study of multivariate distributions with fixed univariate marginal distributions. Indeed, many of the basic results about copulas can be traced to the early work of Wassily Hoeffding. In (Hoeffding 1940, 1941) one finds bivariate "standardized distributions" whose support is contained in the square $[-1/2,1/2]^2$ and whose margins are uniform on the interval $[-1/2,1/2]$. (As Schweizer (1991) opines, "had Hoeffding chosen the unit square $[0,1]^2$ instead of $[-1/2,1/2]^2$ for his normalization, he would have discovered copulas.")

Hoeffding also obtained the basic best-possible bounds inequality for these functions, characterized the distributions ("functional dependence") corresponding to those bounds, and studied measures of dependence that are "scale-invariant," i.e., invariant under strictly increasing transformations. Unfortunately, until recently this work did not receive the attention it deserved, due primarily to the fact the papers were published in relatively obscure German journals at the outbreak of the Second World War. However, they have recently been translated into English and are among Hoeffding's collected papers, recently published by Fisher and Sen (1994). Unaware of Hoeffding's work, Fréchet (1951) independently obtained many of the same results, which has led to the terms such as "Fréchet bounds" and "Fréchet classes." In recognition of the shared responsibility for these important ideas, we will refer to "Fréchet-Hoeffding bounds" and "Fréchet-Hoeffding classes." After Hoeffding, Fréchet, and Sklar, the functions now known as copulas were rediscovered by several other authors. Kimeldorf and Sampson (1975b) referred to them as *uniform representations*, and Galambos (1978) and Deheuvels (1978) called them *dependence functions*.

At the time that Sklar wrote his 1959 paper with the term "copula," he was collaborating with Berthold Schweizer in the development of the theory of *probabilistic metric spaces*, or *PM spaces*. During the period from 1958 through 1976, most of the important results concerning copulas were obtained in the course of the study of PM spaces. Recall that (informally) a metric space consists of a set S and a metric d that measures "distances" between points, say p and q, in S. In a probabilistic metric space, we replace the distance $d(p,q)$ by a distribution function F_{pq}, whose value $F_{pq}(x)$ for any real x is the probability that the distance between p and q is less than x. The first difficulty in the construction of probabilistic metric spaces comes when one tries to find a "probabilistic" analog of the triangle inequality $d(p,r) \leq d(p,q) + d(q,r)$—what is the corresponding relationship among the distribution functions F_{pr}, F_{pq}, and F_{qr} for all p, q, and r in S? Karl Menger (1942) proposed $F_{pr}(x+y) \geq T(F_{pq}(x), F_{qr}(y))$; where T is a *triangle norm* or *t-norm*. Like a copula, a *t*-norm maps $[0,1]^2$ to $[0,1]$, and joins distribution functions. Some *t*-norms are copulas, and conversely, some copulas are *t*-norms. So, in a sense, it was inevitable that copulas would arise in the study of PM spaces. For a thorough treatment of the theory of PM spaces and the history of its development, see (Schweizer and Sklar 1983; Schweizer 1991).

Among the most important results in PM spaces—for the statistician—is the class of Archimedean *t*-norms, those *t*-norms T that satisfy $T(u,u) < u$ for all u in $(0,1)$. Archimedean *t*-norms that are also copulas are called *Archimedean copulas*. Because of their simple forms, the ease with which they can be constructed, and their many nice properties, Archimedean copulas frequently appear in discussions of multivariate distributions—see, for example, (Genest and MacKay 1986a,b; Marshall and Olkin 1988; Joe 1993, 1997). This important class of copulas is the subject of Chapter 4.

We now turn our attention to copulas and dependence. The earliest paper explicitly relating copulas to the study of dependence among random variables appears to be (Schweizer and Wolff 1981). In that paper, Schweizer and Wolff discussed and modified Rényi's (1959) criteria for measures of dependence between pairs of random variables, presented the basic invariance properties of copulas under strictly monotone transformations of random variables (see Theorems 2.4.3 and 2.4.4), and introduced the measure of dependence now known as *Schweizer and Wolff's* σ (see Section 5.3.1). In their words, since

> ... under almost surely increasing transformations of (the random variables), the copula is invariant while the margins may be changed at will, it follows that it is precisely the copula which captures those properties of the joint distribution which are invariant under almost surely strictly increasing transformations. Hence the study of rank statistics—insofar as it is the study of properties invariant under such transformations—may be characterized as the study of copulas and copula-invariant properties.

Of course, copulas appear implicitly in earlier work on dependence by many other authors, too many to list here, so we will mention only two. Foremost is Hoeffding. In addition to studying the basic properties of "standardized distributions" (i.e., copulas), Hoeffding (1940, 1941) used them to study nonparametric measures of association such as Spearman's rho and his "dependence index" Φ^2 (see Section 5.3.1). Deheuvels (1979, 1981a,b,c) used "empirical dependence functions" (i.e., empirical copulas, the sample analogs of copulas—see Section 5.5) to estimate the population copula and to construct various nonparametric tests of independence. Chapter 5 is devoted to an introduction to the role played by copulas in the study of dependence.

Although this book concentrates on the two applications of copulas mentioned by Fisher (1997)—the construction of families of multivariate distributions and the study of dependence—copulas are being exploited in other ways. We mention but one, which we discuss in the final chapter. Through an ingenious definition of a "product" $*$ of copulas, Darsow, Nguyen, and Olsen (1992) have shown that the Chapman-Kolmogorov equations for the transition probabilities in a real stochastic process can be expressed succinctly in terms of the $*$-product of copulas. This new approach to the theory of Markov processes may well be the key to "capturing the Markov property of such processes in a framework as simple and perspicuous as the conventional framework for analyzing Markov chains" (Schweizer 1991).

The study of copulas and the role they play in probability, statistics, and stochastic processes is a subject still in its infancy. There are many open problems and much work to be done.

2 Definitions and Basic Properties

In the Introduction, we referred to copulas as "functions that join or couple multivariate distribution functions to their one-dimensional marginal distribution functions" and as "distribution functions whose one-dimensional margins are uniform." But neither of these statements is a definition—hence we will devote this chapter to giving a precise definition of copulas and to examining some of their elementary properties.

But first we present a glimpse of where we are headed. Consider for a moment a pair of random variables X and Y, with distribution functions $F(x) = P[X \leq x]$ and $G(y) = P[Y \leq y]$, respectively, and a joint distribution function $H(x,y) = P[X \leq x, Y \leq y]$ (we will review definitions of random variables, distribution functions, and other important topics as needed in the course of this chapter). To each pair of real numbers (x,y) we can associate three numbers: $F(x)$, $G(y)$, and $H(x,y)$. Note that each of these numbers lies in the interval $[0,1]$. In other words, each pair (x,y) of real numbers leads to a point $\big(F(x), G(y)\big)$ in the unit square $[0,1] \times [0,1]$, and this ordered pair in turn corresponds to a number $H(x,y)$ in $[0,1]$. We will show that this correspondence, which assigns the value of the joint distribution function to each ordered pair of values of the individual distribution functions, is indeed a function. Such functions are copulas.

To accomplish what we have outlined above, we need to generalize the notion of "nondecreasing" for univariate functions to a concept applicable to multivariate functions. We begin with some notation and definitions. In Sects. 2.1-2.9, we confine ourselves to the two-dimensional case; in Sect. 2.10, we consider n dimensions.

2.1 Preliminaries

The focus of this section is the notion of a "2-increasing" function—a two-dimensional analog of a nondecreasing function of one variable. But first we need to introduce some notation. We will let $\mathbf{R}$ denote the ordinary real line $(-\infty,\infty)$, $\overline{\mathbf{R}}$ denote the extended real line $[-\infty,\infty]$, and $\overline{\mathbf{R}}^2$ denote the extended real plane $\overline{\mathbf{R}} \times \overline{\mathbf{R}}$. A *rectangle* in $\overline{\mathbf{R}}^2$ is the

Cartesian product B of two closed intervals: $B = [x_1,x_2]\times[y_1,y_2]$. The *vertices* of a rectangle B are the points (x_1,y_1), (x_1,y_2), (x_2,y_1), and (x_2,y_2). The *unit square* $\mathbf{I}^2$ is the product $\mathbf{I}\times\mathbf{I}$ where $\mathbf{I} = [0,1]$. A 2-place *real function H* is a function whose domain, Dom$H$, is a subset of $\overline{\mathbf{R}}^2$ and whose range, Ran$H$, is a subset of $\mathbf{R}$.

Definition 2.1.1. Let S_1 and S_2 be nonempty subsets of $\overline{\mathbf{R}}$, and let H be a two-place real function such that Dom$H = S_1\times S_2$. Let $B = [x_1,x_2]\times[y_1,y_2]$ be a rectangle all of whose vertices are in Dom$H$. Then the *H-volume of B* is given by

$$V_H(B) = H(x_2,y_2) - H(x_2,y_1) - H(x_1,y_2) + H(x_1,y_1). \qquad (2.1.1)$$

Note that if we define the first order differences of H on the rectangle B as

$$\Delta_{x_1}^{x_2}H(x,y) = H(x_2,y) - H(x_1,y) \text{ and } \Delta_{y_1}^{y_2}H(x,y) = H(x,y_2) - H(x,y_1),$$

then the H-volume of a rectangle B is the *second order difference* of H on B,

$$V_H(B) = \Delta_{y_1}^{y_2}\Delta_{x_1}^{x_2}H(x,y).$$

Definition 2.1.2. A 2-place real function H is *2-increasing* if $V_H(B) \geq 0$ for all rectangles B whose vertices lie in Dom$H$.

When H is 2-increasing, we will occasionally refer to the H-volume of a rectangle B as the *H-measure of B*. Some authors refer to 2-increasing functions as *quasi-monotone*.

We note here that the statement "H is 2-increasing" neither implies nor is implied by the statement "H is nondecreasing in each argument," as the following two examples illustrate. The verifications are elementary, and are left as exercises.

Example 2.1. Let H be the function defined on $\mathbf{I}^2$ by $H(x,y) = \max(x,y)$. Then H is a nondecreasing function of x and of y; however, $V_H(\mathbf{I}^2) = -1$, so that H is not 2-increasing. ∎

Example 2.2. Let H be the function defined on $\mathbf{I}^2$ by $H(x,y) = (2x-1)(2y-1)$. Then H is 2-increasing, however it is a decreasing function of x for each y in $(0,1/2)$ and a decreasing function of y for each x in $(0,1/2)$. ∎

The following lemmas will be very useful in the next section in establishing the continuity of subcopulas and copulas. The first is a direct consequence of Definitions 2.1.1 and 2.1.2.

Lemma 2.1.3. *Let S_1 and S_2 be nonempty subsets of $\overline{\mathbf{R}}$, and let H be a 2-increasing function with domain $S_1 \times S_2$. Let x_1, x_2 be in S_1 with $x_1 \leq x_2$, and let y_1, y_2 be in S_2 with $y_1 \leq y_2$. Then the function $t \mapsto H(t,y_2) - H(t,y_1)$ is nondecreasing on S_1, and the function $t \mapsto H(x_2,t) - H(x_1,t)$ is nondecreasing on S_2.*

As an immediate application of this lemma, we can show that with an additional hypothesis, a 2-increasing function H is nondecreasing in each argument. Suppose S_1 has a least element a_1 and that S_2 has a least element a_2. We say that a function H from $S_1 \times S_2$ into $\mathbf{R}$ is *grounded* if $H(x,a_2) = 0 = H(a_1,y)$ for all (x,y) in $S_1 \times S_2$. Hence we have

Lemma 2.1.4. *Let S_1 and S_2 be nonempty subsets of $\overline{\mathbf{R}}$, and let H be a grounded 2-increasing function with domain $S_1 \times S_2$. Then H is nondecreasing in each argument.*

Proof. Let a_1, a_2 denote the least elements of S_1, S_2, respectively, and set $x_1 = a_1$, $y_1 = a_2$ in Lemma 2.1.3. $\square$

Now suppose that S_1 has a greatest element b_1 and that S_2 has a greatest element b_2. We then say that a function H from $S_1 \times S_2$ into $\mathbf{R}$ *has margins*, and that the margins of H are the functions F and G given by:

$$\mathrm{Dom}F = S_1, \text{ and } F(x) = H(x,b_2) \text{ for all } x \text{ in } S_1;$$

$$\mathrm{Dom}G = S_2, \text{ and } G(y) = H(b_1,y) \text{ for all } y \text{ in } S_2.$$

Example 2.3. Let H be the function with domain $[-1,1] \times [0,\infty]$ given by

$$H(x,y) = \frac{(x+1)(e^y - 1)}{x + 2e^y - 1}.$$

Then H is grounded because $H(x,0) = 0$ and $H(-1,y) = 0$; and H has margins $F(x)$ and $G(y)$ given by

$$F(x) = H(x,\infty) = (x+1)/2 \text{ and } G(y) = H(1,y) = 1 - e^{-y}. \quad \blacksquare$$

We close this section with an important lemma concerning grounded 2-increasing functions with margins.

Lemma 2.1.5. *Let S_1 and S_2 be nonempty subsets of $\overline{\mathbf{R}}$, and let H be a grounded 2-increasing function, with margins, whose domain is $S_1 \times S_2$. Let (x_1,y_1) and (x_2,y_2) be any points in $S_1 \times S_2$. Then*

$$\left| H(x_2,y_2) - H(x_1,y_1) \right| \leq \left| F(x_2) - F(x_1) \right| + \left| G(y_2) - G(y_1) \right|.$$

Proof. From the triangle inequality, we have

$$\left|H(x_2,y_2)-H(x_1,y_1)\right| \le \left|H(x_2,y_2)-H(x_1,y_2)\right| + \left|H(x_1,y_2)-H(x_1,y_1)\right|.$$

Now assume $x_1 \le x_2$. Because H is grounded, 2-increasing, and has margins, Lemmas 2.1.3 and 2.1.4 yield $0 \le H(x_2,y_2) - H(x_1,y_2) \le F(x_2)-F(x_1)$. An analogous inequality holds when $x_2 \le x_1$, hence it follows that for any x_1,x_2 in S_1, $\left|H(x_2,y_2)-H(x_1,y_2)\right| \le \left|F(x_2)-F(x_1)\right|$. Similarly for any y_1,y_2 in S_2, $\left|H(x_1,y_2)-H(x_1,y_1)\right| \le \left|G(y_2)-G(y_1)\right|$, which completes the proof. □

2.2 Copulas

We are now in a position to define the functions—copulas—that are the subject of this book. To do so, we first define subcopulas as a certain class of grounded 2-increasing functions with margins; then we define copulas as subcopulas with domain $\mathbf{I}^2$.

Definition 2.2.1. A *two-dimensional subcopula* (or *2-subcopula*, or briefly, a *subcopula*) is a function C' with the following properties:

1. $\operatorname{Dom} C' = S_1 \times S_2$, where S_1 and S_2 are subsets of $\mathbf{I}$ containing 0 and 1;
2. C' is grounded and 2-increasing;
3. For every u in S_1 and every v in S_2,

$$C'(u,1)=u \quad \text{and} \quad C'(1,v)=v. \tag{2.2.1}$$

Note that for every (u,v) in $\operatorname{Dom} C'$, $0 \le C'(u,v) \le 1$, so that $\operatorname{Ran} C'$ is also a subset of $\mathbf{I}$.

Definition 2.2.2. A *two-dimensional copula* (or *2-copula*, or briefly, a *copula*) is a 2-subcopula C whose domain is $\mathbf{I}^2$.

Equivalently, a copula is a function C from $\mathbf{I}^2$ to $\mathbf{I}$ with the following properties:

1. For every u, v in $\mathbf{I}$,

$$C(u,0)=0=C(0,v) \tag{2.2.2a}$$

and

$$C(u,1)=u \quad \text{and} \quad C(1,v)=v; \tag{2.2.2b}$$

2. For every u_1,u_2,v_1,v_2 in $\mathbf{I}$ such that $u_1 \le u_2$ and $v_1 \le v_2$,

$$C(u_2,v_2)-C(u_2,v_1)-C(u_1,v_2)+C(u_1,v_1) \ge 0. \tag{2.2.3}$$

PRODUCT COPULA $\Pi(u,v) \approx u.v$

Because $C(u,v) = V_C([0,u] \times [0,v])$, one can think of $C(u,v)$ as an assignment of a number in $\mathbf{I}$ to the rectangle $[0,u] \times [0,v]$. Thus (2.2.3) gives an "inclusion-exclusion" type formula for the number assigned by C to each rectangle $[u_1,u_2] \times [v_1,v_2]$ in $\mathbf{I}^2$ and states that the number so assigned must be nonnegative.

The distinction between a subcopula and a copula (the domain) may appear to be a minor one, but it will be rather important in the next section when we discuss Sklar's theorem. In addition, many of the important properties of copulas are actually properties of subcopulas.

Theorem 2.2.3. *Let C' be a subcopula. Then for every (u,v) in* $\mathrm{Dom}\, C'$,

$$\max(u + v - 1, 0) \leq C'(u,v) \leq \min(u,v). \tag{2.2.4}$$

Proof. Let (u,v) be an arbitrary point in $\mathrm{Dom}\, C'$. Now $C'(u,v) \leq C'(u,1) = u$ and $C'(u,v) \leq C'(1,v) = v$ yield $C'(u,v) \leq \min(u,v)$. Furthermore, $V_{C'}([u,1] \times [v,1]) \geq 0$ implies $C'(u,v) \geq u + v - 1$, which when combined with $C'(u,v) \geq 0$ yields $C'(u,v) \geq \max(u + v - 1, 0)$. □

Because every copula is a subcopula, the inequality in the above theorem holds for copulas. Indeed, the bounds in (2.2.4) are themselves copulas (see Exercise 2.2) and are commonly denoted by $M(u,v) = \min(u,v)$ and $W(u,v) = \max(u + v - 1, 0)$. Thus for every copula C and every (u,v) in $\mathbf{I}^2$,

$$W(u,v) \leq C(u,v) \leq M(u,v). \tag{2.2.5}$$

Inequality (2.2.5) is the copula version of the *Fréchet-Hoeffding bounds* inequality, which we shall encounter later in terms of distribution functions. We refer to M as the *Fréchet-Hoeffding upper bound* and W as the *Fréchet-Hoeffding lower bound*. A third important copula that we will frequently encounter is the *product copula* $\Pi(u,v) = uv$.

The following theorem, which follows directly from Lemma 2.1.5, establishes the continuity of subcopulas—and hence of copulas—via a Lipschitz condition on $\mathbf{I}^2$.

Theorem 2.2.4. *Let C' be a subcopula. Then for every* $(u_1,u_2),(v_1,v_2)$ *in* $\mathrm{Dom}\, C'$,

$$\left| C'(u_2,v_2) - C'(u_1,v_1) \right| \leq \left| u_2 - u_1 \right| + \left| v_2 - v_1 \right|. \tag{2.2.6}$$

Hence C' is uniformly continuous on its domain.

The *sections* of a copula will be employed in the construction of copulas in the next chapter, and will be used in Chapter 5 to provide interpretations of certain dependence properties:

Definition 2.2.5. Let C be a copula, and let a be any number in $\mathbf{I}$. The *horizontal section of C at a* is the function from $\mathbf{I}$ to $\mathbf{I}$ given by

$t \mapsto C(t,a)$; the *vertical section of C at a* is the function from $\mathbf{I}$ to $\mathbf{I}$ given by $t \mapsto C(a,t)$; and the *diagonal section of C* is the function δ_C from $\mathbf{I}$ to $\mathbf{I}$ defined by $\delta_C(t) = C(t,t)$.

The following corollary is an immediate consequence of Lemma 2.1.4 and Theorem 2.2.4.

Corollary 2.2.6. *The horizontal, vertical, and diagonal sections of a copula C are all nondecreasing and uniformly continuous on* $\mathbf{I}$.

Various applications of copulas that we will encounter in later chapters involve the shape of the graph of a copula, i.e., the surface $z = C(u,v)$. It follows from Definition 2.2.2 and Theorem 2.2.4 that the graph of any copula is a continuous surface within the unit cube $\mathbf{I}^3$ whose boundary is the skew quadrilateral with vertices $(0,0,0)$, $(1,0,0)$, $(1,1,1)$, and $(0,1,0)$; and from Theorem 2.2.3 that this graph lies between the graphs of the Fréchet-Hoeffding bounds, i.e., the surfaces $z = M(u,v)$ and $z = W(u,v)$. In Fig. 2.1 we present the graphs of the copulas M and W, as well as the graph of Π, a portion of the hyperbolic paraboloid $z = uv$.

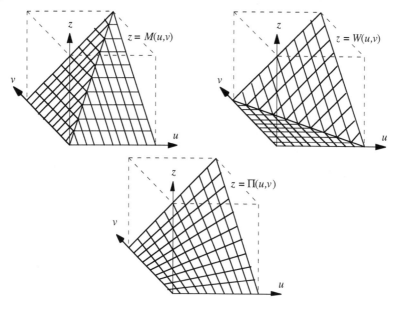

Fig. 2.1. Graphs of the copulas M, Π, and W

A simple but useful way to present the graph of a copula is with a *contour diagram* (Conway 1979), that is, with graphs of its *level sets*—the sets in $\mathbf{I}^2$ given by $C(u,v) = $ a constant, for selected constants in $\mathbf{I}$. In Fig. 2.2 we present the contour diagrams of the copulas M, Π, and W. Note that the points $(t,1)$ and $(1,t)$ are each members of the level set corresponding to the constant t. Hence we do not need to label the level sets in the diagram, as the boundary conditions $C(1,t) = t = C(t,1)$ readily provide the constant for each level set.

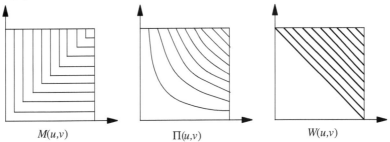

$$M(u,v) \qquad \Pi(u,v) \qquad W(u,v)$$

Fig. 2.2. Contour diagrams of the copulas M, Π, and W

Also note that, given any copula C, it follows from (2.2.5) that for a given t in $\mathbf{I}$ the graph of the level set $\{(u,v) \in \mathbf{I}^2 | C(u,v) = t\}$ must lie in the shaded triangle in Fig. 2.3, whose boundaries are the level sets determined by $M(u,v) = t$ and $W(u,v) = t$.

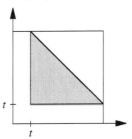

Fig. 2.3. The region that contains the level set $\{(u,v) \in \mathbf{I}^2 | C(u,v) = t\}$

We conclude this section with the two theorems concerning the partial derivatives of copulas. The word "almost" is used in the sense of Lebesgue measure.

Theorem 2.2.7. *Let C be a copula. For any v in $\mathbf{I}$, the partial derivative $\partial C(u,v)/\partial u$ exists for almost all u, and for such v and u,*

$$0 \le \frac{\partial}{\partial u} C(u,v) \le 1. \qquad (2.2.7)$$

Similarly, for any u in $\mathbf{I}$, the partial derivative $\partial C(u,v)/\partial v$ exists for almost all v, and for such u and v,

$$0 \leq \frac{\partial}{\partial v} C(u,v) \leq 1. \tag{2.2.8}$$

Furthermore, the functions $u \mapsto \partial C(u,v)/\partial v$ *and* $v \mapsto \partial C(u,v)/\partial u$ *are defined and nondecreasing almost everywhere on* **I**.

Proof. The existence of the partial derivatives $\partial C(u,v)/\partial u$ and $\partial C(u,v)/\partial v$ is immediate because monotone functions (here the horizontal and vertical sections of the copula) are differentiable almost everywhere. Inequalities (2.2.7) and (2.2.8) follow from (2.2.6) by setting $v_1 = v_2$ and $u_1 = u_2$, respectively. If $v_1 \leq v_2$, then, from Lemma 2.1.3, the function $u \mapsto C(u,v_2) - C(u,v_1)$ is nondecreasing. Hence $\partial\big(C(u,v_2) - C(u,v_1)\big)/\partial u$ is defined and nonnegative almost everywhere on **I**, from which it follows that $v \mapsto \partial C(u,v)/\partial u$ is defined and nondecreasing almost everywhere on **I**. A similar result holds for $u \mapsto \partial C(u,v)/\partial v$. □

Theorem 2.2.8. *Let C be a copula. If* $\partial C(u,v)/\partial v$ *and* $\partial^2 C(u,v)/\partial u \partial v$ *are continous on* $\mathbf{I}^2$ *and* $\partial C(u,v)/\partial u$ *exists for all* $u \in (0,1)$ *when* $v = 0$, *then* $\partial C(u,v)/\partial u$ *and* $\partial^2 C(u,v)/\partial v \partial u$ *exist in* $(0,1)^2$ *and* $\partial^2 C(u,v)/\partial u \partial v$ $= \partial^2 C(u,v)/\partial v \partial u$.

Proof. See (Seeley 1961).

Exercises

2.1   Verify the statements in Examples 2.1 and 2.2.

2.2   Show that $M(u,v) = \min(u,v)$, $W(u,v) = \max(u+v-1,0)$, and $\Pi(u,v) = uv$ are indeed copulas.

2.3   (a) Let C_0 and C_1 be copulas, and let θ be any number in **I**. Show that the weighted arithmetic mean $(1-\theta)C_0 + \theta C_1$ is also a copula. Hence conclude that any convex linear combination of copulas is a copula.
    (b) Show that the geometric mean of two copulas may fail to be a copula. [Hint: Let C be the geometric mean of Π and W, and show that the C-volume of the rectangle $[1/2,3/4] \times [1/2,3/4]$ is negative.]

2.4   *The Fréchet and Mardia families of copulas.*
    (a) Let α, β be in **I** with $\alpha + \beta \leq 1$. Set

$$C_{\alpha,\beta}(u,v) = \alpha M(u,v) + (1-\alpha-\beta)\Pi(u,v) + \beta W(u,v).$$

Show that $C_{\alpha,\beta}$ is a copula. A family of copulas that includes M, Π, and W is called *comprehensive*. This two-parameter comprehensive family is due to Fréchet (1958).

(b) Let θ be in $[-1,1]$, and set

$$C_\theta(u,v) = \frac{\theta^2(1+\theta)}{2}M(u,v) + (1-\theta^2)\Pi(u,v) + \frac{\theta^2(1-\theta)}{2}W(u,v). \quad (2.2.9)$$

Show that C_θ is a copula. This one-parameter comprehensive family is due to Mardia (1970).

2.5 *The Cuadras-Augé family of copulas.* Let θ be in $\mathbf{I}$, and set

$$C_\theta(u,v) = [\min(u,v)]^\theta [uv]^{1-\theta} = \begin{cases} uv^{1-\theta}, & u \le v, \\ u^{1-\theta}v, & u \ge v. \end{cases} \quad (2.2.10)$$

Show that C_θ is a copula. Note that $C_0 = \Pi$ and $C_1 = M$. This family (weighted geometric means of M and Π) is due to Cuadras and Augé (1981).

2.6 Let C be a copula, and let (a,b) be any point in $\mathbf{I}^2$. For (u,v) in $\mathbf{I}^2$, define

$$K_{a,b}(u,v) = V_C\big([a(1-u),u+a(1-u)] \times [b(1-v),v+b(1-v)]\big).$$

Show that $K_{a,b}$ is a copula. Note that $K_{0,0}(u,v) = C(u,v)$. Several special cases will be of interest in Sects. 2.4, 2.7, and 6.4, namely:

$$K_{0,1}(u,v) = u - C(u,1-v),$$
$$K_{1,0}(u,v) = v - C(1-u,v), \text{ and}$$
$$K_{1,1}(u,v) = u + v - 1 + C(1-u,1-v).$$

2.7 Let f be a function from $\mathbf{I}^2$ into $\mathbf{I}$ which is nondecreasing in each variable and has margins given by $f(t,1) = t = f(1,t)$ for all t in $\mathbf{I}$. Prove that f is grounded.

2.8 (a) Show that for any copula C, $\max(2t-1,0) \le \delta_C(t) \le t$ for all t in $\mathbf{I}$.
(b) Show that $\delta_C(t) = \delta_M(t)$ for all t in $\mathbf{I}$ implies $C = M$.
(c) Show $\delta_C(t) = \delta_W(t)$ for all t in $\mathbf{I}$ does not imply that $C = W$.

2.9   The *secondary diagonal section* of C is given by $C(t,1-t)$. Show that $C(t,1-t) = 0$ for all t in $\mathbf{I}$ implies $C = W$.

2.10 Let t be in $[0,1)$, and let C_t be the function from $\mathbf{I}^2$ into $\mathbf{I}$ given by

$$C_t(u,v) = \begin{cases} \max(u+v-1,t), & (u,v) \in [t,1]^2, \\ \min(u,v), & \text{otherwise.} \end{cases}$$

(a) Show that C_t is a copula.

(b) Show that the level set $\{(u,v) \in \mathbf{I}^2 | C_t(u,v) = t\}$ is the set of points in the triangle with vertices $(t,1)$, $(1,t)$, and (t,t), that is, the shaded region in Fig. 2.3. The copula in this exercise illustrates why the term "level set" is preferable to "level curve" for some copulas.

2.11 This exercise shows that the 2-increasing condition (2.2.3) for copulas is not a consequence of simpler properties. Let Q be the function from $\mathbf{I}^2$ into $\mathbf{I}$ given by

$$Q(u,v) = \begin{cases} \min\left(u,v,\dfrac{1}{3},u+v-\dfrac{2}{3}\right), & \dfrac{2}{3} \le u+v \le \dfrac{4}{3}, \\ \max\left(u+v-1,0\right), & \text{otherwise;} \end{cases}$$

that is, Q has the values given in Fig. 2.4 in the various parts of $\mathbf{I}^2$.
(a) Show that for every u,v in $\mathbf{I}$, $Q(u,0) = 0 = Q(0,v)$, $Q(u,1) = u$ and $Q(1,v) = v$; $W(u,v) \le Q(u,v) \le M(u,v)$; and that Q is continuous, satisfies the Lipschitz condition (2.2.6), and is nondecreasing in each variable.
(b) Show that Q fails to be 2-increasing, and hence is not a copula. [Hint: consider the Q-volume of the rectangle $[1/3,2/3]^2$.]

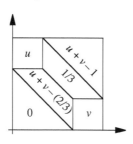

Fig. 2.4. The function Q in Exercise 2.11

2.3 Sklar's Theorem

The theorem in the title of this section is central to the theory of copulas and is the foundation of many, if not most, of the applications of that theory to statistics. Sklar's theorem elucidates the role that copulas play in the relationship between multivariate distribution functions and their univariate margins. Thus we begin this section with a short discussion of distribution functions.

Definition 2.3.1. A *distribution function* is a function F with domain $\overline{\mathbf{R}}$ such that
1. F is nondecreasing,
2. $F(-\infty) = 0$ and $F(\infty) = 1$.

Example 2.4. For any number a in $\mathbf{R}$, the *unit step at a* is the distribution function ε_a given by

$$\varepsilon_a(x) = \begin{cases} 0, & x \in [-\infty, a), \\ 1, & x \in [a, \infty]; \end{cases}$$

and for any numbers a, b in $\mathbf{R}$ with $a < b$, the *uniform distribution on* $[a,b]$ is the distribution function U_{ab} given by

$$U_{ab}(x) = \begin{cases} 0, & x \in [-\infty, a), \\ \dfrac{x-a}{b-a}, & x \in [a, b], \\ 1, & x \in (b, \infty]. \end{cases}$$ ■

Definition 2.3.2. A *joint distribution function* is a function H with domain $\overline{\mathbf{R}}^2$ such that
1. H is 2-increasing,
2. $H(x, -\infty) = H(-\infty, y) = 0$, and $H(\infty, \infty) = 1$.

Thus H is grounded, and because $\mathrm{Dom}H = \overline{\mathbf{R}}^2$, H has margins F and G given by $F(x) = H(x, \infty)$ and $G(y) = H(\infty, y)$. By virtue of Corollary 2.2.6, F and G are distribution functions.

Example 2.5. Let H be the function with domain $\overline{\mathbf{R}}^2$ given by

$$H(x,y) = \begin{cases} \dfrac{(x+1)(e^y - 1)}{x + 2e^y - 1}, & (x,y) \in [-1,1] \times [0, \infty], \\ 1 - e^{-y}, & (x,y) \in (1, \infty] \times [0, \infty], \\ 0, & \text{elsewhere.} \end{cases}$$

It is tedious but elementary to verify that H is 2-increasing and grounded, and that $H(\infty,\infty) = 1$. Hence H is a joint distribution function. The margins of H are the distribution functions F and G given by

$$F = U_{-1,1} \text{ and } G(y) = \begin{cases} 0, & y \in [-\infty,0), \\ 1-e^{-y}, & y \in [0,\infty]. \end{cases}$$

[Cf. Examples 2.3 and 2.4.]   ∎

Note that there is nothing "probabilistic" in these definitions of distribution functions. Random variables are not mentioned, nor is left-continuity or right-continuity. All the distribution functions of one or of two random variables usually encountered in statistics satisfy either the first or the second of the above definitions. Hence any results we derive for such distribution functions will hold when we discuss random variables, regardless of any additional restrictions that may be imposed.

Theorem 2.3.3. Sklar's theorem. *Let H be a joint distribution function with margins F and G. Then there exists a copula C such that for all x,y in $\overline{\mathbf{R}}$,*

$$H(x,y) = C(F(x),G(y)). \tag{2.3.1}$$

If F and G are continuous, then C is unique; otherwise, C is uniquely determined on RanF ×RanG. *Conversely, if C is a copula and F and G are distribution functions, then the function H defined by (2.3.1) is a joint distribution function with margins F and G.*

This theorem first appeared in (Sklar 1959). The name "copula" was chosen to emphasize the manner in which a copula "couples" a joint distribution function to its univariate margins. The argument that we give below is essentially the same as in (Schweizer and Sklar 1974). It requires two lemmas.

Lemma 2.3.4. *Let H be a joint distribution function with margins F and G. Then there exists a unique subcopula C' such that*

    1. Dom C' = RanF ×RanG,

    2. *For all x,y in $\overline{\mathbf{R}}$, $H(x,y) = C'(F(x),G(y))$.*

Proof. The joint distribution H satisfies the hypotheses of Lemma 2.1.5 with $S_1 = S_2 = \overline{\mathbf{R}}$. Hence for any points (x_1,y_1) and (x_2,y_2) in $\overline{\mathbf{R}}^2$,

$$\left|H(x_2,y_2) - H(x_1,y_1)\right| \leq \left|F(x_2) - F(x_1)\right| + \left|G(y_2) - G(y_1)\right|.$$

It follows that if $F(x_1) = F(x_2)$ and $G(y_1) = G(y_2)$, then $H(x_1,y_1) = H(x_2,y_2)$. Thus the set of ordered pairs

$$\left\{ \left((F(x),G(y)),H(x,y) \right) \middle| x,y \in \overline{\mathbf{R}} \right\}$$

defines a 2-place real function C' whose domain is $\text{Ran}F \times \text{Ran}G$. That this function is a subcopula follows directly from the properties of H. For instance, to show that (2.2.2) holds, we first note that for each u in $\text{Ran}F$, there is an x in $\overline{\mathbf{R}}$ such that $F(x) = u$. Thus $C'(u,1) = C'(F(x),G(\infty)) = H(x,\infty) = F(x) = u$. Verifications of the other conditions in Definition 2.2.1 are similar.   □

Lemma 2.3.5. *Let C' be a subcopula. Then there exists a copula C such that $C(u,v) = C'(u,v)$ for all (u,v) in $\text{Dom}\,C'$; i.e., any subcopula can be extended to a copula. The extension is generally non-unique.*

Proof. Let $\text{Dom}\,C' = S_1 \times S_2$. Using Theorem 2.2.4 and the fact that C' is nondecreasing in each place, we can extend C' by continuity to a function C'' with domain $\overline{S}_1 \times \overline{S}_2$, where $\overline{S}_1$ is the closure of S_1 and $\overline{S}_2$ is the closure of S_2. Clearly C'' is also a subcopula. We next extend C'' to a function C with domain $\mathbf{I}^2$. To this end, let (a,b) be any point in $\mathbf{I}^2$, let a_1 and a_2 be, respectively, the greatest and least elements of $\overline{S}_1$ that satisfy $a_1 \le a \le a_2$; and let b_1 and b_2 be, respectively, the greatest and least elements of $\overline{S}_2$ that satisfy $b_1 \le b \le b_2$. Note that if a is in $\overline{S}_1$, then $a_1 = a = a_2$; and if b is in $\overline{S}_2$, then $b_1 = b = b_2$. Now let

$$\lambda_1 = \begin{cases} (a-a_1)/(a_2-a_1), & \text{if } a_1 < a_2, \\ 1, & \text{if } a_1 = a_2; \end{cases}$$

$$\mu_1 = \begin{cases} (b-b_1)/(b_2-b_1), & \text{if } b_1 < b_2, \\ 1, & \text{if } b_1 = b_2; \end{cases}$$

and define

$$C(a,b) = (1-\lambda_1)(1-\mu_1)C''(a_1,b_1) + (1-\lambda_1)\mu_1 C''(a_1,b_2) \\ + \lambda_1(1-\mu_1)C''(a_2,b_1) + \lambda_1\mu_1 C''(a_2,b_2). \tag{2.3.2}$$

Notice that the interpolation defined in (2.3.2) is linear in each place (what we call *bilinear interpolation*) because λ_1 and μ_1 are linear in a and b, respectively.

It is obvious that $\text{Dom}\,C = \mathbf{I}^2$, that $C(a,b) = C''(a,b)$ for any (a,b) in $\text{Dom}\,C''$; and that C satisfies (2.2.2a) and (2.2.2b). Hence we only must show that C satisfies (2.2.3). To accomplish this, let (c,d) be another point in $\mathbf{I}^2$ such that $c \ge a$ and $d \ge b$, and let $c_1, d_1, c_2, d_2, \lambda_2, \mu_2$ be related to c and d as $a_1, b_1, a_2, b_2, \lambda_1, \mu_1$ are related to a and b. In evaluating $V_C(B)$ for the rectangle $B = [a,c] \times [b,d]$, there will be several cases to consider, depending upon whether or not there is a point in $\overline{S}_1$ strictly between a and c, and whether or not there is a point in $\overline{S}_2$

strictly between b and d. In the simplest of these cases, there is no point in $\overline{S}_1$ strictly between a and c, and no point in $\overline{S}_2$ strictly between b and d, so that $c_1 = a_1$, $c_2 = a_2$, $d_1 = b_1$, and $d_2 = b_2$. Substituting (2.3.2) and the corresponding terms for $C(a,d)$, $C(c,b)$ and $C(c,d)$ into the expression given by (2.1.1) for $V_C(B)$ and simplifying yields

$$V_C(B) = V_C([a,c] \times [b,d]) = (\lambda_2 - \lambda_1)(\mu_2 - \mu_1)V_C([a_1,a_2] \times [b_1,b_2]),$$

from which it follows that $V_C(B) \geq 0$ in this case, as $c \geq a$ and $d \geq b$ imply $\lambda_2 \geq \lambda_1$ and $\mu_2 \geq \mu_1$.

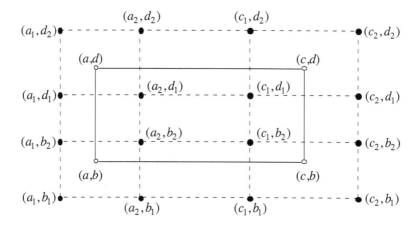

Fig. 2.5. The least simple case in the proof of Lemma 2.3.5

At the other extreme, the least simple case occurs when there is at least one point in $\overline{S}_1$ strictly between a and c, and at least one point in $\overline{S}_2$ strictly between b and d, so that $a < a_2 \leq c_1 < c$ and $b < b_2 \leq d_1 < d$. In this case—which is illustrated in Fig. 2.5—substituting (2.3.2) and the corresponding terms for $C(a,d)$, $C(c,b)$ and $C(c,d)$ into the expression given by (2.1.1) for $V_C(B)$ and rearranging the terms yields

$$\begin{aligned}
V_C(B) = {}& (1 - \lambda_1)\mu_2 V_C([a_1,a_2] \times [d_1,d_2]) + \mu_2 V_C([a_2,c_1] \times [d_1,d_2]) \\
& + \lambda_2 \mu_2 V_C([c_1,c_2] \times [d_1,d_2]) + (1 - \lambda_1)V_C([a_1,a_2] \times [b_2,d_1]) \\
& + V_C([a_2,c_1] \times [b_2,d_1]) + \lambda_2 V_C([c_1,c_2] \times [b_2,d_1]) \\
& + (1 - \lambda_1)(1 - \mu_1)V_C([a_1,a_2] \times [b_1,b_2]) \\
& + (1 - \mu_1)V_C([a_2,c_1] \times [b_1,b_2]) + \lambda_2(1 - \mu_1)V_C([c_1,c_2] \times [b_1,b_2]).
\end{aligned}$$

The right-hand side of the above expression is a combination of nine nonnegative quantities (the C-volumes of the nine rectangles determined by the dashed lines in Fig. 2.5) with nonnegative coefficients, and hence is nonnegative. The remaining cases are similar, which completes the proof.   □

Example 2.6. Let (a,b) be any point in $\mathbf{R}^2$, and consider the following distribution function H:

$$H(x,y) = \begin{cases} 0, & x < a \text{ or } y < b, \\ 1, & x \geq a \text{ and } y \geq b. \end{cases}$$

The margins of H are the unit step functions ε_a and ε_b. Applying Lemma 2.3.4 yields the subcopula C' with domain $\{0,1\}\times\{0,1\}$ such that $C'(0,0) = C'(0,1) = C'(1,0) = 0$ and $C'(1,1) = 1$. The extension of C' to a copula C via Lemma 2.3.5 is the copula $C = \Pi$, i.e., $C(u,v) = uv$. Notice however, that *every* copula agrees with C' on its domain, and thus is an extension of this C'.   ■

We are now ready to prove Sklar's theorem, which we restate here for convenience.

Theorem 2.3.3. Sklar's theorem. *Let H be a joint distribution function with margins F and G. Then there exists a copula C such that for all x,y in $\overline{\mathbf{R}}$,*

$$H(x,y) = C(F(x),G(y)). \tag{2.3.1}$$

If F and G are continuous, then C is unique; otherwise, C is uniquely determined on $\mathrm{Ran}F\times\mathrm{Ran}G$. Conversely, if C is a copula and F and G are distribution functions, then the function H defined by (2.3.1) is a joint distribution function with margins F and G.

Proof. The existence of a copula C such that (2.3.1) holds for all x,y in $\overline{\mathbf{R}}$ follows from Lemmas 2.3.4 and 2.3.5. If F and G are continuous, then $\mathrm{Ran}F = \mathrm{Ran}G = \mathbf{I}$, so that the unique subcopula in Lemma 2.3.4 is a copula. The converse is a matter of straightforward verification.   □

Equation (2.3.1) gives an expression for joint distribution functions in terms of a copula and two univariate distribution functions. But (2.3.1) can be inverted to express copulas in terms of a joint distribution function and the "inverses" of the two margins. However, if a margin is not strictly increasing, then it does not possess an inverse in the usual sense. Thus we first need to define "quasi-inverses" of distribution functions (recall Definition 2.3.1).

Definition 2.3.6. Let F be a distribution function. Then a *quasi-inverse* of F is any function $F^{(-1)}$ with domain $\mathbf{I}$ such that

1. if t is in $\mathrm{Ran}F$, then $F^{(-1)}(t)$ is any number x in $\overline{\mathbf{R}}$ such that $F(x) = t$, i.e., for all t in $\mathrm{Ran}F$,

$$F(F^{(-1)}(t)) = t;$$

2. if t is not in $\mathrm{Ran}F$, then

$$F^{(-1)}(t) = \inf\{x|F(x) \geq t\} = \sup\{x|F(x) \leq t\}.$$

If F is strictly increasing, then it has but a single quasi-inverse, which is of course the ordinary inverse, for which we use the customary notation F^{-1}.

Example 2.7. The quasi-inverses of ε_a, the unit step at a (see Example 2.4) are the functions given by

$$\varepsilon_a^{(-1)}(t) = \begin{cases} a_0, & t=0, \\ a, & t \in (0,1), \\ a_1, & t=1, \end{cases}$$

where a_0 and a_1 are any numbers in $\overline{\mathbf{R}}$ such that $a_0 < a \leq a_1$. ∎

Using quasi-inverses of distribution functions, we now have the following corollary to Lemma 2.3.4.

Corollary 2.3.7. Let H, F, G, and C' be as in Lemma 2.3.4, and let $F^{(-1)}$ and $G^{(-1)}$ be quasi-inverses of F and G, respectively. Then for any (u,v) in $\operatorname{Dom} C'$,

$$C'(u,v) = H(F^{(-1)}(u), G^{(-1)}(v)). \tag{2.3.3}$$

When F and G are continuous, the above result holds for copulas as well and provides a method of constructing copulas from joint distribution functions. We will exploit Corollary 2.3.7 in the next chapter to construct families of copulas, but for now the following examples will serve to illustrate the procedure.

Example 2.8. Recall the distribution function H from Example 2.5:

$$H(x,y) = \begin{cases} \dfrac{(x+1)(e^y-1)}{x+2e^y-1}, & (x,y) \in [-1,1] \times [0,\infty], \\ 1-e^{-y}, & (x,y) \in (1,\infty] \times [0,\infty], \\ 0, & \text{elsewhere.} \end{cases}$$

with margins F and G given by

$$F(x) = \begin{cases} 0, & x < -1, \\ (x+1)/2, & x \in [-1,1], \\ 1, & x > 1, \end{cases} \quad \text{and} \quad G(y) = \begin{cases} 0, & y < 0, \\ 1-e^{-y}, & y \geq 0. \end{cases}$$

Quasi-inverses of F and G are given by $F^{(-1)}(u) = 2u-1$ and $G^{(-1)}(v) = -\ln(1-v)$ for u,v in $\mathbf{I}$. Because $\operatorname{Ran} F = \operatorname{Ran} G = \mathbf{I}$, (2.3.3) yields the copula C given by

$$C(u,v) = \frac{uv}{u+v-uv}. \qquad (2.3.4)$$ ∎

Example 2.9. *Gumbel's bivariate exponential distribution* (Gumbel 1960a). Let H_θ be the joint distribution function given by

$$H_\theta(x,y) = \begin{cases} 1 - e^{-x} - e^{-y} + e^{-(x+y+\theta xy)}, & x \geq 0, y \geq 0, \\ 0, & \text{otherwise;} \end{cases}$$

where θ is a parameter in $[0,1]$. Then the marginal distribution functions are exponentials, with quasi-inverses $F^{(-1)}(u) = -\ln(1-u)$ and $G^{(-1)}(v) = -\ln(1-v)$ for u,v in $\mathbf{I}$. Hence the corresponding copula is

$$C_\theta(u,v) = u + v - 1 + (1-u)(1-v)e^{-\theta \ln(1-u)\ln(1-v)}. \qquad (2.3.5)$$ ∎

Example 2.10. It is an exercise in many mathematical statistics texts to find an example of a bivariate distribution with standard normal margins that is not the standard bivariate normal with parameters $\mu_x = \mu_y = 0$, $\sigma_x^2 = \sigma_y^2 = 1$, and Pearson's product-moment correlation coefficient ρ. With Sklar's theorem and Corollary 2.3.7 this becomes trivial—let C be a copula such as one in either of the preceding examples, and use standard normal margins in (2.3.1). Indeed, if Φ denotes the standard (univariate) normal distribution function and N_ρ denotes the standard bivariate normal distribution function (with Pearson's product-moment correlation coefficient ρ), then *any* copula *except* one of the form

$$C(u,v) = N_\rho(\Phi^{-1}(u), \Phi^{-1}(v))$$

$$= \frac{1}{2\pi\sqrt{1-\rho^2}} \int_{-\infty}^{\Phi^{-1}(u)} \int_{-\infty}^{\Phi^{-1}(v)} \exp\left[\frac{-(s^2 - 2\rho st + t^2)}{2(1-\rho^2)}\right] ds\, dt \qquad (2.3.6)$$

(with $\rho \neq -1$, 0, or 1) will suffice. Explicit constructions using the copulas in Exercises 2.4, 2.12, and 3.11, Example 3.12, and Sect. 3.3.1 can be found in (Kowalski 1973), and one using the copula $C_{1/2}$ from Exercise 2.10 in (Vitale 1978). ∎

We close this section with one final observation. With an appropriate extension of its domain to $\overline{\mathbf{R}}^2$, every copula is a joint distribution function with margins that are uniform on $\mathbf{I}$. To be precise, let C be a copula, and define the function H_C on $\overline{\mathbf{R}}^2$ via

$$H_C(x,y) = \begin{cases} 0, & x<0 \text{ or } y<0, \\ C(x,y), & (x,y) \in \mathbf{I}^2, \\ x, & y>1, x \in \mathbf{I}, \\ y, & x>1, y \in \mathbf{I}, \\ 1, & x>1 \text{ and } y>1. \end{cases}$$

Then H_C is a distribution function both of whose margins are readily seen to be U_{01}. Indeed, it is often quite useful to think of copulas as restrictions to $\mathbf{I}^2$ of joint distribution functions whose margins are U_{01}.

2.4 Copulas and Random Variables

In this book, we will use the term "random variable" in the statistical rather than the probabilistic sense; that is, a random variable is a quantity whose values are described by a (known or unknown) probability distribution function. Of course, all of the results to follow remain valid when a random variable is defined in terms of measure theory, i.e., as a measurable function on a given probability space. But for our purposes it suffices to adopt the descriptions of Wald (1947), "a variable x is called a random variable if for any given value c a definite probability can be ascribed to the event that x will take a value less than c"; and of Gnedenko (1962), "a *random variable* is a variable quantity whose values depend on chance and for which there exists a distribution function." For a detailed discussion of this point of view, see (Menger 1956).

In what follows, we will use capital letters, such as X and Y, to represent random variables, and lowercase letters x, y to represent their values. We will say that F is the *distribution function of the random variable X* when for all x in $\mathbf{R}$, $F(x) = P[X \le x]$. We are defining distribution functions of random variables to be right-continuous—but that is simply a matter of custom and convenience. Left-continuous distribution functions would serve equally as well. A random variable is continuous if its distribution function is continuous.

When we discuss two or more random variables, we adopt the same convention—two or more random variables are the components of a quantity (now a vector) whose values are described by a joint distribution function. As a consequence, we always assume that the collection of random variables under discussion can be defined on a common probability space.

We are now in a position to restate Sklar's theorem in terms of random variables and their distribution functions:

Theorem 2.4.1. *Let X and Y be random variables with distribution functions F and G, respectively, and joint distribution function H. Then* *there exists a copula C such that* (2.3.1) *holds. If F and G are continuous, C is unique. Otherwise, C is uniquely determined on* RanF×RanG.

The copula C in Theorem 2.4.1 will be called the *copula of X and Y*, and denoted C_{XY} when its identification with the random variables X and Y is advantageous.

The following theorem shows that the product copula $\Pi(u,v) = uv$ characterizes independent random variables when the distribution functions are continuous. Its proof follows from Theorem 2.4.1 and the observation that X and Y are independent if and only if $H(x,y) = F(x)G(y)$ for all x,y in $\overline{\mathbf{R}}^2$.

Theorem 2.4.2. *Let X and Y be continuous random variables. Then X and Y are independent if and only if* $C_{XY} = \Pi$.

Much of the usefulness of copulas in the study of nonparametric statistics derives from the fact that for strictly monotone transformations of the random variables, copulas are either invariant or change in predictable ways. Recall that if the distribution function of a random variable X is continuous, and if α is a strictly monotone function whose domain contains RanX, then the distribution function of the random variable $\alpha(X)$ is also continuous. We treat the case of strictly increasing transformations first.

Theorem 2.4.3. *Let X and Y be continuous random variables with copula* C_{XY}. *If α and β are strictly increasing on* RanX *and* RanY, *respectively, then* $C_{\alpha(X)\beta(Y)} = C_{XY}$. *Thus* C_{XY} *is invariant under strictly increasing transformations of X and Y.*

Proof. Let F_1, G_1, F_2, and G_2 denote the distribution functions of X, Y, $\alpha(X)$, and $\beta(Y)$, respectively. Because α and β are strictly increasing, $F_2(x) = P[\alpha(X) \le x] = P[X \le \alpha^{-1}(x)] = F_1(\alpha^{-1}(x))$, and likewise $G_2(y) = G_1(\beta^{-1}(y))$. Thus, for any x,y in $\overline{\mathbf{R}}$,

$$C_{\alpha(X)\beta(Y)}(F_2(x),G_2(y)) = P[\alpha(X) \le x, \beta(Y) \le y]$$
$$= P[X \le \alpha^{-1}(x), Y \le \beta^{-1}(y)]$$
$$= C_{XY}(F_1(\alpha^{-1}(x)), G_1(\beta^{-1}(y)))$$
$$= C_{XY}(F_2(x), G_2(y)).$$

Because X and Y are continuous, $\operatorname{Ran} F_2 = \operatorname{Ran} G_2 = \mathbf{I}$, whence it follows that $C_{\alpha(X)\beta(Y)} = C_{XY}$ on $\mathbf{I}^2$. □

When at least one of α and β is strictly decreasing, we obtain results in which the copula of the random variables $\alpha(X)$ and $\beta(Y)$ is a simple transformation of C_{XY}. Specifically, we have:

Theorem 2.4.4. *Let X and Y be continuous random variables with copula C_{XY}. Let α and β be strictly monotone on* RanX *and* RanY, *respectively.*

1. *If α is strictly increasing and β is strictly decreasing, then*

$$C_{\alpha(X)\beta(Y)}(u,v) = u - C_{XY}(u,1-v).$$

2. *If α is strictly decreasing and β is strictly increasing, then*

$$C_{\alpha(X)\beta(Y)}(u,v) = v - C_{XY}(1-u,v).$$

3. *If α and β are both strictly decreasing, then*

$$C_{\alpha(X)\beta(Y)}(u,v) = u + v - 1 + C_{XY}(1-u,1-v).$$

The proof of Theorem 2.4.4 is left as an exercise. Note that in each case the form of the copula is independent of the particular choices of α and β, and note further that the three forms for $C_{\alpha(X)\beta(Y)}$ that appear in this theorem were first encountered in Exercise 2.6. [Remark: We could be somewhat more general in the preceding two theorems by replacing phrases such as "strictly increasing" by "almost surely strictly increasing"—to allow for subsets of Lebesgue measure zero where the property may fail to hold.]

Although we have chosen to avoid measure theory in our definition of random variables, we will nevertheless need some terminology and results from measure theory in the remaining sections of this chapter and in chapters to come. Each joint distribution function H induces a probability measure on $\mathbf{R}^2$ via $V_H((-\infty,x]\times(-\infty,y]) = H(x,y)$ and a standard extension to Borel subsets of $\mathbf{R}^2$ using measure-theoretic techniques. Because copulas are joint distribution functions (with uniform $(0,1)$ margins), each copula C induces a probability measure on $\mathbf{I}^2$ via $V_C([0,u]\times[0,v]) = C(u,v)$ in a similar fashion—that is, the C-measure of a set is its C-volume V_C. Hence, at an intuitive level, the C-measure of a subset of $\mathbf{I}^2$ is the probability that two uniform $(0,1)$ random variables U and V with joint distribution function C assume values in that subset. C-measures are often called *doubly stochastic measures*, as for any measurable subset S of $\mathbf{I}$, $V_C(S\times\mathbf{I}) = V_C(\mathbf{I}\times S) = \lambda(S)$, where λ denotes ordinary Lebesgue measure on $\mathbf{I}$. The term "doubly stochastic" is taken from matrix theory, where doubly stochastic matrices have nonnegative entries and all row sums and column sums are 1.

For any copula C, let

$$C(u,v) = A_C(u,v) + S_C(u,v),$$

where

$$A_C(u,v) = \int_0^u \int_0^v \frac{\partial^2}{\partial s \partial t} C(s,t)\, dt\, ds \text{ and } S_C(u,v) = C(u,v) - A_C(u,v). \quad (2.4.1)$$

Unlike bivariate distributions in general, the margins of a copula are continuous, hence a copula has no "atoms" (individual points in $\mathbf{I}^2$ whose C-measure is positive).

If $C \equiv A_C$ on $\mathbf{I}^2$—that is, if considered as a joint distribution function, C has a joint density given by $\partial^2 C(u,v)/\partial u \partial v$—then C is *absolutely continuous*, whereas if $C \equiv S_C$ on $\mathbf{I}^2$—that is, if $\partial^2 C(u,v)/\partial u \partial v = 0$ almost everywhere in $\mathbf{I}^2$—then C is *singular*. Otherwise, C has an *absolutely continuous component* A_C and a *singular component* S_C. In this case neither A_C nor S_C is a copula, because neither has uniform $(0,1)$ margins. In addition, the C-measure of the absolutely continuous component is $A_C(1,1)$, and the C-measure of the singular component is $S_C(1,1)$.

Just as the support of a joint distribution function H is the complement of the union of all open subsets of $\mathbf{R}^2$ with H-measure zero, the *support of a copula* is the complement of the union of all open subsets of $\mathbf{I}^2$ with C-measure zero. When the support of C is $\mathbf{I}^2$, we say C has "full support." When C is singular, its support has Lebesgue measure zero (and conversely). However, many copulas that have full support have both an absolutely continuous and a singular component.

Example 2.11. The support of the Fréchet-Hoeffding upper bound M is the main diagonal of $\mathbf{I}^2$, i.e., the graph of $v = u$ for u in $\mathbf{I}$, so that M is singular. This follows from the fact that the M-measure of any open rectangle that lies entirely above or below the main diagonal is zero. Also note that $\partial^2 M/\partial u \partial v = 0$ everywhere in $\mathbf{I}^2$ except on the main diagonal. Similarly, the support of the Fréchet-Hoeffding lower bound W is the secondary diagonal of $\mathbf{I}^2$, i.e., the graph of $v = 1 - u$ for u in $\mathbf{I}$, and thus W is singular as well. ∎

Example 2.12. The product copula $\Pi(u,v) = uv$ is absolutely continuous, because for all (u,v) in $\mathbf{I}^2$,

$$A_\Pi(u,v) = \int_0^u \int_0^v \frac{\partial^2}{\partial s \partial t} \Pi(s,t)\, dt\, ds = \int_0^u \int_0^v 1\, dt\, ds = uv = \Pi(u,v). \quad \blacksquare$$

In Sect. 3.1.1 we will illustrate a general procedure for decomposing a copula into the sum of its absolutely continuous and singular components and for finding the probability mass (i.e., C-measure) of each component.

Exercises

2.12 *Gumbel's bivariate logistic distribution* (Gumbel 1961). Let X and Y be random variables with a joint distribution function given by

$$H(x,y) = (1 + e^{-x} + e^{-y})^{-1}$$

for all x,y in $\overline{\mathbf{R}}$.
(a) Show that X and Y have standard (univariate) logistic distributions, i.e.,

$$F(x) = (1 + e^{-x})^{-1} \text{ and } G(y) = (1 + e^{-y})^{-1}.$$

(b) Show that the copula of X and Y is the copula given by (2.3.4) in Example 2.8.

2.13 *Type B bivariate extreme value distributions* (Johnson and Kotz 1972). Let X and Y be random variables with a joint distribution function given by

$$H_\theta(x,y) = \exp[-(e^{-\theta x} + e^{-\theta y})^{1/\theta}]$$

for all x,y in $\overline{\mathbf{R}}$, where $\theta \geq 1$. Show that the copula of X and Y is given by

$$C_\theta(u,v) = \exp\left(-\left[(-\ln u)^\theta + (-\ln v)^\theta\right]^{1/\theta}\right). \qquad (2.4.2)$$

This parametric family of copulas is known as the *Gumbel-Hougaard* family (Hutchinson and Lai 1990), which we shall see again in Chapter 4.

2.14 Conway (1979) and Hutchinson and Lai (1990) note that Gumbel's bivariate logistic distribution (Exercise 2.12) suffers from the defect that it lacks a parameter, which limits its usefulness in applications. This can be corrected in a number of ways, one of which (Ali et al. 1978) is to define H_θ as

$$H_\theta(x,y) = \left(1 + e^{-x} + e^{-y} + (1-\theta)e^{-x-y}\right)^{-1}$$

for all x,y in $\overline{\mathbf{R}}$, where θ lies in $[-1,1]$. Show that
(a) the margins are standard logistic distributions;

(b) when $\theta = 1$, we have Gumbel's bivariate logistic distribution;

(c) when $\theta = 0$, X and Y are independent; and

(d) the copula of X and Y is given by

$$C_\theta(u,v) = \frac{uv}{1 - \theta(1-u)(1-v)}. \qquad (2.4.3)$$

This is the *Ali-Mikhail-Haq* family of copulas (Hutchinson and Lai 1990), which we will encounter again in Chapters 3 and 4.

2.15 Let X_1 and Y_1 be random variables with continuous distribution functions F_1 and G_1, respectively, and copula C. Let F_2 and G_2 be another pair of continuous distribution functions, and set $X_2 = F_2^{(-1)}(F_1(X_1))$ and $Y_2 = G_2^{(-1)}(G_1(Y_1))$. Prove that
(a) the distribution functions of X_2 and Y_2 are F_2 and G_2, respectively; and
(b) the copula of X_2 and Y_2 is C.

2.16 (a) Let X and Y be continuous random variables with copula C and univariate distribution functions F and G, respectively. The random variables $\max(X,Y)$ and $\min(X,Y)$ are the *order statistics* for X and Y. Prove that the distribution functions of the order statistics are given by

$$P[\max(X,Y) \leq t] = C(F(t),G(t))$$

and

$$P[\min(X,Y) \leq t] = F(t) + G(t) - C(F(t),G(t)),$$

so that when $F = G$,

$$P[\max(X,Y) \leq t] = \delta_C(F(t)) \text{ and}$$
$$P[\min(X,Y) \leq t] = 2F(t) - \delta_C(F(t)).$$

(b) Show that bounds on the distribution functions of the order statistics are given by

$$\max(F(t) + G(t) - 1, 0) \leq P[\max(X,Y) \leq t] \leq \min(F(t),G(t))$$

and

$$\max(F(t),G(t)) \leq P[\min(X,Y) \leq t] \leq \min(F(t) + G(t), 1).$$

2.17 Prove Theorem 2.4.4.

2.5 The Fréchet-Hoeffding Bounds for Joint Distribution Functions

In Sect. 2.2 we encountered the Fréchet-Hoeffding bounds as universal bounds for copulas, i.e., for any copula C and for all u,v in $\mathbf{I}$,

$$W(u,v) = \max(u+v-1,0) \le C(u,v) \le \min(u,v) = M(u,v).$$

As a consequence of Sklar's theorem, if X and Y are random variables with a joint distribution function H and margins F and G, respectively, then for all x,y in $\overline{\mathbf{R}}$,

$$\max(F(x)+G(y)-1,0) \le H(x,y) \le \min(F(x),G(y)) \qquad (2.5.1)$$

Because M and W are copulas, the above bounds are joint distribution functions and are called the *Fréchet-Hoeffding bounds* for joint distribution functions H with margins F and G. Of interest in this section is the following question: What can we say about the random variables X and Y when their joint distribution function H is equal to one of its Fréchet-Hoeffding bounds?

To answer this question, we first need to introduce the notions of nondecreasing and nonincreasing sets in $\overline{\mathbf{R}}^2$.

Definition 2.5.1. A subset S of $\overline{\mathbf{R}}^2$ is *nondecreasing* if for any (x,y) and (u,v) in S, $x < u$ implies $y \le v$. Similarly, a subset S of $\overline{\mathbf{R}}^2$ is *nonincreasing* if for any (x,y) and (u,v) in S, $x < u$ implies $y \ge v$.

Fig. 2.6 illustrates a simple nondecreasing set.

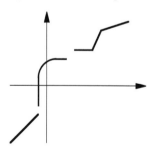

Fig. 2.6. The graph of a nondecreasing set

We will now prove that the joint distribution function H for a pair (X,Y) of random variables is the Fréchet-Hoeffding upper bound (i.e., the copula is M) if and only if the support of H lies in a nondecreasing set. The following proof is based on the one that appears in (Mikusiński, Sherwood and Taylor 1991-1992). But first, we need two lemmas:

Lemma 2.5.2. *Let S be a subset of* $\overline{\mathbf{R}}^2$. *Then S is nondecreasing if and only if for each* (x,y) *in* $\overline{\mathbf{R}}^2$, *either*

1. *for all* (u,v) *in S,* $u \leq x$ *implies* $v \leq y$; *or*   (2.5.2)

2. *for all* (u,v) *in S,* $v \leq y$ *implies* $u \leq x$.   (2.5.3)

Proof. First assume that S is nondecreasing, and that neither (2.5.2) nor (2.5.3) holds. Then there exist points (a,b) and (c,d) in S such that $a \leq x, b > y, d \leq y$, and $c > x$. Hence $a < c$ and $b > d$; a contradiction. In the opposite direction, assume that S is not nondecreasing. Then there exist points (a,b) and (c,d) in S with $a < c$ and $b > d$. For $(x,y) = \left((a+c)/2,(b+d)/2\right)$, neither (2.5.2) nor (2.5.3) holds. □

Lemma 2.5.3. *Let X and Y be random variables with joint distribution function H. Then H is equal to its Fréchet-Hoeffding upper bound if and only if for every* (x,y) *in* $\overline{\mathbf{R}}^2$, *either* $P[X > x, Y \leq y] = 0$ *or* $P[X \leq x, Y > y] = 0$.

Proof: As usual, let F and G denote the margins of H. Then

$$F(x) = P[X \leq x] = P[X \leq x, Y \leq y] + P[X \leq x, Y > y]$$
$$= H(x,y) + P[X \leq x, Y > y],$$

and

$$G(y) = P[Y \leq y] = P[X \leq x, Y \leq y] + P[X > x, Y \leq y]$$
$$= H(x,y) + P[X > x, Y \leq y].$$

Hence $H(x,y) = M(F(x),G(y))$ if and only if $\min(P[X \leq x, Y > y], P[X > x, Y \leq y]) = 0$, from which the desired conclusion follows. □

We are now ready to prove

Theorem 2.5.4. *Let X and Y be random variables with joint distribution function H. Then H is identically equal to its Fréchet-Hoeffding upper bound if and only if the support of H is a nondecreasing subset of* $\overline{\mathbf{R}}^2$.

Proof. Let S denote the support of H, and let (x,y) be any point in $\overline{\mathbf{R}}^2$. Then (2.5.2) holds if and only if $\{(u,v)|u \leq x \text{ and } v > y\} \cap S = \varnothing$; or equivalently, if and only if $P[X \leq x, Y > y] = 0$. Similarly, (2.5.3) holds if and only if $\{(u,v)|u > x \text{ and } v \leq y\} \cap S = \varnothing$; or equivalently, if and only if $P[X > x, Y \leq y] = 0$. The theorem now follows from Lemmas 2.5.2 and 2.5.3. □

Of course, there is an analogous result for the Fréchet-Hoeffding lower bound—its proof is outlined in Exercises 2.18 through 2.20:

Theorem 2.5.5. *Let X and Y be random variables with joint distribution function H. Then H is identically equal to its Fréchet-Hoeffding lower bound if and only if the support of H is a nonincreasing subset of* $\overline{\mathbf{R}}^2$.

When X and Y are continuous, the support of H can have no horizontal or vertical line segments, and in this case it is common to say that "Y is almost surely an increasing function of X" if and only if the copula of X and Y is M; and "Y is almost surely a decreasing function of X" if and only if the copula of X and Y is W. If U and V are uniform $(0,1)$ random variables whose joint distribution function is the copula M, then $P[U = V] = 1$; and if the copula is W, then $P[U + V = 1] = 1$.

Random variables with copula M are often called *comonotonic*, and random variables with copula W are often called *countermonotonic*.

2.6 Survival Copulas

In many applications, the random variables of interest represent the lifetimes of individuals or objects in some population. The probability of an individual living or surviving beyond time x is given by the *survival function* (or *survivor function*, or *reliability function*) $\overline{F}(x) = P[X > x] = 1 - F(x)$, where, as before, F denotes the distribution function of X. When dealing with lifetimes, the natural range of a random variable is often $[0,\infty)$; however, we will use the term "survival function" for $P[X > x]$ even when the range is $\mathbf{R}$.

For a pair (X,Y) of random variables with joint distribution function H, the *joint survival function* is given by $\overline{H}(x,y) = P[X > x, Y > y]$. The margins of $\overline{H}$ are the functions $\overline{H}(x,-\infty)$ and $\overline{H}(-\infty,y)$, which are the univariate survival functions $\overline{F}$ and $\overline{G}$, respectively. A natural question is the following: Is there a relationship between univariate and joint survival functions analogous to the one between univariate and joint distribution functions, as embodied in Sklar's theorem? To answer this question, suppose that the copula of X and Y is C. Then we have

$$\overline{H}(x,y) = 1 - F(x) - G(y) + H(x,y)$$
$$= \overline{F}(x) + \overline{G}(y) - 1 + C(F(x),G(y))$$
$$= \overline{F}(x) + \overline{G}(y) - 1 + C(1 - \overline{F}(x), 1 - \overline{G}(y)),$$

so that if we define a function $\hat{C}$ from $\mathbf{I}^2$ into $\mathbf{I}$ by

$$\hat{C}(u,v) = u + v - 1 + C(1 - u, 1 - v), \qquad (2.6.1)$$

we have

$$\overline{H}(x,y) = \hat{C}(\overline{F}(x), \overline{G}(y)). \qquad (2.6.2)$$

First note that, as a consequence of Exercise 2.6, the function $\hat{C}$ in (2.6.1) is a copula (see also part 3 of Theorem 2.4.4). We refer to $\hat{C}$ as the *survival copula* of X and Y. Secondly, notice that $\hat{C}$ "couples" the joint survival function to its univariate margins in a manner completely analogous to the way in which a copula connects the joint distribution function to its margins.

Care should be taken not to confuse the survival copula $\hat{C}$ with the joint survival function $\overline{C}$ for two uniform $(0,1)$ random variables whose joint distribution function is the copula C. Note that $\overline{C}(u,v) = P[U > u, V > v] = 1 - u - v + C(u,v) = \hat{C}(1 - u, 1 - v)$.

Example 2.13. In Example 2.9, we obtained the copula C_θ in (2.3.5) for Gumbel's bivariate exponential distribution: for θ in $[0,1]$,

$$C_\theta(u,v) = u + v - 1 + (1-u)(1-v)e^{-\theta \ln(1-u)\ln(1-v)}.$$

Just as the survival function for univariate exponentially distributed random variables is functionally simpler than the distribution function, the same is often true in the bivariate case. Employing (2.6.1), we have

$$\hat{C}_\theta(u,v) = uve^{-\theta \ln u \ln v}. \qquad \blacksquare$$

Example 2.14. *A bivariate Pareto distribution* (Hutchinson and Lai 1990). Let X and Y be random variables whose joint survival function is given by

$$\overline{H}_\theta(x,y) = \begin{cases} (1+x+y)^{-\theta}, & x \geq 0, y \geq 0, \\ (1+x)^{-\theta}, & x \geq 0, y < 0, \\ (1+y)^{-\theta}, & x < 0, y \geq 0, \\ 1, & x < 0, y < 0; \end{cases}$$

where $\theta > 0$. Then the marginal survival functions $\overline{F}$ and $\overline{G}$ are

$$\overline{F}(x) = \begin{cases} (1+x)^{-\theta}, & x \geq 0 \\ 1, & x < 0, \end{cases} \text{ and } \overline{G}(y) = \begin{cases} (1+y)^{-\theta}, & y \geq 0 \\ 1, & y < 0, \end{cases}$$

so that X and Y have identical Pareto distributions. Inverting the survival functions and employing the survival version of Corollary 2.3.7 (see Exercise 2.25) yields the survival copula

$$\hat{C}_\theta(u,v) = (u^{-1/\theta} + v^{-1/\theta} - 1)^{-\theta}. \qquad (2.6.3)$$

We shall encounter this family again in Chapter 4. $\blacksquare$

Two other functions closely related to copulas—and survival copulas—are the *dual of a copula* and the *co-copula* (Schweizer and Sklar 1983). The dual of a copula C is the function $\tilde{C}$ defined by $\tilde{C}(u,v) = u + v - C(u,v)$; and the co-copula is the function C^* defined by $C^*(u,v) = 1 - C(1-u, 1-v)$. Neither of these is a copula, but when C

is the copula of a pair of random variables X and Y, the dual of the copula and the co-copula each express a probability of an event involving X and Y. Just as

$$P[X \leq x, Y \leq y] = C(F(x), G(y)) \text{ and } P[X > x, Y > y] = \hat{C}(\overline{F}(x), \overline{G}(y)),$$

we have

$$P[X \leq x \text{ or } Y \leq y] = \tilde{C}(F(x), G(y)), \tag{2.6.4}$$

and

$$P[X > x \text{ or } Y > y] = C^*(\overline{F}(x), \overline{G}(y)). \tag{2.6.5}$$

Other relationships among C, $\hat{C}$, $\tilde{C}$, and C^* are explored in Exercises 2.24 and 2.25.

Exercises

2.18 Prove the "Fréchet-Hoeffding lower bound" version of Lemma 2.5.2: Let S be a subset of $\overline{\mathbf{R}}^2$. Then S is nonincreasing if and only if for each (x,y) in $\overline{\mathbf{R}}^2$, either

   1. for all (u,v) in S, $u \leq x$ implies $v > y$; or

   2. for all (u,v) in S, $v > y$ implies $u \leq x$.

2.19 Prove the "Fréchet-Hoeffding lower bound" version of Lemma 2.5.3: Let X and Y be random variables whose joint distribution function H is equal to its Fréchet-Hoeffding lower bound. Then for every (x,y) in $\overline{\mathbf{R}}^2$, either $P[X > x, Y > y] = 0$ or $P[X \leq x, Y \leq y] = 0$

2.20 Prove Theorem 2.5.5.

2.21 Let X and Y be nonnegative random variables whose survival function is $\overline{H}(x,y) = (e^x + e^y - 1)^{-1}$ for $x,y \geq 0$.
   (a) Show that X and Y are standard exponential random variables.
   (b) Show that the copula of X and Y is the copula given by (2.3.4) in Example 2.8 [cf. Exercise 2.12].

2.22 Let X and Y be continuous random variables whose joint distribution function is given by $C(F(x), G(y))$, where C is the copula of X and Y, and F and G are the distribution functions of X and Y respectively. Verify that (2.6.4) and (2.6.5) hold.

2.23 Let $X_1, Y_1, F_1, G_1, F_2, G_2$, and C be as in Exercise 2.15. Set $X_2 = F_2^{(-1)}(1 - F_1(X_1))$ and $Y_2 = G_2^{(-1)}(1 - G_1(Y_1))$. Prove that
(a) The distribution functions of X_2 and Y_2 are F_2 and G_2, respectively; and
(b) The copula of X_2 and Y_2 is $\hat{C}$.

2.24 Let X and Y be continuous random variables with copula C and a common univariate distribution function F. Show that the distribution and survival functions of the order statistics (see Exercise 2.16) are given by

Order statistic	Distribution function	Survival function
$\max(X,Y)$	$\delta(F(t))$	$\delta^*(\overline{F}(t))$
$\min(X,Y)$	$\tilde{\delta}(F(t))$	$\hat{\delta}(\overline{F}(t))$

where $\delta, \hat{\delta}, \tilde{\delta}$, and δ^* denote the diagonal sections of $C, \hat{C}, \tilde{C}$, and C^*, respectively.

2.25 Show that under composition $\circ$, the set of operations of forming the survival copula, the dual of a copula, and the co-copula of a given copula, along with the identity (i.e., "$\wedge$", "$\sim$", "$*$", and "i") yields the dihedral group (e.g., $C^{**} = C$, so $* \circ * = i$; $\hat{C}^* = \tilde{C}$, so $\wedge \circ * = \sim$, etc.):

$\circ$	i	$\wedge$	$\sim$	$*$
i	i	$\wedge$	$\sim$	$*$
$\wedge$	$\wedge$	i	$*$	$\sim$
$\sim$	$\sim$	$*$	i	$\wedge$
$*$	$*$	$\sim$	$\wedge$	i

2.26 Prove the following "survival" version of Corollary 2.3.7: Let $\overline{H}, \overline{F}, \overline{G}$, and $\hat{C}$ be as in (2.6.2), and let $\overline{F}^{(-1)}$ and $\overline{G}^{(-1)}$ be quasi-inverses of $\overline{F}$ and $\overline{G}$, respectively. Then for any (u,v) in $\mathbf{I}^2$,

$$\hat{C}(u,v) = \overline{H}(\overline{F}^{(-1)}(u), \overline{G}^{(-1)}(v)).$$

2.7 Symmetry

If X is a random variable and a is a real number, we say that X is *symmetric about* a if the distribution functions of the random variables $X-a$ and $a-X$ are the same, that is, if for any x in $\mathbf{R}$, $P[X-a \leq x] = P[a-X \leq x]$. When X is continuous with distribution function F, this is equivalent to

$$F(a+x) = \overline{F}(a-x) \qquad (2.7.1)$$

[when F is discontinuous, (2.7.1) holds only at the points of continuity of F].

Now consider the bivariate situation. What does it mean to say that a pair (X,Y) of random variables is "symmetric" about a point (a,b)? There are a number of ways to answer this question, and each answer leads to a different type of bivariate symmetry.

Definition 2.7.1. Let X and Y be random variables and let (a,b) be a point in $\mathbf{R}^2$.

1. (X,Y) is *marginally symmetric* about (a,b) if X and Y are symmetric about a and b, respectively.

2. (X,Y) is *radially symmetric* about (a,b) if the joint distribution function of $X-a$ and $Y-b$ is the same as the joint distribution function of $a-X$ and $b-Y$.

3. (X,Y) is *jointly symmetric* about (a,b) if the following four pairs of random variables have a common joint distribution: $(X-a,Y-b)$, $(X-a,b-Y)$, $(a-X,Y-b)$, and $(a-X,b-Y)$.

When X and Y are continuous, we can express the condition for radial symmetry in terms of the joint distribution and survival functions of X and Y in a manner analogous to the relationship in (2.7.1) between univariate distribution and survival functions:

Theorem 2.7.2. *Let X and Y be continuous random variables with joint distribution function H and margins F and G, respectively. Let (a,b) be a point in $\mathbf{R}^2$. Then (X,Y) is radially symmetric about (a,b) if and only if*

$$H(a+x,b+y) = \overline{H}(a-x,b-y) \text{ for all } (x,y) \text{ in } \mathbf{R}^2. \qquad (2.7.2)$$

The term "radial" comes from the fact that the points $(a+x,b+y)$ and $(a-x,b-y)$ that appear in (2.7.2) lie on rays emanating in opposite directions from (a,b). Graphically, Theorem 2.7.2 states that regions such as those shaded in Fig. 2.7(a) always have equal H-volume.

Example 2.15. The bivariate normal distribution with parameters μ_x, μ_y, σ_x^2, σ_y^2, and ρ is radially symmetric about the point (μ_x,μ_y). The proof is straightforward (but tedious)—evaluate double integrals of the joint density over the shaded regions in Fig. 2.7(a). ∎

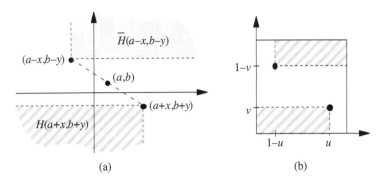

Fig. 2.7. Regions of equal probability for radially symmetric random variables

Example 2.16. The bivariate normal is a member of the family of elliptically contoured distributions. The densities for such distributions have contours that are concentric ellipses with constant eccentricity. Well-known members of this family, in addition to the bivariate normal, are bivariate Pearson type II and type VII distributions (the latter including bivariate t and Cauchy distributions as special cases). Like the bivariate normal, elliptically contoured distributions are radially symmetric.   ■

It is immediate that joint symmetry implies radial symmetry and easy to see that radial symmetry implies marginal symmetry (setting $x = \infty$ in (2.7.2) yields (2.7.1); similarly for $y = \infty$). Indeed, joint symmetry is a very strong condition—it is easy to show that jointly symmetric random variables must be uncorrelated when the requisite second-order moments exist (Randles and Wolfe 1979). Consequently, we will focus on radial symmetry, rather than joint symmetry, for bivariate distributions.

Because the condition for radial symmetry in (2.7.2) involves both the joint distribution and survival functions, it is natural to ask if copulas and survival copulas play a role in radial symmetry. The answer is provided by the next theorem.

Theorem 2.7.3. *Let X and Y be continuous random variables with joint distribution function H, marginal distribution functions F and G, respectively, and copula C. Further suppose that X and Y are symmetric about a and b, respectively. Then (X,Y) is radially symmetric about (a,b), i.e., H satisfies (2.7.2), if and only if $C = \hat{C}$, i.e., if and only if C satisfies the functional equation*

$$C(u,v) = u + v - 1 + C(1-u, 1-v) \text{ for all } (u,v) \text{ in } \mathbf{I}^2. \quad (2.7.3)$$

Proof. Employing (2.6.2) and (2.7.1), the theorem follows from the following chain of equivalent statements:

$H(a+x,b+y) = \overline{H}(a-x,b-y)$ for all (x,y) in $\mathbf{R}^2$

$\Leftrightarrow C(F(a+x),G(b+y)) = \hat{C}(\overline{F}(a-x),\overline{G}(b-y))$ for all (x,y) in $\mathbf{R}^2$,

$\Leftrightarrow C(F(a+x),G(b+y)) = \hat{C}(F(a+x),G(b+y))$ for all (x,y) in $\mathbf{R}^2$,

$\Leftrightarrow C(u,v) = \hat{C}(u,v)$ for all (u,v) in $\mathbf{I}^2$.   □

Geometrically, (2.7.3) states that for any (u,v) in $\mathbf{I}^2$, the rectangles $[0,u]\times[0,v]$ and $[1-u,1]\times[1-v,1]$ have equal C-volume, as illustrated in Fig. 2.7(b).

Another form of symmetry is exchangeability—random variables X and Y are *exchangeable* if the vectors (X,Y) and (Y,X) are identically distributed. Hence if the joint distribution function of X and Y is H, then $H(x,y) = H(y,x)$ for all x,y in $\overline{\mathbf{R}}^2$. Clearly exchangeable random variables must be identically distributed, i.e., have a common univariate distribution function. For identically distributed random variables, exchangeability is equivalent to the symmetry of their copula as expressed in the following theorem, whose proof is straightforward.

Theorem 2.7.4. *Let X and Y be continuous random variables with joint distribution function H, margins F and G, respectively, and copula C. Then X and Y are exchangeable if and only if F = G and C(u,v) = C(v,u) for all (u,v) in $\mathbf{I}^2$.*

When $C(u,v) = C(v,u)$ for all (u,v) in $\mathbf{I}^2$, we will say simply that C is *symmetric*.

Example 2.17. Although identically distributed independent random variables must be exchangeable (because the copula Π is symmetric), the converse is of course not true—identically distributed exchangeable random variables need not be independent. To show this, simply choose for the copula of X and Y *any* symmetric copula except Π, such as one from Example 2.8, 2.9 (or 2.13), or from one of the families in Exercises 2.4 and 2.5.   ∎

There are other bivariate symmetry concepts. See (Nelsen 1993) for details.

2.8 Order

The Fréchet-Hoeffding bounds inequality—$W(u,v) \le C(u,v) \le M(u,v)$ for every copula C and all u,v in $\mathbf{I}$—suggests a partial order on the set of copulas:

Definition 2.8.1. If C_1 and C_2 are copulas, we say that C_1 *is smaller than* C_2 (or C_2 *is larger than* C_1), and write $C_1 \prec C_2$ (or $C_2 \succ C_1$) if $C_1(u,v) \leq C_2(u,v)$ for all u,v in $\mathbf{I}$.

In other words, the Fréchet-Hoeffding lower bound copula W is smaller than every copula, and the Fréchet-Hoeffding upper bound copula M is larger than every copula. This point-wise partial ordering of the set of copulas is called the *concordance ordering* and will be important in Chapter 5 when we discuss the relationship between copulas and dependence properties for random variables (at which time the reason for the name of the ordering will become apparent). It is a partial order rather than a total order because not every pair of copulas is comparable.

Example 2.18. The product copula Π and the copula obtained by averaging the Fréchet-Hoeffding bounds are not comparable. If we let $C(u,v) = [W(u,v)+M(u,v)]/2$, then $C(1/4,1/4) > \Pi(1/4,1/4)$ and $C(1/4,3/4) < \Pi(1/4,3/4)$, so that neither $C \prec \Pi$ nor $\Pi \prec C$ holds. ■

However, there are families of copulas that are totally ordered. We will call a totally ordered parametric family $\{C_\theta\}$ of copulas *positively ordered* if $C_\alpha \prec C_\beta$ whenever $\alpha \leq \beta$; and *negatively ordered* if $C_\alpha \succ C_\beta$ whenever $\alpha \leq \beta$.

Example 2.19. The Cuadras-Augé family of copulas (2.2.10), introduced in Exercise 2.5, is positively ordered, as for $0 \leq \alpha \leq \beta \leq 1$ and u,v in $(0,1)$,

$$\frac{C_\alpha(u,v)}{C_\beta(u,v)} = \left(\frac{uv}{\min(u,v)}\right)^{\beta-\alpha} \leq 1$$

and hence $C_\alpha \prec C_\beta$. ■

Exercises

2.27 Let X and Y be continuous random variables symmetric about a and b with marginal distribution functions F and G, respectively, and with copula C. Is (X,Y) is radially symmetric (or jointly symmetric) about (a,b) if C is
(a) a member of the Fréchet family in Exercise 2.4?
(b) a member of the Cuadras-Augé family in Exercise 2.5?

2.28. Suppose X and Y are identically distributed continuous random variables, each symmetric about a. Show that "exchangeability" does not imply "radial symmetry," nor does "radial symmetry" imply "exchangeability."

2.29 Prove the following analog of Theorem 2.7.2 for jointly symmetric random variables: Let X and Y be continuous random variables with joint distribution function H and margins F and G, respectively. Let (a,b) be a point in $\mathbf{R}^2$. Then (X,Y) is jointly symmetric about (a,b) if and only if

$$H(a+x,b+y) = F(a+x) - H(a+x,b-y) \text{ for all } (x,y) \text{ in } \overline{\mathbf{R}}^2$$

and

$$H(a+x,b+y) = G(b+y) - H(a-x,b+y) \text{ for all } (x,y) \text{ in } \overline{\mathbf{R}}^2.$$

2.30 Prove the following analog of Theorem 2.7.3 for jointly symmetric random variables: Let X and Y be continuous random variables with joint distribution function H, marginal distribution functions F and G, respectively, and copula C. Further suppose that X and Y are symmetric about a and b, respectively. Then (X,Y) is jointly symmetric about (a,b), i.e., H satisfies the equations in Exercise 2.28, if and only if C satisfies

$$C(u,v) = u - C(u,1-v) \text{ and } C(u,v) = v - C(1-u,v) \qquad (2.8.1)$$

for all (u,v) in $\mathbf{I}^2$. [Cf. Exercise 2.6 and Theorem 2.4.4].

2.31 (a) Show that $C_1 \prec C_2$ if and only if $\overline{C}_1 \prec \overline{C}_2$.
      (b) Show that $C_1 \prec C_2$ if and only if $\hat{C}_1 \prec \hat{C}_2$.

2.32 Show that the Ali-Mikhail-Haq family of copulas (2.4.3) from Exercise 2.14 is positively ordered.

2.33 Show that the Mardia family of copulas (2.2.9) from Exercise 2.4 is neither positively nor negatively ordered. [Hint: evaluate C_0, $C_{1/4}$, and $C_{1/2}$ at $(u,v) = (3/4,1/4)$.]

2.9 Random Variate Generation

One of the primary applications of copulas is in simulation and Monte Carlo studies. In this section, we will address the problem of generating a sample from a specified joint distribution. Such samples can then be used to study mathematical models of real-world systems, or for statistical studies, such as the comparison of a new statistical method with competitors, robustness properties, or the agreement of asymptotic with small sample results.

We assume that the reader is familiar with various procedures used to generate independent uniform variates and with algorithms for using those variates to obtain samples from a given univariate distribution. One such method is the inverse distribution function method. To obtain an observation x of a random variable X with distribution function F:

1. Generate a variate u that is uniform on $(0,1)$;
2. Set $x = F^{(-1)}(u)$, where $F^{(-1)}$ is any quasi-inverse of F (see Definition 2.3.6).

For a discussion and for alternative methods, see (Johnson 1987) or (Devroye 1986).

There are a variety of procedures used to generate observations (x,y) of a pair or random variables (X,Y) with a joint distribution function H. In this section, we will focus on using the copula as a tool. By virtue of Sklar's theorem, we need only generate a pair (u,v) of observations of uniform $(0,1)$ random variables (U,V) whose joint distribution function is C, the copula of X and Y, and then transform those uniform variates via the algorithm such as the one in the preceding paragraph. One procedure for generating such of a pair (u,v) of uniform $(0,1)$ variates is the conditional distribution method. For this method, we need the conditional distribution function for V given $U = u$, which we denote $c_u(v)$:

$$c_u(v) = P[V \leq v | U = u] = \lim_{\Delta u \to 0} \frac{C(u + \Delta u, v) - C(u,v)}{\Delta u} = \frac{\partial C(u,v)}{\partial u} \qquad (2.9.1)$$

[Recall from Theorem 2.2.7 that the function $v \mapsto \partial C(u,v)/\partial u$, which we are now denoting $c_u(v)$, exists and is nondecreasing almost everywhere in $\mathbf{I}$].

1. Generate two independent uniform $(0,1)$ variates u and t;
2. Set $v = c_u^{(-1)}(t)$, where $c_u^{(-1)}$ denotes a quasi-inverse of c_u.
3. The desired pair is (u,v).

As with univariate distributions, there are many other algorithms—see (Johnson 1987) or (Devroye 1986) for details.

Example 2.20. Let X and Y be random variables whose joint distribution function H is

$$H(x,y) = \begin{cases} \dfrac{(x+1)(e^y - 1)}{x + 2e^y - 1}, & (x,y) \in [-1,1] \times [0,\infty], \\ 1 - e^{-y}, & (x,y) \in (1,\infty] \times [0,\infty], \\ 0, & \text{elsewhere.} \end{cases}$$

[recall Examples 2.5 and 2.8]. The copula C of X and Y is

$$C(u,v) = \frac{uv}{u + v - uv},$$

and so the conditional distribution function c_u and its inverse $c_u^{(-1)}$ are given by

$$c_u(v) = \frac{\partial}{\partial u} C(u,v) = \left(\frac{v}{u+v-uv} \right)^2 \quad \text{and} \quad c_u^{-1}(t) = \frac{u\sqrt{t}}{1-(1-u)\sqrt{t}}.$$

Thus an algorithm to generate random variates (x,y) is:

1. Generate two independent uniform $(0,1)$ variates u and t;

2. Set $v = \dfrac{u\sqrt{t}}{1-(1-u)\sqrt{t}}$,

3. Set $x = 2u - 1$ and $y = -\ln(1 - v)$ [See Example 2.8 for the inverses of the marginal distribution functions.]

4. The desired pair is (x,y).   ∎

Survival copulas can also be used in the conditional distribution function method to generate random variates from a distribution with a given survival function. Recall [see part 3 of Theorem 2.4.4 and (2.6.1)] that if the copula C is the distribution function of a pair (U,V), then the corresponding survival copula $\hat{C}(u,v) = u+v-1+C(1-u,1-v)$ is the distribution function of the pair $(1-U,1-V)$. Also note that if U is uniform on $(0,1)$, so is the random variable $1-U$. Hence we have the following algorithm to generate a pair (U,V) whose distribution function is the copula C, given $\hat{C}$:

1. Generate two independent uniform $(0,1)$ variates u and t;

2. Set $v = \hat{c}_u^{(-1)}(t)$, where $\hat{c}_u^{(-1)}$ denotes a quasi-inverse of $\hat{c}_u(v) = \partial \hat{C}(u,v)/\partial u$.

3. The desired pair is (u,v).

In the next chapter we will be presenting methods that can be used to construct families of copulas. For many of those families, we will also indicate methods for generating random samples from the distributions that correspond to those copulas.

2.10 Multivariate Copulas

In this section, we extend the results of the preceding sections to the multivariate case. Although many of the definitions and theorems have analogous multivariate versions, not all do, so one must proceed with care. In the interest of clarity, we will restate most of the definitions and theorems in their multivariate versions. We will omit the proofs of theorems for which the proof is similar to that in the bivariate case. Many of the theorems in this section (with proofs) may be found in (Schweizer and Sklar 1983) or the references contained therein.

Some new notation will be advantageous here. For any positive integer n, we let $\overline{\mathbf{R}}^n$ denote the extended n-space $\overline{\mathbf{R}} \times \overline{\mathbf{R}} \times \cdots \times \overline{\mathbf{R}}$. We will use vector notation for points in $\overline{\mathbf{R}}^n$, e.g., $\mathbf{a} = (a_1, a_2, \cdots, a_n)$, and we will write $\mathbf{a} \le \mathbf{b}$ when $a_k \le b_k$ for all k; and $\mathbf{a} < \mathbf{b}$ when $a_k < b_k$ for all k. For $\mathbf{a} \le \mathbf{b}$, we will let $[\mathbf{a},\mathbf{b}]$ denote the n-box $B = [a_1,b_1] \times [a_2,b_2] \times \cdots \times [a_n,b_n]$, the Cartesian product of n closed intervals. The *vertices* of an n-box B are the points $\mathbf{c} = (c_1, c_2, \cdots, c_n)$ where each c_k is equal to either a_k or b_k. The unit n-cube $\mathbf{I}^n$ is the product $\mathbf{I} \times \mathbf{I} \times \cdots \times \mathbf{I}$. An *n-place real function H* is a function whose domain, Dom$H$, is a subset of $\overline{\mathbf{R}}^n$ and whose range, Ran$H$, is a subset of $\mathbf{R}$. Note that the unit "2-cube" is the unit square $\mathbf{I}^2$, and a "2-box" is a rectangle $[x_1,x_2] \times [y_1,y_2]$ in $\overline{\mathbf{R}}^2$.

Definition 2.10.1. Let S_1, $S_2, \cdots S_n$ be nonempty subsets of $\overline{\mathbf{R}}$, and let H be an n-place real function such that Dom$H = S_1 \times S_2 \times \cdots \times S_n$. Let $B = [\mathbf{a},\mathbf{b}]$ be an n-box all of whose vertices are in Dom$H$. Then the H-*volume of B* is given by

$$V_H(B) = \sum \mathrm{sgn}(\mathbf{c}) H(\mathbf{c}), \qquad (2.10.1)$$

where the sum is taken over all vertices $\mathbf{c}$ of B, and sgn($\mathbf{c}$) is given by

$$\mathrm{sgn}(\mathbf{c}) = \begin{cases} 1, & \text{if } c_k = a_k \text{ for an even number of } k\text{'s,} \\ -1, & \text{if } c_k = a_k \text{ for an odd number of } k\text{'s.} \end{cases}$$

Equivalently, the H-volume of an n-box $B = [\mathbf{a},\mathbf{b}]$ is the nth order difference of H on B

$$V_H(B) = \Delta_{\mathbf{a}}^{\mathbf{b}} H(\mathbf{t}) = \Delta_{a_n}^{b_n} \Delta_{a_{n-1}}^{b_{n-1}} \cdots \Delta_{a_2}^{b_2} \Delta_{a_1}^{b_1} H(\mathbf{t}),$$

where we define the n first order differences of an n-place function (such as H) as

$$\Delta_{a_k}^{b_k} H(\mathbf{t}) = H(t_1, \cdots, t_{k-1}, b_k, t_{k+1}, \cdots, t_n) - H(t_1, \cdots, t_{k-1}, a_k, t_{k+1}, \cdots, t_n).$$

Example 2.21. Let H be a 3-place real function with domain $\overline{\mathbf{R}}^3$, and let B be the 3-box $[x_1,x_2] \times [y_1,y_2] \times [z_1,z_2]$. The H-volume of B is

$$V_H(B) = H(x_2,y_2,z_2) - H(x_2,y_2,z_1) - H(x_2,y_1,z_2) - H(x_1,y_2,z_2)$$
$$+ H(x_2,y_1,z_1) + H(x_1,y_2,z_1) + H(x_1,y_1,z_2) - H(x_1,y_1,z_1). \qquad \blacksquare$$

Definition 2.10.2. An n-place real function H is *n-increasing* if $V_H(B) \ge 0$ for all n-boxes B whose vertices lie in Dom H.

Suppose that the domain of an n-place real function H is given by Dom$H = S_1 \times S_2 \times \cdots \times S_n$ where each S_k has a least element a_k. We say that H is *grounded* if $H(\mathbf{t}) = 0$ for all $\mathbf{t}$ in $\mathrm{Dom}\,H$ such that $t_k = a_k$ for at least one k. If each S_k is nonempty and has a greatest element b_k, then we say that H *has margins*, and the *one-dimensional margins* of H are the functions H_k given by $\mathrm{Dom}\,H_k = S_k$ and

$$H_k(x) = H(b_1,\cdots,b_{k-1},x,b_{k+1},\cdots,b_n) \text{ for all } x \text{ in } S_k. \quad (2.10.2)$$

Higher dimensional margins are defined by fixing fewer places in H.

Example 2.22. Let H be the function with domain $[-1,1]\times[0,\infty]\times[0,\pi/2]$ given by

$$H(x,y,z) = \frac{(x+1)(e^y-1)\sin z}{x+2e^y-1}.$$

Then H is grounded because $H(x,y,0) = 0$, $H(x,0,z) = 0$, and $H(-1,y,z) = 0$; H has one-dimensional margins $H_1(x)$, $H_2(y)$, and $H_3(z)$ given by

$$H_1(x) = H(x,\infty,\pi/2) = (x+1)/2, \ H_2(y) = H(1,y,\pi/2) = 1-e^{-y},$$
$$\text{and } H_3(z) = H(1,\infty,z) = \sin z;$$

and H has two-dimensional margins $H_{1,2}(x,y)$, $H_{2,3}(y,z)$, and $H_{1,3}(x,z)$ given by

$$H_{1,2}(x,y) = H(x,y,\pi/2) = \frac{(x+1)(e^y)-1)}{x+2e^y-1},$$
$$H_{2,3}(y,z) = H(1,y,z) = (1-e^{-y})\sin z, \text{ and}$$
$$H_{1,3}(x,z) = H(x,\infty,z,) = \frac{(x+1)\sin z}{2}. \qquad ■$$

In the sequel, one-dimensional margins will be simply "margins," and for $k \geq 2$, we will write "k-margins" for k-dimensional margins.

Lemma 2.10.3. *Let $S_1, S_2,\cdots S_n$ be nonempty subsets of $\overline{\mathbf{R}}$, and let H be a grounded n-increasing function with domain $S_1\times S_2\times\cdots\times S_n$. Then H is nondecreasing in each argument, that is, if $(t_1,\cdots,t_{k-1},x,t_{k+1},\cdots,t_n)$ and $(t_1,\cdots,t_{k-1},y,t_{k+1},\cdots,t_n)$ are in $\mathrm{Dom}\,H$ and $x < y$, then $H(t_1,\cdots,t_{k-1},x,t_{k+1},\cdots,t_n) \leq H(t_1,\cdots,t_{k-1},y,t_{k+1},\cdots,t_n)$.*

The following lemma, which is the n-dimensional analog of Lemma 2.1.5, is needed to show that n-copulas are uniformly continuous, and in the proof of the n-dimensional version of Sklar's theorem. Its proof, however, is somewhat more complicated than that of Lemma 2.1.5; see (Schweizer and Sklar 1983) for details.

Lemma 2.10.4. *Let $S_1, S_2,\cdots S_n$ be nonempty subsets of $\overline{\mathbf{R}}$, and let H be a grounded n-increasing function with margins whose domain is*

$S_1 \times S_2 \times \cdots \times S_n$. *Let* $\mathbf{x} = (x_1, x_2, \cdots, x_n)$ *and* $\mathbf{y} = (y_1, y_2, \cdots, y_n)$ *be any points in* $S_1 \times S_2 \times \cdots \times S_n$. *Then*

$$\left| H(\mathbf{x}) - H(\mathbf{y}) \right| \leq \sum_{k=1}^{n} \left| H_k(x_k) - H_k(y_k) \right|.$$

We are now in a position to define n-dimensional subcopulas and copulas. The definitions are analogous to Definitions 2.2.1 and 2.2.2.

Definition 2.10.5. An n-*dimensional subcopula* (or n-*subcopula*) is a function C' with the following properties:

1. $\operatorname{Dom} C' = S_1 \times S_2 \times \cdots \times S_n$, where each S_k is a subset of $\mathbf{I}$ containing 0 and 1;
2. C' is grounded and n-increasing;
3. C' has (one-dimensional) margins $C'_k, k = 1, 2, \cdots, n$, which satisfy

$$C'_k(u) = u \text{ for all } u \text{ in } S_k. \tag{2.10.3}$$

Note that for every $\mathbf{u}$ in $\operatorname{Dom} C'$, $0 \leq C'(\mathbf{u}) \leq 1$, so that $\operatorname{Ran} C'$ is also a subset of $\mathbf{I}$.

Definition 2.10.6. An n-*dimensional copula* (or n-*copula*) is an n-subcopula C whose domain is $\mathbf{I}^n$.

Equivalently, an n-copula is a function C from $\mathbf{I}^n$ to $\mathbf{I}$ with the following properties:

1. For every $\mathbf{u}$ in $\mathbf{I}^n$,

$$C(\mathbf{u}) = 0 \text{ if at least one coordinate of } \mathbf{u} \text{ is } 0, \tag{2.10.4a}$$

and

$$\text{if all coordinates of } \mathbf{u} \text{ are 1 except } u_k, \text{ then } C(\mathbf{u}) = u_k; \tag{2.10.4b}$$

2. For every $\mathbf{a}$ and $\mathbf{b}$ in $\mathbf{I}^n$ such that $\mathbf{a} \leq \mathbf{b}$,

$$V_C([\mathbf{a}, \mathbf{b}]) \geq 0. \tag{2.10.4c}$$

It is easy to show (see Exercise 2.34) that for any n-copula $C, n \geq 3$, each k-margin of C is a k-copula, $2 \leq k < n$.

Example 2.23. (a) Let $C(u, v, w) = w \cdot \min(u, v)$. Then C is a 3-copula, as it is easily seen that C satisfies (2.10.4a) and (2.10.4b), and the C-volume of the 3-box $B = [a_1, b_1] \times [a_2, b_2] \times [a_3, b_3]$ (where $a_k \leq b_k$) is

$$V_C(B) = \Delta_{a_3}^{b_3} \Delta_{a_2}^{b_2} \Delta_{a_1}^{b_1} C(u, v, w) = (b_3 - a_3) \Delta_{a_2}^{b_2} \Delta_{a_1}^{b_1} \min(u, v) \geq 0.$$

The 2-margins of C are the 2-copulas $C_{1,2}(u,v) = C(u,v,1) = 1 \cdot \min(u,v)$ $= M(u,v)$, $C_{1,3}(u,w) = C(u,1,w) = w \cdot \min(u,1) = \Pi(u,w)$, and $C_{2,3}(v,w) = C(1,v,w) = w \cdot \min(1,v) = \Pi(v,w)$.

(b) Let $C(u,v,w) = \min(u,v) - \min(u,v,1-w)$. The verification that C is a 3-copula is somewhat tedious. Here the 2-margins are $C_{1,2}(u,v) = M(u,v)$, $C_{1,3}(u,w) = u - \min(u,1-w) = W(u,w)$, and $C_{2,3}(v,w) = v - \min(v,1-w) = W(v,w)$. ∎

A consequence of Lemma 2.10.4 is the uniform continuity of n-subcopulas (and hence n-copulas):

Theorem 2.10.7. *Let C' be an n-subcopula. Then for every* $\mathbf{u}$ *and* $\mathbf{v}$ *in* $\mathrm{Dom}\, C'$,

$$\left| C'(\mathbf{v}) - C'(\mathbf{u}) \right| \le \sum_{k=1}^{n} \left| v_k - u_k \right|. \tag{2.10.5}$$

Hence C' is uniformly continuous on its domain.

We are now in a position to state the n-dimensional version of Sklar's theorem. To do so, we first define n-dimensional distribution functions:

Definition 2.10.8. An *n-dimensional distribution function* is a function H with domain $\overline{\mathbf{R}}^n$ such that
1. H is n-increasing,
2. $H(\mathbf{t}) = 0$ for all $\mathbf{t}$ in $\overline{\mathbf{R}}^n$ such that $t_k = -\infty$ for at least one k, and
$H(\infty,\infty,\cdots,\infty) = 1$.

Thus H is grounded, and because $\mathrm{Dom}\, H = \overline{\mathbf{R}}^n$, it follows from Lemma 2.10.3 that the one-dimensional margins, given by (2.10.2), of an n-dimensional distribution function are distribution functions, which for $n \ge 3$ we will denote by $F_1, F_2, \cdots, F_n$.

Theorem 2.10.9. Sklar's theorem in n-dimensions. *Let H be an n-dimensional distribution function with margins $F_1, F_2, \cdots, F_n$. Then there exists an n-copula C such that for all $\mathbf{x}$ in $\overline{\mathbf{R}}^n$,*

$$H(x_1, x_2, \cdots, x_n) = C\big(F_1(x_1), F_2(x_2), \cdots, F_n(x_n)\big). \tag{2.10.6}$$

If $F_1, F_2, \cdots, F_n$ are all continuous, then C is unique; otherwise, C is uniquely determined on $\mathrm{Ran}\, F_1 \times \mathrm{Ran}\, F_2 \times \cdots \times \mathrm{Ran}\, F_n$. *Conversely, if C is an n-copula and $F_1, F_2, \cdots, F_n$ are distribution functions, then the function H defined by (2.10.6) is an n-dimensional distribution function with margins $F_1, F_2, \cdots, F_n$.*

The proof of Theorem 2.10.9 proceeds as in the case of two dimensions—one first proves the n-dimensional versions of Lemma 2.3.4 (which is straightforward) and then Lemma 2.3.5, the "extension lemma." The proof of the n-dimensional extension lemma, in which one shows that every n-subcopula can be extended to an n-copula, proceeds via a "multilinear interpolation" of the subcopula to a copula similar to two-dimensional version in (2.3.2). The proof in the n-dimensional case, however, is somewhat more involved (Moore and Spruill 1975; Deheuvels 1978; Sklar 1996).

Corollary 2.10.10. *Let* $H, C, F_1, F_2, \cdots, F_n$ *be as in Theorem* 2.10.9, *and let* $F_1^{(-1)}, F_2^{(-1)}, \cdots, F_n^{(-1)}$ *be quasi-inverses of* $F_1, F_2, \cdots, F_n$, *respectively. Then for any* **u** *in* $\mathbf{I}^n$,

$$C(u_1, u_2, \cdots, u_n) = H\left(F_1^{(-1)}(u_1), F_2^{(-1)}(u_2), \cdots, F_n^{(-1)}(u_n) \right). \qquad (2.10.7)$$

Of course, the n-dimensional version of Sklar's theorem for random variables (again defined on a common probability space) is similar to Theorem 2.4.1:

Theorem 2.10.11. *Let* $X_1, X_2, \cdots, X_n$ *be random variables with distribution functions* $F_1, F_2, \cdots, F_n$, *respectively, and joint distribution function* H. *Then there exists an* n-copula C *such that* (2.10.6) *holds. If* $F_1, F_2, \cdots, F_n$ *are all continuous,* C *is unique. Otherwise,* C *is uniquely determined on* $\mathrm{Ran}\, F_1 \times \mathrm{Ran}\, F_2 \times \cdots \times \mathrm{Ran}\, F_n$.

The extensions of the 2-copulas M, Π, and W to n dimensions are denoted M^n, Π^n, and W^n (a superscript on the name of a copula will denote dimension rather than exponentiation), and are given by:

$$M^n(\mathbf{u}) = \min(u_1, u_2, \cdots, u_n);$$
$$\Pi^n(\mathbf{u}) = u_1 u_2 \cdots u_n; \qquad (2.10.8)$$
$$W^n(\mathbf{u}) = \max(u_1 + u_2 + \cdots + u_n - n + 1, 0).$$

The functions M^n and Π^n are n-copulas for all $n \geq 2$ (Exercise 2.34), whereas the function W^n fails to be an n-copula for any $n > 2$ (Exercise 2.36). However, we do have the following n-dimensional version of the Fréchet-Hoeffding bounds inequality first encountered in (2.2.5). The proof follows directly from Lemmas 2.10.3 and 2.10.4.

Theorem 2.10.12. *If* C' *is any* n-subcopula, *then for every* **u** *in* $\mathrm{Dom}\, C'$,

$$W^n(\mathbf{u}) \leq C'(\mathbf{u}) \leq M^n(\mathbf{u}). \qquad (2.10.9)$$

Although the Fréchet-Hoeffding lower bound W^n is never a copula for $n > 2$, the left-hand inequality in (2.10.9) is "best-possible," in the sense that for any $n \geq 3$ and any $\mathbf{u}$ in $\mathbf{I}^n$, there is an n-copula C such that $C(\mathbf{u}) = W^n(\mathbf{u})$:

Theorem 2.10.13. *For any $n \geq 3$ and any $\mathbf{u}$ in $\mathbf{I}^n$, there exists an n-copula C (which depends on $\mathbf{u}$) such that*

$$C(\mathbf{u}) = W^n(\mathbf{u}).$$

Proof (Sklar 1998). Let $\mathbf{u} = (u_1, u_2, \cdots, u_n)$ be a (fixed) point in $\mathbf{I}^n$ other than $\mathbf{0} = (0,0,\cdots,0)$ or $\mathbf{1} = (1,1,\cdots,1)$. There are two cases to consider.

1. Suppose $0 < u_1 + u_2 + \cdots + u_n \leq n - 1$. Consider the set of points $\mathbf{v} = (v_1, v_2, \cdots, v_n)$ where each v_k is 0, 1, or $t_k = \min\{(n-1)u_k/(u_1 + u_2 + \cdots + u_n), 1\}$. Define an n-place function C' on these points by $C'(\mathbf{v}) = W^n(\mathbf{v})$. It is straightforward to verify that C' satisfies the conditions in Definition 2.10.5 and hence is an n-subcopula. Now extend C' to an n-copula C via a "multilinear interpolation" similar to (2.3.2). Then for each $\mathbf{x}$ in the n-box $[\mathbf{0}, \mathbf{t}]$, $\mathbf{t} = (t_1, t_2, \cdots, t_n)$ (which includes $\mathbf{u}$), $C(\mathbf{x}) = W^n(\mathbf{x}) = 0$.

2. Suppose $n - 1 < u_1 + u_2 + \cdots + u_n < n$, and consider the set of points $\mathbf{v} = (v_1, v_2, \cdots, v_n)$ where now each v_k is 0, 1, or $s_k = 1 - (1 - u_k)/[n - (u_i + u_2 + \cdots + u_n)]$. Define an n-place function C' on these points by $C'(\mathbf{v}) = W^n(\mathbf{v})$, and extend to an n-copula C as before. Let $\mathbf{s} = (s_1, s_2, \cdots, s_n)$, then for each $\mathbf{x}$ in the n-box $[\mathbf{s}, \mathbf{1}]$ (which includes $\mathbf{u}$), we have $C(\mathbf{x}) = W^n(\mathbf{x}) = x_1 + x_2 + \cdots + x_n - n + 1$. □

The n-copulas M^n and Π^n have characterizations similar to the characterizations of M and Π given in Theorems 2.4.2 and 2.5.4.

Theorem 2.10.14. *For $n \geq 2$, let $X_1, X_2, \cdots, X_n$ be continuous random variables. Then*

1. *$X_1, X_2, \cdots, X_n$ are independent if and only if the n-copula of $X_1, X_2, \cdots, X_n$ is Π^n, and*

2. *each of the random variables $X_1, X_2, \cdots, X_n$ is almost surely a strictly increasing function of any of the others if and only if the n-copula of $X_1, X_2, \cdots, X_n$ is M^n.*

Exercises

2.34 (a) Show that the $(n-1)$-margins of an n-copula are $(n-1)$-copulas. [Hint: consider n-boxes of the form $[a_1,b_1]\times\cdots$ $\times[a_{k-1},b_{k-1}]\times[0,1]\times[a_{k+1},b_{k+1}]\times\cdots\times[a_n,b_n].]$
(b) Show that if C is an n-copula, $n \geq 3$, then for any k, $2 \leq k < n$, all $\binom{n}{k}$ k-margins of C are k-copulas.

2.35 Let M^n and Π^n be the functions defined in (2.10.4), and let $[\mathbf{a},\mathbf{b}]$ be an n-box in $\mathbf{I}^n$. Prove that

$$V_{M^n}([\mathbf{a},\mathbf{b}]) = \max\left(\min(b_1,b_2,\cdots,b_n) - \max(a_1,a_2,\cdots,a_n),0\right)$$

and

$$V_{\Pi^n}([\mathbf{a},\mathbf{b}]) = (b_1 - a_1)(b_2 - a_2)\cdots(b_n - a_n),$$

and hence conclude that M^n and Π^n are n-copulas for all $n \geq 2$.

2.36 Show that

$$V_{W^n}([\mathbf{1/2},\mathbf{1}]) = 1-(n/2),$$

where $\mathbf{1} = (1,1,\cdots,1)$ and $\mathbf{1/2} = (1/2,1/2,\cdots,1/2)$, and hence W^n fails to be an n-copula whenever $n > 2$.

2.37 Let $X_1,X_2,\cdots,X_n$ be continuous random variables with copula C and distribution functions $F_1,F_2,\cdots,F_n$, respectively. Let $X_{(1)}$ and $X_{(n)}$ denote the extreme *order statistics* for $X_1,X_2,\cdots,X_n$ (i.e., $X_{(1)}$ $= \min(X_1,X_2,\cdots,X_n)$ and $X_{(n)} = \max(X_1,X_2,\cdots,X_n)$) [cf. Exercise 2.16]. Prove that the distribution functions $F_{(1)}$ and $F_{(n)}$, respectively, of $X_{(1)}$ and $X_{(n)}$ satisfy

$$\max\left(F_1(t),F_2(t),\cdots,F_n(t)\right) \leq F_{(1)}(t) \leq \min\left(\sum_{k=1}^{n} F_k(t),1\right)$$

and

$$\max\left(\sum_{k=1}^{n} F_k(t) - n + 1,0\right) \leq F_{(n)}(t) \leq \min\left(F_1(t),F_2(t),\cdots,F_n(t)\right).$$

3 Methods of Constructing Copulas

If we have a collection of copulas, then, as a consequence of Sklar's theorem, we automatically have a collection of bivariate or multivariate distributions with whatever marginal distributions we desire. Clearly this can be useful in modeling and simulation. Furthermore, by virtue of Theorem 2.4.3, the nonparametric nature of the dependence between two random variables is expressed by the copula. Thus the study of concepts and measures of nonparametric dependence is a study of properties of copulas—a topic we will pursue in Chapter 5. For this study, it is advantageous to have a variety of copulas at our disposal.

In this chapter, we present and illustrate several general methods of constructing bivariate copulas. In the inversion method, we exploit Sklar's theorem, via Corollary 2.3.7, to produce copulas directly from joint distribution functions. Using geometric methods, we construct singular copulas whose support lies in a specified set and copulas with sections given by simple functions such as polynomials. We also discuss three geometrically motivated construction procedures that yield copulas known as ordinal sums, shuffles of M, and convex sums. In the algebraic method, we construct copulas from relationships involving the bivariate and marginal distributions functions—our examples concern cases in which the algebraic relationship is a ratio. We conclude this chapter with a study of problems associated with the construction of multivariate copulas. Another general method, yielding bivariate and multivariate Archimedean copulas, will be presented in the next chapter.

A note on notation is in order. Throughout this chapter, we will repeatedly use α, β, and θ as subscripts to denote parameters in families of copulas, i.e., C_θ represents a member of a one-parameter family, and $C_{\alpha,\beta}$ represents a member of a two-parameter family. The particular family to which C_θ or $C_{\alpha,\beta}$ belongs will be clear from the context of the particular example or exercise.

3.1 The Inversion Method

In Sect. 2.3, we presented several simple examples of this procedure: given a bivariate distribution function H with continuous margins F and G, "invert" via (2.1.3) to obtain a copula:

$$C(u,v) = H(F^{(-1)}(u), G^{(-1)}(v)).\qquad(3.1.1)$$

With this copula, new bivariate distributions with arbitrary margins, say F' and G', can be constructed using Sklar's theorem: $H'(x,y) = C(F'(x), G'(y))$. Of course, this can be done equally as well using survival functions from (2.6.2) (recall that $\hat{C}$ *is* a copula):

$$\hat{C}(u,v) = \overline{H}(\overline{F}^{(-1)}(u), \overline{G}^{(-1)}(v)).\qquad(3.1.2)$$

where $\overline{F}^{(-1)}$ denotes a quasi-inverse of $\overline{F}$, defined analogously to $F^{(-1)}$ in Definition 2.3.6; or equivalently, $\overline{F}^{(-1)}(t) = F^{(-1)}(1-t)$.

We will now illustrate this procedure to find the copulas for the Marshall-Olkin system of bivariate exponential distributions and for the uniform distribution on a circle.

3.1.1 The Marshall-Olkin Bivariate Exponential Distribution

The univariate exponential distribution plays a central role in mathematical statistics because it is the distribution of waiting time in a standard Poisson process. The following bivariate exponential distribution, first described by Marshall and Olkin (1967a,b), plays a similar role in a two-dimensional Poisson process.

Consider a two-component system—such as a two engine aircraft, or a desktop computer with both a CPU (central processing unit) and a co-processor. The components are subject to "shocks," which are always "fatal" to one or both of the components. For example, one of the two aircraft engines may fail, or a massive explosion could destroy both engines simultaneously; or the CPU or the co-processor could fail, or a power surge could eliminate both simultaneously. Let X and Y denote the lifetimes of the components 1 and 2, respectively. As is often the case in dealing with lifetimes, we will find the survival function $\overline{H}(x,y) = P[X > x, Y > y]$, the probability that component 1 survives beyond time x and that component 2 survives beyond time y.

The "shocks" to the two components are assumed to form three independent Poisson processes with (positive) parameters λ_1, λ_2, and λ_{12}; depending on whether the shock kills only component 1, only component 2, or both components simultaneously. The times Z_1, Z_2, and Z_{12} of occurrence of these three shocks are independent exponential random variables with parameters λ_1, λ_2, and λ_{12}, respectively. So $X = \min(Z_1, Z_{12})$, $Y = \min(Z_2, Z_{12})$, and hence for all $x, y \geq 0$,

$$\overline{H}(x,y) = P[Z_1 > x]P[Z_2 > y]P[Z_{12} > \max(x,y)]$$
$$= \exp\left[-\lambda_1 x - \lambda_2 y - \lambda_{12}\max(x,y)\right].$$

The marginal survival functions are $\overline{F}(x) = \exp(-(\lambda_1 + \lambda_{12})x)$ and $\overline{G}(y) = \exp(-(\lambda_2 + \lambda_{12})y)$; and hence X and Y are exponential random variables with parameters $\lambda_1 + \lambda_{12}$ and $\lambda_2 + \lambda_{12}$, respectively.

In order to find the survival copula $\hat{C}(u,v)$ for this distribution, we first express $\overline{H}(x,y)$ in terms of $\overline{F}(x)$ and $\overline{G}(y)$, recalling that $\overline{H}(x,y) = \hat{C}(\overline{F}(x), \overline{G}(y))$. To accomplish this, we first replace $\max(x,y)$ by $x + y - \min(x,y)$, so that

$$\overline{H}(x,y) = \exp(-(\lambda_1 + \lambda_{12})x - (\lambda_2 + \lambda_{12})y + \lambda_{12}\min(x,y))$$
$$= \overline{F}(x)\overline{G}(y)\min\{\exp(\lambda_{12}x),\exp(\lambda_{12}y)\}.$$

Now set $u = \overline{F}(x)$ and $v = \overline{G}(y)$, and for convenience let $\alpha = \lambda_{12}/(\lambda_1 + \lambda_{12})$ and $\beta = \lambda_{12}/(\lambda_2 + \lambda_{12})$. Then $\exp(\lambda_{12}x) = u^{-\alpha}$ and $\exp(\lambda_{12}y) = v^{-\beta}$, thus, using (3.1.2), the survival copula $\hat{C}$ is given by

$$\hat{C}(u,v) = uv\min(u^{-\alpha},v^{-\beta}) = \min(u^{1-\alpha}v, uv^{1-\beta}).$$

Note that because λ_1, λ_2, λ_{12} are positive, α and β satisfy $0 < \alpha, \beta < 1$. Hence the survival copulas for the Marshall-Olkin bivariate exponential distribution yield a two parameter family of copulas given by

$$C_{\alpha,\beta}(u,v) = \min(u^{1-\alpha}v, uv^{1-\beta}) = \begin{cases} u^{1-\alpha}v, & u^\alpha \geq v^\beta, \\ uv^{1-\beta}, & u^\alpha \leq v^\beta. \end{cases} \qquad (3.1.3)$$

This family is known both as the *Marshall-Olkin family* and the *Generalized Cuadras-Augé family*. Note that when $\alpha = \beta = \theta$, (3.1.3) reduces to the Cuadras-Augé family in Exercise 2.5, corresponding to the case in which $\lambda_1 = \lambda_2$, i.e., the case in which X and Y are exchangeable. The parameter range can be extended to $0 \leq \alpha, \beta \leq 1$ (Exercise 3.1) and indeed, $C_{\alpha,0} = C_{0,\beta} = \Pi$ and $C_{1,1} = M$.

It is interesting to note that, although the copulas in this family have full support (for $0 < \alpha, \beta < 1$), they are neither absolutely continuous nor singular, but rather have both absolutely continuous and singular components $A_{\alpha,\beta}$ and $S_{\alpha,\beta}$, respectively. Because

$$\frac{\partial^2}{\partial u \partial v}C_{\alpha,\beta}(u,v) = \begin{cases} (1-\alpha)u^{-\alpha}, & u^\alpha > v^\beta, \\ (1-\beta)v^{-\beta}, & u^\alpha < v^\beta, \end{cases}$$

the mass of the singular component must be concentrated on the curve $u^{\alpha} = v^{\beta}$ in $\mathbf{I}^2$. Evaluating the double integral in (2.4.1) yields, for $u^{\alpha} < v^{\beta}$,

$$A_{\alpha,\beta}(u,v) = uv^{1-\beta} - \frac{\alpha\beta}{\alpha + \beta - \alpha\beta}\left(u^{\alpha}\right)^{(\alpha+\beta-\alpha\beta)/\alpha\beta},$$

and a similar result when $u^{\alpha} > v^{\beta}$. Thus we have

$$A_{\alpha,\beta}(u,v) = C_{\alpha,\beta}(u,v) - \frac{\alpha\beta}{\alpha + \beta - \alpha\beta}\left[\min\left(u^{\alpha},v^{\beta}\right)\right]^{(\alpha+\beta-\alpha\beta)/\alpha\beta},$$

and consequently

$$S_{\alpha,\beta}(u,v) = \frac{\alpha\beta}{\alpha + \beta - \alpha\beta}\left[\min\left(u^{\alpha},v^{\beta}\right)\right]^{\frac{\alpha+\beta-\alpha\beta}{\alpha\beta}} = \int_0^{\min\left(u^{\alpha},v^{\beta}\right)} t^{\frac{1}{\alpha}+\frac{1}{\beta}-2}\,dt.$$

Hence the $C_{\alpha,\beta}$-measure of the singular component of $C_{\alpha,\beta}$ is given by $S_{\alpha,\beta}(1,1) = \alpha\beta/(\alpha + \beta - \alpha\beta)$. In other words, if U and V are uniform [0,1] random variables whose joint distribution function is the copula $C_{\alpha,\beta}$, then $P[U^{\alpha} = V^{\beta}] = \alpha\beta/(\alpha + \beta - \alpha\beta)$.

   In Fig. 3.1 we have scatterplots for two simulations of Marshall-Olkin copulas, each using 500 pairs of points with the algorithm in Exercise 3.4. The one on the left is for $(\alpha,\beta) = (1/2,3/4)$, the one on the right is for $(\alpha,\beta) = (1/3,1/4)$. The singular component is clearly visible in each case.

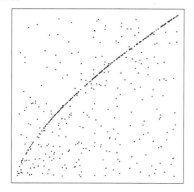 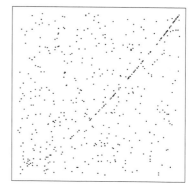

Fig. 3.1. Scatterplots for Marshall-Olkin copulas, $(\alpha,\beta) = (1/2,3/4)$, $(1/3,1/4)$

The Marshall-Olkin family of copulas, endowed with non-exponential margins, has been employed in a variety of applications. See (Hutchinson and Lai 1990) for details and references.

3.1.2 The Circular Uniform Distribution

Let (X,Y) denote the coordinates of a point chosen "at random" on the unit circle, where by "at random," we mean that if the polar coordinates of the point are $(1,\Theta)$, then the random variable Θ is uniformly distributed on the interval $[0,2\pi)$.

The joint distribution function $H(x,y)$ of X and Y may be determined by making use of the fact that $X = \cos\Theta$ and $Y = \sin\Theta$. However, it is more instructive to use a geometric approach. We will show that for points (x,y) within the square $[-1,1]^2$ (which contains the support of this distribution), H is given by

$$H(x,y) = \begin{cases} \dfrac{3}{4} - \dfrac{\arccos x + \arccos y}{2\pi}, & x^2 + y^2 \leq 1, \\[2mm] 1 - \dfrac{\arccos x + \arccos y}{\pi}, & x^2 + y^2 > 1,\ x,y \geq 0, \\[2mm] 1 - \dfrac{\arccos x}{\pi}, & x^2 + y^2 > 1,\ x < 0 \leq y, \\[2mm] 1 - \dfrac{\arccos y}{\pi}, & x^2 + y^2 > 1,\ y < 0 \leq x, \\[2mm] 0, & x^2 + y^2 > 1,\ x,y < 0. \end{cases} \qquad (3.1.4)$$

Of course, outside this square H will be equal to zero or to one of its margins, which are found below.

Suppose (x,y) is a point on or inside the unit circle. Then $2\pi H(x,y)$ is the arc length of that portion of the circle shown in white within the gray region in part (a) of Fig. 3.2. By using the symmetry of the circle and the arcs whose lengths are given by $\arccos x$ and $\arccos y$, we have $2\pi H(x,y) = 3\pi/2 - \arccos x - \arccos y$.

When (x,y) is outside the circle but in the first quadrant portion of $[-1,1]^2$, as shown in part (b) of Fig. 3.2, we have $2\pi H(x,y) = 2\pi - 2(\arccos x + \arccos y)$. The values of $H(x,y)$ for (x,y) in the other regions can be found similarly.

Using a derivation similar to the one above, the margins $F(x)$ and $G(y)$ are readily found—for x and y in $[-1,1]$ they are given by the functions in the third and fourth lines, respectively, in the displayed equation (3.1.4) for H.

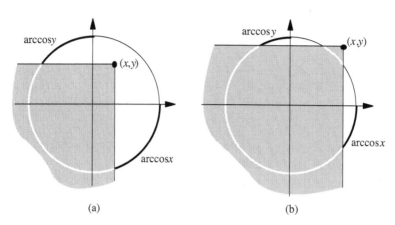

Fig. 3.2. Two cases for the joint distribution function $H(x,y)$ in (3.1.4)

Because it will now be easy to express $H(x,y)$ in terms of $F(x)$ and $G(y)$, and thus to find the copula, the only remaining task is to find the image of the circle $x^2 + y^2 = 1$ under the transformation $x = F^{(-1)}(u)$, $y = G^{(-1)}(v)$. Substitution yields $\sin^2[\pi(u-1/2)] = \cos^2[\pi(v-1/2)]$. But if α and β are in $[-\pi/2, \pi/2]$ such that $\sin^2 \alpha = \cos^2 \beta$, then $|\alpha| + |\beta| = \pi/2$. Hence the image of the unit circle, i.e., the support of the copula of X and Y, is $|u-1/2| + |v-1/2| = 1/2$. The graph of this set is the square whose vertices are the midpoints of the four sides of $\mathbf{I}^2$, as illustrated on part (a) of Fig. 3.3. Thus the copula of the coordinates X and Y of a point chosen at random on the unit circle is given by

$$C(u,v) = \begin{cases} M(u,v), & |u-v| > \dfrac{1}{2}, \\[2mm] W(u,v), & |u+v-1| > \dfrac{1}{2}, \\[2mm] \dfrac{u+v}{2} - \dfrac{1}{4}, & \text{otherwise.} \end{cases} \qquad (3.1.5)$$

In part (b) of Fig. 3.3, we have written the values of C in each of the five regions of $\mathbf{I}^2$.

Note that $\partial^2 C/\partial u \partial v = 0$ almost everywhere in $\mathbf{I}^2$, hence C is singular. It is easy to see that C is symmetric, and that it also satisfies the functional equation $C = \hat{C}$ for radial symmetry. Indeed, it satisfies the functional equations (2.8.1) in Exercise 2.30 for joint symmetry (when endowed with symmetric margins, as is the case for the circular uniform distribution).

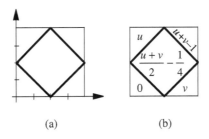

(a)   (b)

Fig. 3.3. The copula of the circular uniform distribution and its support

We conclude this section with two examples of bivariate singular distributions constructed from this copula.

Example 3.1. *A singular bivariate distribution whose margins are Cauchy distributions.* To obtain this joint distribution function, we endow the copula C from (3.1.5) with standard Cauchy margins: $F(x) = 1/2 + (\arctan x)/\pi$ for all real x; and similarly for $G(y)$. The expression for H is quite similar to (3.1.4). However, the support of H is the image of the square $|u-1/2|+|v-1/2|=1/2$ under the transformation $u = F(x)$, $v = G(y)$. This yields $|xy| = 1$, so that the support of this bivariate distribution consists of the four branches of the two rectangular hyperbolas $xy = 1$ and $xy = -1$. ∎

Example 3.2. *A singular bivariate distribution whose margins are normal.* This example is similar to the preceding one, but with $F = G = \Phi$, the standard normal distribution function. The support now lies on the four branches of the curve in the plane given by $|\Phi(x)-1/2|+|\Phi(y)-1/2|=1/2$, which is similar in appearance to $|xy| = 1$. Note that like the distributions in Example 2.10, we have a bivariate distribution with standard normal margins that is not a standard bivariate normal. Also note that this bivariate normal distribution does not possess a density. ∎

Exercises

3.1 Show that when either of the parameters α or β is equal to 0 or 1, the function $C_{\alpha,\beta}$ given in (3.1.3) is a copula.

3.2 Show that a version of the Marshall-Olkin bivariate distribution with Pareto margins (see Example 2.14) has joint survival functions given by

$$\overline{H}(x,y) = (1+x)^{-\theta_1}(1+y)^{-\theta_2}\left[\max(1+x,1+y)\right]^{-\theta_{12}},$$

for $x,y \geq 0$, where θ_1, θ_2, and θ_{12} are positive parameters.

3.3   Prove the following generalization of the Marshall-Olkin family (3.1.3) of copulas: Suppose that a and b are increasing functions defined on $\mathbf{I}$ such that $a(0) = b(0) = 0$ and $a(1) = b(1) = 1$. Further suppose that the functions $u \mapsto a(u)/u$ and $v \mapsto b(v)/v$ are both decreasing on $(0,1]$. Then the function C defined on $\mathbf{I}^2$ by

$$C(u,v) = \min\big(va(u), ub(v)\big)$$

is a copula (Marshall 1996). Note that (3.1.3) is the special case $a(u) = u^{1-\alpha}$ and $b(v) = v^{1-\beta}$. The symmetric case $(a = b)$ is studied in detail in (Durante 2005).

3.4   (a) Show that the following algorithm (Devroye 1987) generates random variates (x,y) from the Marshall-Olkin bivariate exponential distribution with parameters λ_1, λ_2, and λ_{12}:
   1. Generate three independent uniform $(0,1)$ variates r, s, t;
   2. Set $x = \min\left(\dfrac{-\ln r}{\lambda_1}, \dfrac{-\ln t}{\lambda_{12}}\right)$, $y = \min\left(\dfrac{-\ln s}{\lambda_2}, \dfrac{-\ln t}{\lambda_{12}}\right)$;
   3. The desired pair is (x,y).
   (b) Show that $u = \exp[-(\lambda_1 + \lambda_{12})x]$ and $v = \exp[-(\lambda_2 + \lambda_{12})y]$ are uniform $(0,1)$ variates whose joint distribution function is a Marshall-Olkin copula given by (3.1.3).

3.5   Let (X,Y) be random variables with the circular uniform distribution. Find the distribution of $\max(X,Y)$.

3.6   *Raftery's bivariate exponential distribution.* Raftery (1984, 1985) described the following bivariate distribution. Let Z_1, Z_2 and Z_3 be three mutually independent exponential random variables with parameter $\lambda > 0$, and let J be a Bernoulli random variable, independent of the Z's, with parameter θ in $(0,1)$. Set

$$X = (1-\theta)Z_1 + JZ_3,$$
$$Y = (1-\theta)Z_2 + JZ_3.$$

Show that
(a) for $x,y \geq 0$, the joint survival function of X and Y is given by

$$\overline{H}(x,y) = \exp[-\lambda(x \vee y)]$$
$$+ \frac{1-\theta}{1+\theta} \exp\left[\frac{-\lambda}{1-\theta}(x+y)\right]\left\{1 - \exp\left[\lambda \frac{1+\theta}{1-\theta}(x \wedge y)\right]\right\}$$

where $x \vee y = \max(x,y)$ and $x \wedge y = \min(x,y)$;

(b) X and Y are exponential with parameter λ;

(c) the survival copula of X and Y is given by

$$\hat{C}_\theta(u,v) = M(u,v) + \frac{1-\theta}{1+\theta}(uv)^{1/(1-\theta)}\left\{1 - [\max(u,v)]^{-(1+\theta)/(1-\theta)}\right\};$$

(d) $\hat{C}_\theta$ is absolutely continuous, $\hat{C}_0 = \Pi$, and $\hat{C}_1 = M$.

3.2 Geometric Methods

In the previous section, we illustrated how Sklar's theorem could be used to invert joint distribution functions to find copulas. In this section we will, in essence, return to the *definition* of a copula—as given in the paragraph following Definition 2.2.2—as our tool for the construction. That is, without reference to distribution functions or random variables,

we will construct grounded 2-increasing functions on $\mathbf{I}^2$ with uniform margins, utilizing some information of a geometric nature, such as a description of the support or the shape of the graphs of horizontal, vertical, or diagonal sections. We will also examine the "ordinal sum" construction, wherein the members of a set of copulas are scaled and translated in order to construct a new copula; the "shuffles of M," which are constructed from the Fréchet-Hoeffding upper bound; and the "convex sum" construction, a continuous analog of convex linear combinations.

3.2.1 Singular Copulas with Prescribed Support

In this section we illustrate, with three examples, the use of the definition of a copula to construct singular copulas whose support lies in a given set. In the first two examples, the support consists of line segments, and in the third, arcs of circles.

Example 3.3. Let θ be in $[0,1]$, and suppose that probability mass θ is uniformly distributed on the line segment joining $(0,0)$ to $(\theta,1)$, and probability mass $1-\theta$ is uniformly distributed on the line segment joining $(\theta,1)$ to $(1,0)$, as illustrated in part (a) of Fig. 3.4. Note the two limiting cases: when $\theta = 1$, the support is the main diagonal of $\mathbf{I}^2$, and the resulting copula is M; and when $\theta = 0$, the support is the secondary diagonal of $\mathbf{I}^2$, resulting in W. That is, if we let C_θ denote the copula with support as illustrated in part (a) of Fig. 3.4, then $C_1 = M$ and $C_0 = W$.

Using that fact that the support of C_θ lies on the two line segments, we can now find an expression for $C_\theta(u,v)$ by computing the C_θ-volume of appropriate rectangles in $\mathbf{I}^2$. Because the graph of the support divides $\mathbf{I}^2$ into three regions, we have three cases to consider, depending upon where in $\mathbf{I}^2$ the point (u,v) lies.

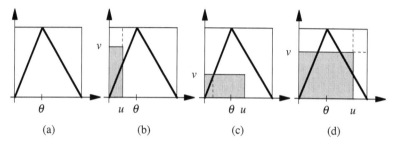

Fig. 3.4. Computing C_θ in Example 3.3

Suppose that $u \leq \theta v$, that is, (u,v) lies in the region above the first segment of the support, as illustrated in part (b) of Fig. 3.4. Then the C_θ-volume of the rectangle $[0,u]\times[0,v]$ is the same as the C_θ-volume of the rectangle $[0,u]\times[0,1]$—and $V_{C_\theta}([0,u]\times[0,1]) = u$ implies $C_\theta(u,v) = u$.

Now suppose $u > \theta v$ and $u < 1 - (1-\theta)v$, that is, (u,v) lies in the region below both segments of the support, as illustrated in part (c) of the figure. Then $C_\theta(u,v) = C_\theta(\theta v,v) = \theta v$, since the C_θ-volume of the rectangle $[\theta v,u]\times[0,v]$ is zero.

Finally, suppose that $u \geq 1 - (1-\theta)v$, so that (u,v) lies in the region above the second segment of the support, as illustrated in part (d) of the figure. We could proceed as we did in the first two cases, but here it will be advantageous to note that the C_θ-volume of any rectangle that does not intersect one of the line segments must be zero. Thus the C_θ-volume of the rectangle $[u,1]\times[v,1]$ is zero—and $V_{C_\theta}([u,1]\times[v,1]) = 0$ implies $C_\theta(u,v) = u + v - 1$. Hence we have

$$C_\theta(u,v) = \begin{cases} u, & 0 \le u \le \theta v \le \theta, \\ \theta v, & 0 \le \theta v < u < 1 - (1-\theta)v, \\ u+v-1, & \theta \le 1 - (1-\theta)v \le u \le 1. \end{cases} \qquad (3.2.1)$$

∎

In several earlier examples (2.10, 2.17, and 3.2), we saw how copulas can be used to construct "counterexamples," that is, examples to show that certain statements do not hold for all joint distributions. In Exercise 3.7, we will see how a member of the family of distributions in (3.2.1) can be similarly employed.

Example 3.4. Again let θ be in $[0,1]$, and suppose the probability mass is uniformly distributed on two line segments, one joining $(0,\theta)$ to $(\theta,0)$ (with mass θ), and the other joining $(\theta,1)$ to $(1,\theta)$ (with mass $1-\theta$), as illustrated in part (a) of Fig. 3.5. These copulas have an interesting probabilistic interpretation. Let $\oplus$ denote "addition mod 1," that is, $x \oplus y = x+y-\lfloor x+y \rfloor$, where $\lfloor t \rfloor$ denotes the integer part of t. If U and V are uniform $(0,1)$ random variables such that $U \oplus V = \theta$ with probability 1, then the support of the joint distribution of U and V lies on the line segments in part (a) of Fig. 3.5, and their copula C_θ is this joint distribution function. In passing we note the limiting cases: $C_0 = C_1 = W$.

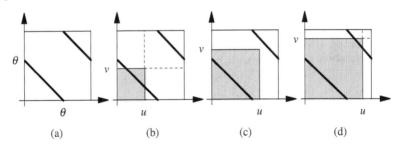

Fig. 3.5. Computing C_θ in Example 3.4

As in the preceding example, we will find an expression for $C_\theta(u,v)$ by considering the regions of $\mathbf{I}^2$ in which (u,v) may lie. If (u,v) is in the rectangle $[0,\theta] \times [\theta,1]$, then $V_{C_\theta}([0,u] \times [0,v]) = V_{C_\theta}([0,u] \times [0,1]) = u$, which yields $C_\theta(u,v) = u$. Similarly if (u,v) is in $[\theta,1] \times [0,\theta]$, then $C_\theta(u,v) = v$. Now suppose (u,v) is in $[0,\theta]^2$ but with $u + v \ge \theta$, as illustrated in part (b) of the figure. Because $V_{C_\theta}([u,1] \times [v,1]) = 1-\theta$, it follows that $C_\theta(u,v) = u+v-\theta$. If (u,v) is in $[\theta,1]^2$ with $u+v < 1+\theta$, as in part (c) if the figure, clearly $C_\theta(u,v) = \theta$. Finally, for (u,v) in $[\theta,1]^2$ with $u+v \geq 1+\theta$, as in part (d) of the figure, $V_{C_\theta}([u,1] \times [v,1]) = 0$ and hence $C_\theta(u,v) = u+v-1$. Thus C_θ is given by

$$C_\theta(u,v) = \begin{cases} \max(0,u+v-\theta), & (u,v) \in [0,\theta]^2, \\ \max(\theta,u+v-1), & (u,v) \in (\theta,1]^2, \\ M(u,v), & \text{otherwise.} \end{cases} \qquad (3.2.2) \quad \blacksquare$$

Example 3.5. Is it possible to find a copula C whose support consists of the two quarter circles shown in part (a) of Fig. 3.6? The upper quarter circle is given by $u^2 + v^2 = 2u$, and the lower one by $u^2 + v^2 = 2v$. Because the support is symmetric with respect to the diagonal, we will construct a symmetric copula, i.e., a C for which $C(u,v) = C(v,u)$. As in the earlier examples, it is easy to show that if (u,v) is in the region above the upper arc, then $C(u,v) = u$; and if (u,v) is in the region below the lower arc, then $C(u,v) = v$. Hence for $u^2 + v^2 > 2\min(u,v)$, we have $C(u,v) = M(u,v)$.

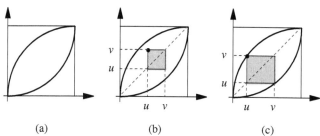

(a)   (b)   (c)

Fig. 3.6. Computing C_θ in Example 3.5

Now assume $u \leq v$, and that $u^2 + v^2 \leq 2u$, so that (u,v) lies above the main diagonal but below the upper arc, as shown in part (b) of the figure. Then $V_C([u,v] \times [u,v]) = 0$, which implies that $C(u,v) + C(v,u) = C(u,u) + C(v,v)$, or equivalently, $C(u,v) = (1/2)[\delta(u) + \delta(v)]$, where δ is the diagonal section of C. Now consider the situation when the point (u,v) lies on the upper quarter circle $u^2 + v^2 = 2u$. By continuity, $u = C(u,v) = (1/2)[\delta(u) + \delta(v)]$ so that $\delta(u) + \delta(v) = 2u = u^2 + v^2$. A solution to this equation is given by $\delta(u) = u^2$, which leads to $C(u,v) = \min(u,v,(u^2 + v^2)/2)$. Of course, it is necessary here to verify that C is 2-increasing on all of $\mathbf{I}^2$, which is readily done. $\blacksquare$

A word of caution is in order. The general problem of determining just what curves in $\mathbf{I}^2$ can serve as the support of a copula is a difficult one. For example, there is no copula whose support consists of the portions of the parabolas $v = u^2$ and $u = v^2$ in $\mathbf{I}^2$. See (Kamiński et al. 1987-1988; Sherwood and Taylor 1988) for a discussion of the general problem.

3.2.2 Ordinal Sums

The copula W for the Fréchet-Hoeffding lower bound is, of course, singular, and the support of W is the secondary diagonal of $\mathbf{I}^2$, the line segment with slope -1 connecting $(0,1)$ to $(1,0)$ Now recall Example 3.4 in the preceding section—in which the support (see part (a) of either Fig. 3.5 or 3.7) of the copula consisted of two line segments each with slope -1. One can view this support as consisting of two copies of the support of W, scaled to fit the subsquares $[0,\theta]^2$ and $[\theta,1]^2$. This illustrates the idea behind the ordinal sum construction.

Let $\{J_i\}$ denote a *partition* of $\mathbf{I}$, that is, a (possibly infinite) collection of closed, non-overlapping (except at common endpoints) nondegenerate intervals $J_i = [a_i,b_i]$ whose union is $\mathbf{I}$. Let $\{C_i\}$ be a collection of copulas with the same indexing as $\{J_i\}$. Then the *ordinal sum of* $\{C_i\}$ *with respect to* $\{J_i\}$ is the copula C given by

$$C(u,v) = \begin{cases} a_i + (b_i - a_i)C_i\left(\dfrac{u-a_i}{b_i-a_i}, \dfrac{v-a_i}{b_i-a_i}\right), & (u,v) \in J_i^2, \\ M(u,v), & \text{otherwise.} \end{cases}$$

To obtain the graph of the support of an ordinal sum, "paste" onto $\mathbf{I}^2$ appropriately scaled copies of the copulas C_i over the squares J_i^2, as illustrated in part (b) of Fig. 3.7. Note that because the support of an ordinal sum is contained in the shaded portion of the square, an ordinal sum must agree with M on the unshaded portion.

Example 3.6. The ordinal sum of $\{W,W\}$ with respect to $\{[0,\theta],[\theta,1]\}$ is the copula C_θ from Example 3.4. Note that for (u,v) in $[0,\theta]^2$,

$$\theta W\left(\frac{u}{\theta}, \frac{v}{\theta}\right) = \max(0, u+v-\theta),$$

and for (u,v) in $[\theta,1]^2$,

$$\theta + (1-\theta)W\left(\frac{u-\theta}{1-\theta}, \frac{v-\theta}{1-\theta}\right) = \max(\theta, u+v-1). \qquad \blacksquare$$

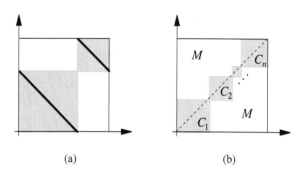

(a)   (b)

Fig. 3.7. Ordinal sums

The following theorem characterizes copulas that have an ordinal
sum representation.

Theorem 3.2.1. *Let C be a copula. Then C is an ordinal sum if and
only if there exists a t in (0,1) such that $C(t,t) = t$.*

Proof. Assume there exists a t in (0,1) such that $C(t,t) = t$. Let C_1 and
C_2 be the functions defined by $C_1(u,v) = C(tu,tv)/t$ and $C_2(u,v) = \left[C(t+(1-t)u,t+(1-t)v)-t\right]/(1-t)$ for u,v in **I**. It is easy to verify that
C_1 and C_2 are copulas, and that C is the ordinal sum of $\{C_1,C_2\}$ with
respect to $\{[0,t],[t,1]\}$. The converse is trivial.   □

If U and V are uniform (0,1) random variables whose joint distribution function is the copula C, then the following conditions are equivalent to the condition $C(t,t) = t$ for some t in (0,1) in the above theorem:
1. $P[\max(U,V) \le t] = t$ for some t in (0,1);
2. $P[\min(U,V) \le t] = t$ for some t in (0,1);
3. $P[\max(U,V) \le t] = P[\min(U,V) \le t]$ for some t in (0,1);
4. $P[(U-t)(V-t) \ge 0] = 1$ for some t in (0,1).

Exercises

3.7   Let U and V be uniform (0,1) random variables whose joint distribution function is the copula C_θ from (3.2.1) in Example 3.3
with $\theta = 1/2$; i.e.,

$$C_{1/2}(u,v) = \begin{cases} u, & 0 \le u \le v/2 \le 1/2, \\ v/2, & 0 \le v/2 < u < 1-v/2, \\ u+v-1, & 1/2 \le 1-v/2 \le u \le 1. \end{cases}$$

(a) Show that $P[V = 1-|2U-1|] = 1$ and $\text{Cov}(U,V) = 0$, so that two random variables can be uncorrelated although one can be predicted perfectly from the other.

(b) Show that $C_{1/2}$ is not symmetric, so that two random variables can be identically distributed and uncorrelated but not exchangeable.

(c) Show that $P[V-U > 0] = 2/3$, so that two random variables can be identically distributed, however their difference need not be symmetric about zero.

(d) Let $X = 2U-1$ and $Y = 2V-1$, so that X and Y are uniform on $(-1,1)$. Show that $P[X+Y > 0] = 2/3$, so that two random variables can each be symmetric about zero, but their sum need not be.

3.8 Let (α,β) be a point in $\mathbf{I}^2$ such that $\alpha > 0$, $\beta > 0$, and $\alpha + \beta < 1$. Suppose that probability mass α is uniformly distributed on the line segment joining (α,β) to $(0,1)$, that probability mass β is uniformly distributed on the line segment joining (α,β) to $(1,0)$, and that probability mass $1-\alpha-\beta$ is uniformly distributed on the line segment joining (α,β) to $(1,1)$, as shown in part (a) of Fig. 3.8. Show that the copula with this support is given by

$$C_{\alpha,\beta}(u,v) = \begin{cases} u - \dfrac{\alpha}{1-\beta}(1-v), & (u,v) \in \Delta_1, \\ v - \dfrac{\beta}{1-\alpha}(1-u), & (u,v) \in \Delta_2, \\ 0, & \text{otherwise,} \end{cases}$$

where Δ_1 is the triangle with vertices (α,β), $(0,1)$, and $(1,1)$; and Δ_2 is the triangle with vertices (α,β), $(1,0)$, and $(1,1)$. Note the limiting cases: $C_{0,0} = M$; $C_{\alpha,1-\alpha} = W$, and when $\alpha = 0$ or $\beta = 0$ we have one-parameter families of copulas similar to the one in Example 3.3.

3.9 Let U and V be uniform $(0,1)$ random variables such that $V = U \oplus \theta$ with probability 1, where θ is a constant in $(0,1)$. Show that if C_θ denotes the copula of U and V, then
(a) the support of C_θ is the set illustrated in part (b) of Fig. 3.8; and
(b) C_θ is given by

$$C_\theta(u,v) = \begin{cases} \min(u,v-\theta), & (u,v) \in [0,1-\theta] \times [\theta,1], \\ \min(u+\theta-1,v), & (u,v) \in [1-\theta,1] \times [0,\theta], \\ W(u,v), & \text{otherwise.} \end{cases}$$

For a related problem that leads to the same copula, see (Marshall 1989).

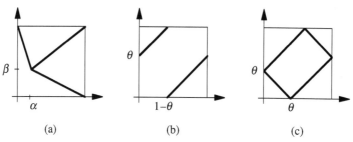

(a)   (b)   (c)

Fig. 3.8. The supports of the copulas in Exercises 3.8, 3.9, and 3.10

3.10 Let θ be in **I**, and let U and V be random variables whose probability mass is uniformly distributed on the boundary of the rectangle in $\mathbf{I}^2$ with vertices $(\theta,0)$, $(0,\theta)$, $(1-\theta,1)$, and $(1,1-\theta)$, as illustrated in part (c) of Fig. 3.8 (Ferguson 1995).
(a) Show that the copula C_θ of U and V is given by

$$C_\theta(u,v) = \begin{cases} M(u,v), & |v-u| \ge \theta, \\ W(u,v), & |u+v-1| \ge 1-\theta, \\ (u+v-\theta)/2, & \text{otherwise.} \end{cases}$$

(b) Show that C_θ satisfies the functional equation $C = \hat{C}$ associated with radial symmetry.
(c) Show that this family is negatively ordered.
Note that $C_0 = M$, $C_1 = W$, and $C_{1/2}$ is the copula of random variables with the circular uniform distribution (see Sect. 3.1.2).

3.11. (a) Show that a singular copula can have support that is symmetric with respect to the diagonal $v = u$ in $\mathbf{I}^2$ yet fail to be a symmetric copula, i.e., $C(u,v) \neq C(v,u)$. [Hint: Let C_θ be a member of the family in Exercise 3.9, and consider $C = (1/3)C_{1/3} + (2/3)C_{2/3}$.]
(b) Show that a singular copula can have support that is radially symmetric with respect to the center of $\mathbf{I}^2$ (i.e., if (u,v) is in the support, so is $(1-u,1-v)$), yet fail to satisfy the functional equation $C = \hat{C}$ in (2.7.3) associated with radial symmetry.

3.12 *Vaswani's Bivariate Normal Distribution.* Vaswani (1947) described the following bivariate distribution: Let Φ denote the standard normal distribution function, let T be a uniform $(0,1)$ random variable, and set

$$X = \Phi^{-1}(T),$$

$$Y = \begin{cases} -\Phi^{-1}(T+1/2), & \text{if } T \in [0,1/2), \\ -\Phi^{-1}(T-1/2), & \text{if } T \in (1/2,1]. \end{cases}$$

(a) Show that X and Y are standard normal random variables.

(b) Show that $\Phi(X) \oplus \Phi(Y) = 1/2$ with probability 1.

(c) Show that the copula of X and Y is given by (3.2.2) with $\theta = 1/2$, or equivalently, the ordinal sum of $\{W,W\}$ with respect to $\{[0,1/2],[1/2,1]\}$.

3.13 Kimeldorf and Sampson (1975a) described the following absolutely continuous joint distribution: Let θ be in $[0,1)$, and let X and Y be random variables whose joint density function h_θ is given by

$$h_\theta(x,y) = \begin{cases} \beta, & (x,y) \in \bigcup_{i=1}^{\lfloor \beta \rfloor} J_i^2, \\ \dfrac{\beta}{\beta - \lfloor \beta \rfloor}, & (x,y) \in \left(\lfloor \beta \rfloor / \beta, 1\right]^2, \\ 0, & \text{otherwise,} \end{cases}$$

where $\beta = (1+\theta)/(1-\theta)$, $\lfloor \beta \rfloor$ represents the integer part of β, and $J_i = [(i-1)/\beta, i/\beta]$. Show that
(a) X and Y are uniform $(0,1)$ random variables, and that
(b) the joint distribution function of X and Y is the ordinal sum of $\{\Pi, \Pi, \cdots, \Pi\}$ with respect to $\{J_i\}_{i=1}^{\lfloor \beta \rfloor} \cup \{[\lfloor \beta \rfloor / \beta, 1]\}$.

3.2.3 Shuffles of *M*

The copulas in Example 3.4 and Exercise 3.9 have similar support—each is a collection of line segments with slope $+1$ or -1 (see Figs. 3.5(a) and 3.8(b)). Among the copulas whose supports share this property are the "shuffles of M." The support of a shuffle of M can be described informally as follows (Mikusiński et al. 1992):

The mass distribution for a shuffle of M can be obtained by (1) placing the mass for M on $\mathbf{I}^2$, (2) cutting $\mathbf{I}^2$ vertically into a finite number of strips, (3) shuffling the strips with perhaps some of them flipped around their vertical axes of symmetry, and then (4) reassembling them to form the square again. The resulting mass distribution will correspond to a copula called a *shuffle of M*.

Formally, a shuffle of M is determined by a positive integer n, a finite partition $\{J_i\} = \{J_1, J_2, \cdots, J_n\}$ of $\mathbf{I}$ into n closed subintervals, a permutation π on $S_n = \{1, 2, \cdots, n\}$, and a function $\omega\colon S_n \to \{-1, 1\}$ where $\omega(i)$ is -1 or 1 according to whether or not the strip $J_i \times \mathbf{I}$ is flipped. We denote permutations by the vector of images $(\pi(1), \pi(2), \cdots, \pi(n))$. The resulting shuffle of M may then be unambiguously denoted by $M(n, \{J_i\}, \pi, \omega)$, where n is the number of connected components in its support. We will assume that all shuffles of M are given in this form. A shuffle of M with $\omega \equiv 1$, i.e., for which none of the strips are flipped, is a *straight shuffle*, and a shuffle of M with $\omega \equiv -1$ is called a *flipped shuffle*. We will also write $\mathbf{I}_n$ for $\{J_i\}$ when it is a *regular* partition of $\mathbf{I}$, i.e., when the width of each subinterval J_i is $1/n$.

Example 3.7. The copula from Exercise 3.9 is the straight shuffle given by $M(2, \{[0, 1-\theta], [1-\theta, 1]\}, (2, 1), 1)$, and the copula from Example 3.4 is the flipped shuffle of M given by $M(2, \{[0, \theta], [\theta, 1]\}, (1, 2), -1)$. ∎

Although shuffles of M are rather simple objects, they can be surprisingly useful. As an example, consider the following question: What is the "opposite" of independence, for random variables? One possible answer is that X and Y should be as "dependent" as possible. We say that X and Y are *mutually completely dependent* (Lancaster 1963) if there exists a one-to-one function φ such that $P[Y = \varphi(X)] = 1$, i.e., X and Y are almost surely invertible functions of one another. As noted in (Kimeldorf and Sampson 1978), mutual complete dependence implies complete predictability of either random variable from the other; whereas independence implies complete unpredictability.

Now suppose that the copula of X and Y is a shuffle of M. Then, because the support of any shuffle is the graph of a one-to-one function, it follows that X and Y are mutually completely dependent (the converse, however, is not true—there are mutually completely dependent random variables with more complex copulas, see Exercise 3.16). But, as we will now show (using shuffles of M), there are mutually completely dependent random variables whose joint distribution functions are arbitrarily close to the joint distribution function of independent random variables with the same marginals. As noted in (Mikusiński et al. 1991), this implies that in practice, the behavior of any pair of independent continuous random variables can be approximated so closely by a pair of mutually completely dependent continuous random variables that it would be impossible, experimentally, to distinguish one pair from the other. In view of Sklar's theorem, we need only prove that that the product copula Π can be approximated arbitrarily closely by shuffles of M. The proof of the following theorem is adapted from those in (Kimeldorf and Sampson 1978; Mikusiński et al. 1991).

Theorem 3.2.2. *For any $\varepsilon > 0$, there exists a shuffle of M, which we denote C_ε, such that*

$$\sup_{u,v \in \mathbf{I}} \left| C_\varepsilon(u,v) - \Pi(u,v) \right| < \varepsilon.$$

Proof: Let m be an integer such that $m \geq 4/\varepsilon$. Then as a consequence of Theorem 2.2.4, for any copula C and for u, v, s, t in $\mathbf{I}$,

$$\left| C(u,v) - C(s,t) \right| < \frac{\varepsilon}{2} \text{ whenever } \left| u - s \right| < \frac{1}{m} \text{ and } \left| v - t \right| < \frac{1}{m}.$$

The integer m now determines C_ε, a shuffle of M, in the following manner: Let $n = m^2$, and let $\{J_i\}$ be the regular partition $\mathbf{I}_n$ of $\mathbf{I}$ into n subintervals of equal width. Let π be the permutation of S_n given by $\pi(m(j-1)+k) = m(k-1) + j$ for $k,j = 1,2,\cdots,m$. Let ω be arbitrary, and set $C_\varepsilon = M(n, \{J_i\}, \pi, \omega)$. The effect of this permutation is to redistribute the probability mass of M so that there is mass $1/n$ in each of the n subsquares of $\mathbf{I}^2$. Figure 3.9 illustrates such a shuffle of M when $m = 3$ (i.e., $n = 9$) in which π is the permutation $(1,4,7,2,5,8,3,6,9)$.

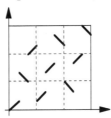

Fig. 3.9. The example of a shuffle of M in the proof of Theorem 3.2.2

Because $V_{C_\varepsilon}([0,p/m] \times [0,q/m]) = V_\Pi([0,p/m] \times [0,q/m]) = pq/n$ for $p,q = 0,1, \cdots,m$, it follows that $C_\varepsilon(p/m,q/m) = \Pi(p/m,q/m)$ for $p,q = 0,1,\cdots,m$.

Now let (u,v) be a point in $\mathbf{I}^2$. Then there exists a pair (p,q) of integers with $p,q \in \{0,1, \cdots ,m\}$ such that $|u - p/m| < 1/m$ and $|v - q/m| < 1/m$. Hence

$$
\begin{aligned}
|C_\varepsilon(u,v) - \Pi(u,v)| &\leq |C_\varepsilon(u,v) - C_\varepsilon(p/m,q/m)| \\
&\quad + |C_\varepsilon(p/m,q/m) - \Pi(p/m,q/m)| \\
&\quad + |\Pi(p/m,q/m) - \Pi(u,v)| \\
&< \frac{\varepsilon}{2} + 0 + \frac{\varepsilon}{2} = \varepsilon,
\end{aligned}
$$

which completes the proof.   □

The preceding theorem can be greatly generalized—the copula Π can be replaced by *any copula whatsoever*. That is to say, if C is any copula, then C can be approximated arbitrarily closely—uniformly—by certain shuffles of M. The proof uses the same permutations, but with partitions $\{J_i\}$ of $\mathbf{I}$ in which the widths of the subintervals are determined by the C-volumes of n $(= m^2)$ nonoverlapping subrectangles of dimension $(1/m) \times (1/m)$ in $\mathbf{I}^2$. To be precise, the width of J_i is $V_C([(k-1)/m,k/m] \times [(j-1)/m,j/m])$, where $i = m(j-1) + k$, $k,j = 1,2,\cdots,m$. See (Mikusiński et al. 1991) for details.

As a consequence, the shuffles of M are dense in the set of all copulas endowed with the sup norm (as in Theorem 3.2.2). Thus we have the phenomenon, alluded to earlier, in which the limit of a sequence of random variables is independence but at each step in the limiting process, each component of a pair of random variables in the sequence is almost surely an invertible function of the other. For a discussion and references, see (Vitale 1990).

When we possess information about the values of a copula at points in the interior of $\mathbf{I}^2$, the Fréchet-Hoeffding bounds (2.2.4) can often be narrowed. In the following theorem, we show that if the value of a copula is specified at a single interior point of $\mathbf{I}^2$, then the bounds can be narrowed to certain shuffles of M.

In the proof of the following theorem, we let x^+ denote the *positive part* of x, i.e., $x^+ = \max(x,0)$.

Theorem 3.2.3. *Let C be a copula, and suppose $C(a,b) = \theta$, where (a,b) is in $(0,1)^2$ and θ satisfies $\max(a + b - 1,0) \leq \theta \leq \min(a,b)$. Then*

$$
C_L(u,v) \leq C(u,v) \leq C_U(u,v), \tag{3.2.3}
$$

where C_U and C_L are the copulas given by

$$C_U = M(4,\{[0,\theta],[\theta,a],[a,a+b-\theta],[a+b-\theta,1]\},(1,3,2,4),1)$$

and

$$C_L = M(4,\{[0,a-\theta],[a-\theta,a],[a,1-b+\theta],[1-b+\theta,1]\},(4,2,3,1),-1).$$

Because $C_L(a,b) = C_U(a,b) = \theta$, the bounds are best-possible.

Proof. Although the shuffle notation is useful to see the geometric structure of C_U and C_L, the positive part notation will be more useful in the proof. The shuffles C_U and C_L are given explicitly by

$$C_U(u,v) = \min\left(u,v,\theta+(u-a)^+ +(v-b)^+\right)$$

and

$$C_L(u,v) = \max\left(0,u+v-1,\theta-(a-u)^+ -(b-v)^+\right),$$

and the supports of C_U and C_L are illustrated in Fig. 3.10 (the solid line segments with slope ± 1) for the case $(a,b) = (0.6,0.3)$ and $\theta = 0.2$.

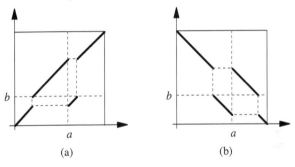

(a)   (b)

Fig. 3.10. The supports of (a) C_U and (b) C_L for a copula C with $C(a,b) = \theta$.

If $u \geq a$, then $0 \leq C(u,v) - C(a,v) \leq u - a$; and if $u < a$, then $0 \leq C(a,v) - C(u,v) \leq a - u$; from which it follows that

$$-(a-u)^+ \leq C(u,v)-C(a,v) \leq (u-a)^+.$$

Similarly, $-(b-v)^+ \leq C(a,v)-C(a,b) \leq (v-b)^+$, and adding yields

$$-(a-u)^+ -(b-v)^+ \leq C(u,v)-C(a,b) \leq (u-a)^+ +(v-b)^+.$$

Because $C(a,b) = \theta$, we have

$$\theta-(a-u)^+ -(b-v)^+ \leq C(u,v) \leq \theta+(u-a)^+ +(v-b)^+.$$

Incorporating the Fréchet-Hoeffding bounds yields (3.2.3). Noting that shuffles are copulas and that $C_U(a,b) = C_L(a,b) = \theta$ completes the proof.   □

For an application of shuffles of M to the joint distribution of (U,V) when $U+V$ and $U-V$ are independent, see (Dall'Aglio 1997a,b).

We close this section with the observation that copulas other than M can be "shuffled," as illustrated in the following example from (Mikusiński et al. 1991).

Example 3.8. Let C be an arbitrary copula, let $n = 2$, $\omega \equiv 1$, let π be given by $(2,1)$ and, for θ in $(0,1)$, consider the partition $\{[0,\theta],[\theta,1]\}$. Equivalently, let the mass distribution for C be sliced vertically at $u = \theta$, and the resulting two strips interchanged. Let C_θ denote this straight "shuffle of C," which we could also denote as $C(2,\{[0,\theta],[\theta,1]\},(2,1),1)$. If U and V are random variables whose joint distribution function is C, then C_θ is the joint distribution function of the pair $(U \oplus \theta, V)$ [cf. Exercise 3.9]. Explicitly, we have

$$
\begin{aligned}
C_\theta(u,v) &= P[U \oplus \theta \leq u, V \leq v] \\
&= \begin{cases} P[U \in (1-\theta, 1-\theta+u], V \leq v], & u \leq \theta, \\ P[U \in (0, u-\theta] \cup (1-\theta, 1), V \leq v], & u > \theta, \end{cases} \\
&= \begin{cases} C(1-\theta+u, v) - C(1-\theta, v), & u \leq \theta, \\ v - C(1-\theta, v) + C(u-\theta, v), & u > \theta. \end{cases}
\end{aligned}
$$ ■

3.2.4 Convex Sums

In Exercise 2.3, it was shown that if $\{C_\theta\}$ is a finite collection of copulas, then any convex linear combination of the copulas in $\{C_\theta\}$ is also a copula. Convex sums are the extension of this idea to infinite collections of copulas indexed by a continuous parameter θ.

We now consider the parameter θ as an observation of a continuous random variable Θ with distribution function Λ. If we set

$$C'(u,v) = \int_{\mathbf{R}} C_\theta(u,v)\, d\Lambda(\theta), \tag{3.2.4}$$

then it is easy to show (Exercise 3.17) that C' is a copula, which we call the *convex sum of* $\{C_\theta\}$ *with respect to* Λ. In this context, Λ is often referred to as the *mixing distribution* of the family $\{C_\theta\}$. When the distribution function of Θ has a parameter, say α, then we have

$$C'_\alpha(u,v) = \int_{\mathbf{R}} C_\theta(u,v)\, d\Lambda_\alpha(\theta). \tag{3.2.5}$$

Example 3.9. Let $\{C_\theta\}$ be the family of copulas from Exercise 3.10, i.e., let

$$C_\theta(u,v) = \begin{cases} M(u,v), & |v-u| \geq \theta, \\ W(u,v), & |u+v-1| \geq 1-\theta, \\ (u+v-\theta)/2, & \text{elsewhere,} \end{cases}$$

for θ in **I**; and let the mixing distribution Λ_α be given by $\Lambda_\alpha(\theta) = \theta^\alpha$, where $\alpha > 0$. Using (3.2.5) and evaluating the integral (over **I**) yields the family $\{C'_\alpha\}$ of convex sums of $\{C_\theta\}$ given by

$$C'_\alpha(u,v) = W(u,v) + \frac{1}{2(\alpha+1)}\left\{\left[1 - |u+v-1|\right]^{\alpha+1} - |u-v|^{\alpha+1}\right\}.$$

Note that $C'_1 = \Pi$, and the limits $C'_0 = M$, $C'_\infty = W$. It is elementary but tedious to show that each C'_α is absolutely continuous as well, so that with this mixing distribution, the convex sums of a family $\{C_\theta\}$ of singular copulas form a comprehensive family $\{C'_\alpha\}$ of absolutely continuous copulas. For convex sums of $\{C_\theta\}$ with other mixing distributions, see (Ferguson 1995). ∎

Example 3.10. Let C be an arbitrary copula, let θ be in $(0,1)$, and let $\{C_\theta\}$ be the family of "shuffles of C" from Example 3.8, i.e.

$$C_\theta(u,v) = \begin{cases} C(1-\theta+u,v) - C(1-\theta,v), & u \leq \theta, \\ v - C(1-\theta,v) + C(u-\theta,v), & u > \theta. \end{cases}$$

If the mixing distribution is uniform on $(0,1)$, i.e., $\Lambda = U_{01}$, then elementary integration in (3.2.4) yields

$$\int_0^1 C_\theta(u,v)\, d\theta = uv = \Pi(u,v).$$

Mikusiński et al. (1991) describe this result as follows:

> Visually one can see this by imagining that the unit square endowed with the mass distribution for C is wrapped around a circular cylinder so that the left and right edges meet. If one spins the cylinder at a constant rate one will see the uniform distribution associated with Π. ∎

We conclude this section with the following example from (Marshall and Olkin 1988; Joe 1993), in which convex sums lead to copulas constructed from Laplace transforms of distribution functions.

Example 3.11. The representation (3.2.4) can be extended by replacing C_θ by more general bivariate distribution functions. For example, set

$$H(u,v) = \int_0^\infty F^\theta(u)G^\theta(v)\,d\Lambda(\theta), \qquad (3.2.6)$$

that is, let H be a convex sum (or mixture) of powers of distribution functions F and G (which are *not* the margins of H, indeed, at this point H may not even be a joint distribution function), and assume $\Lambda(0) = 0$. Let $\psi(t)$ denote the Laplace transform of the mixing distribution Λ, i.e., let

$$\psi(t) = \int_0^\infty e^{-\theta t}\,d\Lambda(\theta).$$

Note that $\psi(-t)$ is the moment generating function of Λ. Now let F and G be the distribution functions given by $F(u) = \exp[-\psi^{-1}(u)]$ and $G(v) = \exp[-\psi^{-1}(v)]$ for u,v in $\mathbf{I}$, then (3.2.6) becomes

$$H(u,v) = \int_0^\infty \exp\!\left[-\theta\!\left(\psi^{-1}(u)+\psi^{-1}(v)\right)\right] d\Lambda(\theta)$$
$$= \psi\!\left(\psi^{-1}(u)+\psi^{-1}(v)\right),$$

which, as Marshall and Olkin (1988) show, is a bivariate distribution function. Furthermore, because $\psi^{-1}(1) = 0$, its margins are uniform, whence H is a copula—that is, when ψ is the Laplace transform of a distribution function, then the function C defined on $\mathbf{I}^2$ by

$$C(u,v) = \psi\!\left(\psi^{-1}(u)+\psi^{-1}(v)\right) \qquad (3.2.7)$$

is a copula. However, the right side of (3.2.7) is a copula for a broader class of functions than Laplace transforms—these copulas are called *Archimedean*, and are the subject of the next chapter. ∎

Exercises

3.14 (a) Show that the graph of every shuffle of M is piecewise planar.
(b) Show that every copula whose graph is piecewise planar is singular.
(c) Show that the converses of (a) and (b) are false.

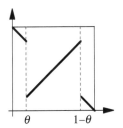

Fig. 3.11. The support of the shuffle of M in Exercise 3.15

3.15 *Mutually completely dependent uncorrelated random variables.*
Let θ be in $[0,1/2]$, and let C_θ be the shuffle of M given by
$M(3,\{[0,\theta],[\theta,1-\theta],[1-\theta,1]\},(3,2,1),\omega)$ where $\omega(1) = \omega(3) = -1$,
$\omega(2) = +1$. (See Fig. 3.11.)
(a) Show that C_θ is also given by

$$C_\theta(u,v) = \begin{cases} M(u,v) - \theta, & (u,v) \in [\theta, 1-\theta]^2, \\ W(u,v), & \text{otherwise.} \end{cases}$$

(b) Show that if C_θ is the joint distribution function of U and V
with $\theta = (2 - \sqrt[3]{4})/4 \cong 0.103$, then $\mathrm{Cov}(U,V) = 0$, that is, U and V
are mutually completely dependent uncorrelated uniform $(0,1)$
random variables.
(c) Let Φ denote the standard normal distribution function. Show
that if $C_\theta(\Phi(x),\Phi(y))$ is the joint distribution function of X and Y
with θ near $\Phi(-1.538) \cong 0.062$, then $\mathrm{Cov}(X,Y) = 0$, that is, X and Y
are mutually completely dependent uncorrelated standard normal
random variables (Melnick and Tennenbein 1982). Also note that
X and Y are exchangeable, and that $X+Y$ is not normal, as
$P[X+Y=0] = 2\theta$.

3.16 Let X be a standard normal random variable, and define Y by

$$Y = \begin{cases} X, & \text{if } \lfloor X \rfloor \text{ is even}, \\ -X, & \text{if } \lfloor X \rfloor \text{ is odd}. \end{cases}$$

Show that X and Y are mutually completely dependent but that the
copula of X and Y is not a shuffle of M.

3.17 Show that the function C' given by (3.2.4) is a copula.

3.18 Let $\{C_t\}$ be the family of copulas from Exercise 2.10, i.e., for t in **I** let

$$C_t(u,v) = \begin{cases} \max(u+v-1,t), & (u,v) \in [t,1]^2, \\ \min(u,v), & \text{otherwise;} \end{cases}$$

and let the mixing distribution Λ_α be given by $\Lambda_\alpha(t) = t^\alpha$, where $\alpha > 0$. Show that the convex sum C'_α of $\{C_t\}$ is given by

$$C'_\alpha(u,v) = M(u,v) - \frac{1}{\alpha+1}\left\{[M(u,v)]^{\alpha+1} - [W(u,v)]^{\alpha+1}\right\}.$$

Show that $C'_0 = W$ and $C'_\infty = M$.

3.19 Show that when $\psi(s) = (1+s)^{-1/\theta}$, i.e., when ψ is the Laplace transform of a gamma distribution with parameters $\alpha = 1/\theta, \beta = 1$; then the construction (3.2.7) generates the survival copulas for the bivariate Pareto distribution in Example 2.14 (Joe 1993).

3.20 Show that the copulas C'_α in Example 3.9 satisfy the functional equation $C = \hat{C}$ associated with radial symmetry.

3.2.5 Copulas with Prescribed Horizontal or Vertical Sections

Just how "simple" can the expression for a copula be? For example, the product copula $\Pi(u,v) = uv$ is linear in both u and v—are there other copulas which are linear in at least one variable? Are there "simple" copulas given by low degree polynomials in u or v? These questions lead us to a study of the sections (recall Definition 2.2.5) of a copula, i.e., the functions $u \mapsto C(u,v)$ and $v \mapsto C(u,v)$. These sections have several statistical interpretations—one of which is the following. When U and V are uniform $(0,1)$ random variables with a joint distribution function C, the sections are proportional to conditional distribution functions. For example, for u_0 in $(0,1)$,

$$P[V \leq v | U \leq u_0] = P[U \leq u_0, V \leq v]/P[U \leq u_0] = C(u_0,v)/u_0. \quad (3.2.8)$$

Furthermore, several of the dependence concepts for random variables that we will encounter in Chapter 5 have geometric interpretations in terms of sections of their copula.

Copulas with Linear Sections

We begin by looking for copulas that are linear in one variable — say u, that is, copulas of the form $C(u,v) = a(v)u + b(v)$ for all (u,v) in $\mathbf{I}^2$. The functions a and b are readily found from the boundary conditions (2.2.2a) and (2.2.2b) in Definition 2.2.2. Thus

$$0 = C(0,v) = b(v) \text{ and } v = C(1,v) = a(v),$$

whence there is only one copula with linear vertical (or horizontal) sections, namely Π.

Copulas with Quadratic Sections

Are there copulas with quadratic sections in, say, u? If so, then C will be given by $C(u,v) = a(v)u^2 + b(v)u + c(v)$ for appropriate functions a, b, and c. Again employing the boundary conditions, we obtain

$$0 = C(0,v) = c(v) \text{ and } v = C(1,v) = a(v) + b(v).$$

If we let $a(v) = -\psi(v)$, then $b(v) = v - a(v) = v + \psi(v)$, and we have

$$C(u,v) = uv + \psi(v)u(1-u) \tag{3.2.9}$$

where ψ is a function such that C is 2-increasing and $\psi(0) = \psi(1) = 0$ (so that $C(u,0) = 0$ and $C(u,1) = u$).

Example 3.12. *The Farlie-Gumbel-Morgenstern family of copulas.* Suppose that C is symmetric and has quadratic sections in u. Then C satisfies (3.2.9) and $C(u,v) = uv + \psi(u)v(1-v)$. Consequently, $\psi(v) = \theta v(1-v)$ for some parameter θ, so that

$$C_\theta(u,v) = uv + \theta uv(1-u)(1-v). \tag{3.2.10}$$

The C_θ-volume of a rectangle $[u_1,u_2] \times [v_1,v_2]$ is given, after some simplification, by

$$V_{C_\theta}\big([u_1,u_2] \times [v_1,v_2]\big) = (u_2 - u_1)(v_2 - v_1)[1 + \theta(1 - u_1 - u_2)(1 - v_1 - v_2)].$$

Because $(1 - u_1 - u_2)(1 - v_1 - v_2)$ is in $[-1,1]$ for all u_1, u_2, v_1, v_2 in $\mathbf{I}$, it follows that C_θ is 2-increasing, and hence a copula, if and only if θ is in $[-1,1]$.

This family is known as the *Farlie-Gumbel-Morgenstern family* (often abbreviated "FGM") and contains as members all copulas with quadratic sections in both u and v. The family was discussed by Morgenstern (1956), Gumbel (1958), and Farlie (1960); however, it seems that the earliest publication with the basic functional form (3.2.10) is Eyraud (1938). Additional properties of the FGM family are explored in Exercises 3.21 and 3.22. Primarily because of their simple analytical form, FGM distributions have been widely used in modeling, for tests of association, and in studying the efficiency of nonparametric procedures. For extensive lists of applications and references, see (Conway 1983; Hutchinson and Lai 1990). ∎

However, FGM copulas can only model relatively weak dependence. Using the algorithm in Exercise 3.2.3, we have simulated 500 observations from the two extreme members ($\theta = 1$ and $\theta = -1$) of this family (see Exercise 3.21(b)). The scatterplots appear in Fig. 3.12.

We now return to the question of choosing ψ so that the function C in (3.2.9) is a copula. Answers are provided by the following theorem and corollary, due to (Quesada Molina and Rodríguez Lallena 1995):

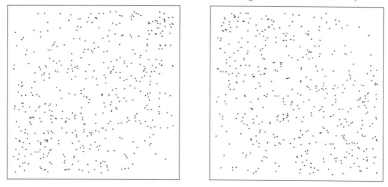

Fig. 3.12. Scatterplots for FGM copulas with $\theta = 1$ (left) and -1 (right)

Theorem 3.2.4. *Let ψ be a function with domain* **I**, *and let C be given by (3.2.9) for u,v in* **I**. *Then C is a copula if and only if:*
1. $\psi(0) = \psi(1) = 0$;   (3.2.11)
2. $\psi(v)$ *satisfies the Lipschitz condition*

$$\left| \psi(v_2) - \psi(v_1) \right| \le \left| v_2 - v_1 \right|$$ (3.2.12)

for all v_1, v_2 in **I**. *Furthermore, C is absolutely continuous.*

Proof: As noted earlier, the boundary conditions $C(u,0) = 0$ and $C(u,1) = u$ are equivalent to (3.2.11). Furthermore, C is 2-increasing if and only if

$$V_C\left([u_1,u_2] \times [v_1,v_2]\right) =$$
$$(u_2 - u_1)\left\{(v_2 - v_1) + \left[\psi(v_2) - \psi(v_1)\right](1 - u_1 - u_2)\right\} \ge 0.$$

If $u_1 = u_2, v_1 = v_2$, or if $u_1 + u_2 = 1$, then $V_C([u_1,u_2] \times [v_1,v_2]) = 0$.
So for $u_1 < u_2$ and $v_1 < v_2$, we have

$$\frac{\psi(v_2) - \psi(v_1)}{v_2 - v_1} \leq \frac{1}{u_2 + u_1 - 1} \text{ if } u_1 + u_2 > 1,$$

and

$$\frac{\psi(v_2) - \psi(v_1)}{v_2 - v_1} \geq \frac{1}{u_2 + u_1 - 1} \text{ if } u_1 + u_2 < 1.$$

However, $\inf\{1/(u_1 + u_2 - 1)|0 \leq u_1 \leq u_2 \leq 1, u_1 + u_2 > 1\} = 1$ and $\sup\{1/(u_1 + u_2 - 1)|0 \leq u_1 \leq u_2 \leq 1, u_1 + u_2 < 1\} = -1$, and hence C is 2-increasing if and only if

$$-1 \leq \frac{\psi(v_2) - \psi(v_1)}{v_2 - v_1} \leq 1$$

for v_1, v_2 in **I** such that $v_1 < v_2$, which is equivalent to (3.2.12). Lastly, the absolute continuity of C follows from the absolute continuity of ψ (with $|\psi'(v)| \leq 1$ almost everywhere on **I**), a condition equivalent to (3.2.12). □

Of course, copulas with quadratic sections in v can be obtained by exchanging the roles of u and v in (3.2.9) and Theorem 3.2.4. The following corollary, whose proof is left as an exercise, summarizes the salient properties of the function ψ for copulas with quadratic sections.

Corollary 3.2.5. *The function C defined by (3.2.9) is a copula if and only if ψ satisfies the following three properties:*

  1. *$\psi(v)$ is absolutely continuous on **I**;*

  2. *$|\psi'(v)| \leq 1$ almost everywhere on **I**;*

  3. *$|\psi(v)| \leq \min(v, 1-v)$ for all v in **I**.*

Furthermore, C is absolutely continuous.

So, to construct copulas with quadratic sections in u, we need only choose functions ψ satisfying the three properties in Corollary 3.2.5—that is, continuous piecewise differentiable functions whose graphs lie in the shaded region in Fig. 3.13, and whose derivatives (where they exist) do not exceed 1 in absolute value. One example of such functions is $\psi(v) = \theta v(1-v)$ for θ in $[-1,1]$, and this leads to the Farlie-Gumbel-Morgenstern family presented in Example 3.12. Other examples of functions that lead to parametric families of copulas are considered in Exercise 3.25. See (Quesada Molina and Rodríguez Lallena 1995) for a discussion of these families and further examples.

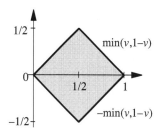

Fig. 3.13. Bounds for the graph of $\psi(v)$

Copulas with Cubic Sections

In a similar fashion, these ideas can be extended to construct copulas whose horizontal or vertical sections are cubic polynomials. The development is quite similar to that for copulas with quadratic sections, and hence we will present the results without proof. The proofs can be found in (Nelsen et al. 1997).

Let C be a copula with cubic sections in u. Then C is given by $C(u,v)$ $= a(v)u^3 + b(v)u^2 + c(v)u + d(v)$ for appropriate functions a, b, c, and d. Again employing the boundary conditions, we obtain $0 = C(0,v) = d(v)$ and $v = C(1,v) = a(v) + b(v) + c(v)$, so that $c(v) = v - a(v) - b(v)$. If we let $\alpha(v) = -a(v) - b(v)$ and $\beta(v) = -2a(v) - b(v)$, then we have

$$C(u,v) = uv + u(1-u)\left[\alpha(v)(1-u) + \beta(v)u\right] \qquad (3.2.13)$$

where α and β are functions such that $\alpha(0) = \alpha(1) = \beta(0) = \beta(1) = 0$ (so that $C(u,0) = 0$ and $C(u,1) = u$) and for which C is 2-increasing.

The requisite conditions for C in (3.2.13) to be 2-increasing (and hence a copula) are given in the next theorem, whose proof is similar to that of Theorem 3.2.4.

Theorem 3.2.6. *Let α, β be two functions from $\mathbf{I}$ to $\mathbf{R}$ satisfying $\alpha(0) = \alpha(1) = \beta(0) = \beta(1) = 0$, and let C be the function defined by (3.2.13). Then C is a copula if and only if for every u_1, u_2, v_1, v_2 in $\mathbf{I}$ such that $u_1 < u_2, v_1 < v_2$, we have*

$$\left[(1-u_1)^2 + (1-u_2)^2 + u_1 u_2 - 1\right]\frac{\alpha(v_2) - \alpha(v_1)}{v_2 - v_1} - $$
$$\left[u_1^2 + u_2^2 + (1-u_1)(1-u_2) - 1\right]\frac{\beta(v_2) - \beta(v_1)}{v_2 - v_1} \geq -1.$$

But this theorem is hardly a manageable result. However, the following lemma is easily established:

Lemma 3.2.7. *Let α, β, and C be as in Theorem 3.2.6. Then C is a copula if and only if*

1. *$\alpha(v)$ and $\beta(v)$ are absolutely continuous and*

2. *$1 + \alpha'(v)(1 - 4u + 3u^2) + \beta'(v)(2u - 3u^2) \geq 0$ for all u in $\mathbf{I}$ and almost all v in $\mathbf{I}$.*

With this lemma we can establish the following theorem, which will be used in the sequel to construct copulas with cubic sections. Theorems 3.2.8 and 3.2.10 both refer to a set S, the union of the set of points in the square $[-1,2] \times [-2,1]$ and the set of points in and on the ellipse in $\mathbf{R}^2$ whose equation is $x^2 - xy + y^2 - 3x + 3y = 0$. The graph of S is given in Fig. 3.14.

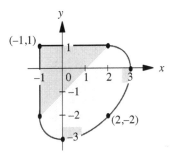

Fig. 3.14. The set S in Theorems 3.2.8 and 3.2.10

Theorem 3.2.8. *Let α, β, and C be as in Theorem 3.2.6. Then C is a copula if and only if*

1. *$\alpha(v)$ and $\beta(v)$ are absolutely continuous and*

2. *for almost every v in $\mathbf{I}$, the point $(\alpha'(v), \beta'(v))$ lies in S. In other words, for almost every v in $\mathbf{I}$, either*

$$-1 \leq \alpha'(v) \leq 2 \quad and \quad -2 \leq \beta'(v) \leq 1$$

or

$$[\alpha'(v)]^2 - \alpha'(v)\beta'(v) + [\beta'(v)]^2 - 3\alpha'(v) + 3\beta'(v) \leq 0.$$

Moreover, C is absolutely continuous.

The next theorem gives necessary and sufficient conditions for a copula with cubic sections to be associated with either a radially symmetric or jointly symmetric pair of random variables (recall Theorem 2.7.3 and Exercise 2.29). Its proof is left as an exercise.

Theorem 3.2.9. *Let α, β, and C be as in Theorem 3.2.6. Then*

1. *The survival copula $\hat{C}$ associated with C is given by*

$$\hat{C}(u,v) = uv + u(1-u)\big[\beta(1-v)(1-u) + \alpha(1-v)u\big]; \qquad (3.2.14)$$

2. $C = \hat{C}$, *i.e. C satisfies* (2.7.3), *if and only if* $\alpha(v) = \beta(1-v)$ *for all* v *in* **I**; *and*

3. *C satisfies* (2.8.1) *if and only if* $\alpha(v) = \beta(1-v) = -\alpha(1-v) = -\beta(v)$ *for all* v *in* **I**.

Example 3.13. *Iterated Farlie-Gumbel-Morgenstern Distributions.* Let C_θ be a member of the FGM family (3.2.10). The joint survival function (not the survival copula) $\overline{C}_\theta(u,v) = 1 - u - v + C_\theta(u,v)$ associated with C_θ is

$$\overline{C}_\theta(u,v) = (1-u)(1-v)(1+\theta uv). \qquad (3.2.15)$$

Noting that both the joint distribution function uv and the joint survival function $(1-u)(1-v)$ for independence appear in both (3.2.10) and (3.2.15), Kotz and Johnson (1977) "iterated" the FGM distribution by replacing the " $(1-u)(1-v)$ " term in (3.2.10) by $\overline{C}_\theta$ (but with a new parameter, say φ) in (3.2.15) to obtain *Kotz and Johnson's iterated FGM*:

$$C_{\theta,\varphi}(u,v) = uv + \theta uv(1-u)(1-v)[1+\varphi uv]. \qquad (3.2.16)$$

When $C_{\theta,\varphi}$ is written in the form (3.2.13), $\alpha(v) = \theta v(1-v)$ and $\beta(v) = \theta v(1-v)(1+\varphi v)$.

In a similar vein, Lin (1987) iterated the FGM by replacing the "uv" term in (3.2.15) by C_θ from (3.2.10) (but again with a new parameter, say φ) and then solved for $C_{\theta,\varphi}$ to obtain *Lin's iterated FGM*:

$$C_{\theta,\varphi}(u,v) = uv + \theta uv(1-u)(1-v)[1+\varphi(1-u)(1-v)]; \qquad (3.2.17)$$

which are also the survival copulas for the Kotz and Johnson family given by (3.2.16). ∎

In Example 3.12, we saw that the Farlie-Gumbel-Morgenstern family (3.2.10) had as its members all the copulas with quadratic sections in both u and v. Note that the iterated FGM copulas in Example 3.13 have cubic sections in both u and v. We now examine the extension of these families to find all copulas with cubic sections in both u and v. That is, we seek copulas $C(u,v)$ which can be written both as (3.2.13) and as

$$C(u,v) = uv + v(1-v)\big[\gamma(u)(1-v) + \chi(u)v\big] \qquad (3.2.18)$$

where α, β, γ and χ satisfy the hypotheses of Theorem 3.2.6. As a consequence of Theorem 3.2.8, we have

Theorem 3.2.10. *Suppose that C has cubic sections in both u and v, i.e., C is given by both* (3.2.13) *and* (3.2.18). *Then*

$$C(u,v) = uv + uv(1-u)(1-v)\big[A_1 v(1-u) + \\ A_2(1-v)(1-u) + B_1 uv + B_2 u(1-v)\big],$$ (3.2.19)

where A_1, A_2, B_1, B_2 *are real constants such that the points* (A_2, A_1), (B_1, B_2), (B_1, A_1), *and* (A_2, B_2) *all lie in S.*

Note that when C has cubic sections in both u and v, the functions α, β, γ and χ in (3.2.13) and (3.2.18) are given by

$$\alpha(v) = v(1-v)\big[A_1 v + A_2(1-v)\big],$$

$$\beta(v) = v(1-v)\big[B_1 v + B_2(1-v)\big],$$

$$\gamma(u) = u(1-u)\big[B_2 u + A_2(1-u)\big], \text{ and}$$

$$\chi(u) = u(1-u)\big[B_1 u + A_1(1-u)\big].$$

Example 3.14. The two families of iterated FGM copulas in Example 3.13 have cubic sections in both u and v, and hence can be written in the form (3.2.19). For the Johnson and Kotz family of copulas given by (3.2.16), $A_1 = A_2 = B_2 = \theta$ and $B_1 = \theta(1+\varphi)$; and for the Lin family of copulas in (3.2.17), $A_1 = B_1 = B_2 = \theta$ and $A_2 = \theta(1+\varphi)$. If we let $\xi = \theta(1+\varphi)$, then the ranges of the parameters θ and ξ for both families are: $\theta \in [-1,1]$ and $-1 - \theta \le \xi \le [3 - \theta + (9 - 6\theta - 3\theta^2)^{1/2}]/2$. ∎

As an immediate consequence of Theorem 3.2.9, we have

Corollary 3.2.11. *Suppose that C has cubic sections in both u and v, i.e., C is given by* (3.2.19) *in Theorem 3.2.10. Then*

1. *C is symmetric, i.e.,* $C(u,v) = C(v,u)$, *if and only if* $A_1 = B_2$;

2. $C = \hat{C}$, *i.e. C satisfies* (2.7.3), *if and only if* $A_1 = B_2$ *and* $A_2 = B_1$;

3. *C satisfies* (2.8.1) *if and only if* $A_1 = B_2 = -A_2 = -B_1$.

The next two examples show the ease with which the procedures outlined in the above theorems can be use to construct families of copulas.

Example 3.15. If we set $A_1 = B_2 = a - b$ and $A_2 = B_1 = a + b$ in (3.2.19), we obtain a two-parameter family of copulas each of which is symmetric and satisfies the functional equation $C = \hat{C}$ associated with radial symmetry. From Theorem 3.2.10, the constants a and b satisfy $b \in [-1,2]$; $|a| \le b + 1$ for $b \in [-1,1/2]$ and $|a| \le (6b - 3b^2)^{1/2}$ for $b \in [1/2,2]$. Explicitly we have

$$C_{a,b}(u,v) = uv + uv(1-u)(1-v)[a + b(1-2u)(1-2v)] \quad (3.2.20)$$

for u,v in **I**. Several subfamilies are of interest.

1. When $b = 0$ the copulas given by (3.2.20) are the FGM family.

2. When $a = 3\theta$ and $b = 5\theta^2$, we obtain a family constructed by Sarmanov (1974) from the first two Legendre polynomials. Applications of this family are discussed in (Lee 1996).

3. When $a = 0$, we have a family in which each copula satisfies the functional equations (2.8.1) associated with joint symmetry.

Additional subfamilies of (3.2.20) are considered in Exercises 3.39 and 4.9. ■

Example 3.16. Theorem 3.2.10 can also be used to construct families of asymmetric copulas. For example, if we set $A_1 = A_2 = a$ and $B_1 = B_2 = b$ for $-1 \leq a,b \leq 1$; $a \neq b$, we obtain a family of asymmetric copulas with cubic sections in u but quadratic sections in v (when $a = b$ we have the FGM family)—see Exercise 3.30. If we set $A_1 = a$, $A_2 = B_1 = B_2 = b$ where $|b| \leq 1$, $[b - 3 - (9 + 6b - 3b^2)^{1/2}]/2 \leq a \leq 1$, $a \neq b$, then we have the following family of asymmetric copulas with cubic sections in both u and v:

$$C(u,v) = uv + uv(1-u)(1-v)[(a-b)v(1-u)+b].$$ ■

The ideas developed in this section can be extended to investigate copulas whose horizontal or vertical sections are higher degree polynomials or other simple functions such as hyperbolas (see Exercise 3.26).

3.2.6 Copulas with Prescribed Diagonal Sections

We now turn to the construction of copulas having a prescribed diagonal section. Recall from Sects. 2.2 and 2.6 that the diagonal section of a copula C is the function δ_C from **I** to **I** defined by $\delta_C(t) = C(t,t)$, and that the diagonal section of the dual of C (see Sect. 2.6) is the function $\tilde{\delta}_C(t)$ from **I** to **I** given by $\tilde{\delta}_C(t) = 2t - \delta_C(t)$. Diagonal sections are of interest because if X and Y are random variables with a common distribution function F and copula C, then the distribution functions of the order statistics $\max(X,Y)$ and $\min(X,Y)$ are $\delta_C(F(t))$ and $\tilde{\delta}_C(F(t))$, respectively [see Exercises 2.16 and 2.24].

As a consequence of Theorem 2.2.4 and Exercise 2.8, it follows that if δ is the diagonal section of a copula, then

$$\delta(1) = 1;\qquad\qquad (3.2.21a)$$

$$0 \leq \delta(t_2) - \delta(t_1) \leq 2(t_2 - t_1) \text{ for all } t_1, t_2 \text{ in } \mathbf{I} \text{ with } t_1 \leq t_2; \quad (3.2.21b)$$

and

$$\delta(t) \le t \text{ for all } t \text{ in } \mathbf{I}. \qquad (3.2.21c)$$

In this section, *any* function $\delta : \mathbf{I} \to \mathbf{I}$ that satisfies (3.2.21a)-(3.2.21c) will be called simply a *diagonal*, while we will refer to the function $\delta_C(t) = C(t,t)$ as the *diagonal section* of C.

Now suppose that δ is any diagonal. Is there a copula C whose diagonal section is δ? The answer is provided in the following theorem.

Theorem 3.2.12. *Let δ be any diagonal, and set*

$$C(u,v) = \min\big(u,v,(1/2)[\delta(u)+\delta(v)]\big). \qquad (3.2.22)$$

Then C is a copula whose diagonal section is δ.

The proof of the above theorem, which is a somewhat technical but straightforward verification of the fact that C is 2-increasing, can be found in (Fredricks and Nelsen 1997a). Copulas of the form given by $C(u,v)$ in (3.2.22) are called *diagonal copulas*.

Example 3.17. (a) Let $\delta(t) = t$, the diagonal section of M. The diagonal copula constructed from this diagonal is M, as it must be as the only copula whose diagonal section is the identity is M [see Exercise 2.8].

(b) Let $\delta(t) = \max(0,2t-1)$, the diagonal section of W. The diagonal copula constructed from this diagonal is not W but rather the shuffle of M given by $M(2,\mathbf{I}_2,(2,1),1)$, i.e., the copula $C_{1/2}$ from Exercise 3.9.

(c) Let $\delta(t) = t^2$, the diagonal section of Π. The diagonal copula constructed from this diagonal is the singular copula from Example 3.5 that assigns the probability mass to two quarter circles in $\mathbf{I}^2$ with radius 1, one centered at (0,1) and one centered at (1,0). ∎

Because a diagonal copula is constructed from a function that can be the distribution function of the maximum of two random variables, it is natural to suspect that diagonal copulas are related to joint distributions of order statistics. Such is indeed the case.

Theorem 3.2.13. *Suppose X and Y are continuous random variables with copula C and a common marginal distribution function. Then the joint distribution function of $\max(X,Y)$ and $\min(X,Y)$ is the Fréchet-Hoeffding upper bound if and only if C is a diagonal copula.*

Proof. By virtue of Sklar's theorem, we need only prove the following: Suppose U and V are random variables whose joint distribution function is the copula C. Then the joint distribution function of $\max(U,V)$ and $\min(U,V)$ is the Fréchet-Hoeffding upper bound if and only if C is a diagonal copula.

Let $H(z,\tilde{z})$ be the joint distribution function of $Z = \max(U,V)$ and $\tilde{Z} = \min(U,V)$. Recall that δ_C and $\tilde{\delta}_C$ are the distribution functions of Z and $\tilde{Z}$, respectively. Then, setting $\delta_C = \delta$,

$$H(z,\tilde{z}) = P[\max(U,V) \le z, \min(U,V) \le \tilde{z})$$
$$= \begin{cases} \delta(z), & z \le \tilde{z}, \\ C(z,\tilde{z}) + C(\tilde{z},z) - \delta(\tilde{z}), & z \ge \tilde{z}. \end{cases}$$

Assume C is a diagonal copula, i.e., $C(u,v) = \min(u,v,(1/2)[\delta(u)+\delta(v)])$. Then if $z \ge \tilde{z}$, $H(z, \tilde{z}) = 2C(z, \tilde{z}) - \delta(\tilde{z}) = \min(2\tilde{z} - \delta(\tilde{z}),\delta(z)) = \min(\tilde{\delta}(\tilde{z}),\delta(z))$. If $z < \tilde{z}$, then $\delta(z) = \min(\tilde{\delta}(\tilde{z}),\delta(z))$ since $\delta(z) \le \delta(\tilde{z}) \le \tilde{z} \le \tilde{\delta}(\tilde{z})$. Thus $H(z,\tilde{z}) = M(\tilde{\delta}(\tilde{z}),\delta(z))$.

In the opposite direction, assume $H(z,\tilde{z}) = M(\tilde{\delta}(\tilde{z}),\delta(z))$, where again δ denotes the diagonal section of C. Assuming that C is symmetric, we will show that C must be a diagonal copula [for the proof in the general case, see (Fredricks and Nelsen 1997b)]. If $z > \tilde{z}$, then $2C(z,\tilde{z}) - \delta(\tilde{z}) = M(\tilde{\delta}(\tilde{z}),\delta(z)) = \min(2\tilde{z}-\delta(\tilde{z}),\delta(z))$, and hence $C(z,\tilde{z}) = \min(\tilde{z},(1/2)[\delta(z)+\delta(\tilde{z})])$. By symmetry, $C(z,\tilde{z}) = \min(z,(1/2)[\delta(z)+\delta(\tilde{z})])$ when $z \le \tilde{z}$. Thus $C(u,v) = \min(u,v,(1/2)[\delta(u)+\delta(v)])$ for all u,v in $\mathbf{I}^2$.   □

There are other ways to construct copulas with prescribed diagonal sections. For example, Bertino (1977) shows that if δ is a diagonal then

$$B_\delta(u,v) = \begin{cases} u - \inf_{u \le t \le v}[t - \delta(t)], & u \le v, \\ v - \inf_{v \le t \le u}[t - \delta(t)], & v \le u, \end{cases} \qquad (3.2.23)$$

is a copula whose diagonal section is δ. For a thorough treatment of the properties of Bertino copulas, including a characterization similar to Theorem 3.2.13, see (Fredricks and Nelsen 2002).

Exercises

3.21 (a) Show that the arithmetic mean of two Farlie-Gumbel-Morgenstern copulas is again a Farlie-Gumbel-Morgenstern copula, i.e., if C_α and C_β are given by (3.2.10), then the arithmetic mean of C_α and C_β is $C_{(\alpha+\beta)/2}$ (see Exercise 2.3).

(b) Show that each FGM copula is a weighted arithmetic mean of the two extreme members of the family, i.e., for all θ in $[-1,1]$,

$$C_\theta(u,v) = \frac{1-\theta}{2}C_{-1}(u,v) + \frac{1+\theta}{2}C_{+1}(u,v).$$

3.22 Show that each member of the Farlie-Gumbel-Morgenstern family of copulas is absolutely continuous and satisfies the condition $C = \hat{C}$ for radial symmetry. Also show that the FGM family is positively ordered.

3.23 Show that the following algorithm (Johnson 1986) generates random variates (u,v) from an FGM distribution with parameter θ:
  1. Generate two independent uniform $(0,1)$ variates u, t;
  2. Set $a = 1 + \theta(1 - 2u)$; $b = \sqrt{a^2 - 4(a-1)t}$;
  3. Set $v = 2t/(b + a)$;
  4. The desired pair is (u,v).

3.24 Prove Corollary 3.2.4.

3.25 Show that for each of the following choices of ψ, the function C given by (3.2.9) is a copula (Quesada Molina and Rodríguez Lallena 1995):
  (a) $\psi(v) = \min\{\alpha v, \beta(1-v)\}$ (or $-\min\{\alpha v, \beta(1-v)\}$) for α, β in $\mathbf{I}$;
  (b) $\psi(v) = (\theta/\pi)\sin(\pi v)$ for θ in $[-1,1]$;
  (c) $\psi(v) = (\theta/2\pi)\sin(2\pi v)$ for θ in $[-1,1]$;
  (d) $\psi(v) = \theta[\zeta(v) + \zeta(1 - v)]$ for θ in $[-1,1]$, where ζ is the piecewise linear function whose graph connects $(0,0)$ to $(1/4,1/4)$ to $(1/2,0)$ to $(1,0)$.
  (e) $\psi(v) = \theta\zeta(v) - (1-\theta)\zeta(1-v)$ for θ in $\mathbf{I}$, where ζ is the piecewise linear function in part (d).

3.26 Show that a family of copulas with hyperbolic sections in both u and v is the Ali-Mikhail-Haq family, first encountered in Exercise 2.14. Are there other such families?

3.27 Prove Theorem 3.2.9.

3.28 Let C be a copula with cubic sections in u, i.e., let C be given by (3.2.13) where α and β satisfy Theorem 3.2.7. Prove that for all v in $\mathbf{I}$, $\max(-v,3(v-1)) \leq \alpha(v) \leq \min(1-v,3v)$ and $\max(-3v, v-1) \leq \beta(v) \leq \min(3(1-v),v)$; that is, the graphs of α and β lie in the shaded regions shown in Fig. 3.15.

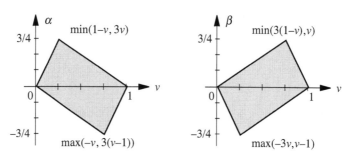

Fig. 3.15. Regions for α and β in Exercise 3.28

3.29 Prove Corollary 3.2.11.

3.30 Let C be a copula with cubic sections in u and v, i.e., let C be given by (3.2.19) in Theorem 3.2.10. Show that

(a) if $A_1 = A_2$ and $B_1 = B_2$, then the sections in v are actually quadratic rather than cubic;

(b) if $A_1 = B_1$ and $A_2 = B_2$, then the sections in u are quadratic rather than cubic;

(c) if $A_1 = A_2 = B_1 = B_2 = \theta$, then (3.2.19) degenerates to the Farlie-Gumbel-Morgenstern family (3.2.10).

3.31 Let C be any symmetric copula whose diagonal section is δ. Show that $C(u,v) \leq \min(u,v,(1/2)[\delta(u)+\delta(v)])$ for all u,v in $\mathbf{I}^2$.

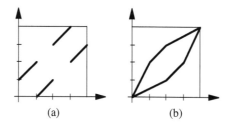

Fig. 3.16. The supports of the copulas in Exercises 3.32 and 3.33

3.32 Let $\delta(t) = \min(\max(0,2t - 1/2),\max(1/2,2t - 1))$, a piecewise linear function whose graph connects $(0,0)$ to $(1/4,0)$ to $(1/2,1/2)$ to $(3/4,1/2)$ to $(1,1)$. Show that the diagonal copula constructed from this diagonal is the straight shuffle of M given by $M(4,\mathbf{I}_4,(2,1,4,3),1\}$, whose support is illustrated in part (a) of Fig. 3.16.

3.33 Let δ_θ be a convex combination of the diagonal sections of M and W, i.e., $\delta_\theta(t) = \theta\delta_M(t) + (1-\theta)\delta_W(t)$, where θ is in $\mathbf{I}$. Let C_θ be the diagonal copula constructed from δ_θ. Show that the support of C_θ is the hexagon in $\mathbf{I}^2$ with vertices $(0,0)$, $(1/2,\omega)$, $(1-\omega,1/2), (1,1), (1/2,1-\omega)$, and $(\omega,1/2)$, where $\omega = \theta/(4-2\theta)$, as illustrated in part (b) of Fig. 3.16 for the case $\theta = 2/3$.

3.34 Let C be a copula whose support is a subset of the union of the two diagonals of $\mathbf{I}^2$. Show that

$$C(u,v) = \begin{cases} \delta\big(\min(u,v)\big), & u+v \leq 1, \\ |u-v| + \delta\big(\max(u,v)\big), & u+v > 1, \end{cases}$$

where δ is a diagonal such that $t - \delta(t) = (1-t) - \delta(1-t)$ for t in $\mathbf{I}$, i.e., the graph of $y = t - \delta(t)$ is symmetric with respect to $t = 1/2$.

3.3 Algebraic Methods

In this section, we will construct two well-known families of copulas, the Plackett and Ali-Mikhail-Haq families, to illustrate the procedure for using an algebraic relationship between the joint distribution function and its univariate margins to find copulas. In both cases, the algebraic relationship concerns an "odds" ratio—in the first case we generalize 2×2 contingency tables, and in the second case we work with survival odds ratios.

3.3.1 Plackett Distributions

A measure of "association" or "dependence" in 2×2 contingency tables is the *cross product ratio*, or *odds ratio*, which we will denote by θ. To illustrate, consider the 2×2 table in Fig. 3.17. For convenience, we have labeled the categories for each variable as "low" and "high." If the observed counts in the four categories are a, b, c, and d, as shown, then the cross product ratio is the positive real number θ given by $\theta = ad/bc$.

		Column variable: low	Column variable: high	
Row variable	low	a	b	$a{+}b$
	high	c	d	$c{+}d$
		$a{+}c$	$b{+}d$	

Fig. 3.17. A 2×2 contingency table

The value $\theta = 1$ corresponds to independence, for when $\theta = 1$, $ad = bc$, which implies that each "observed" entry (such as a) is equal to its "expected value" under independence (here $(a+b)(a+c)/n$, where $n = a+b+c+d$). When $\theta > 1$, the observations are concentrated in the "low-low" and "high-high" cells; and when θ is between 0 and 1, the observations are concentrated in the "low-high" and "high-low" cells.

With simple algebra, θ is the ratio of the "odds" for the rows given the column, or equivalently, for the columns given the row:

$$\theta = \frac{a/c}{b/d} = \frac{\dfrac{a}{a+c}\Big/\dfrac{c}{a+c}}{\dfrac{b}{b+d}\Big/\dfrac{d}{b+d}} \quad \text{and} \quad \theta = \frac{a/b}{c/d} = \frac{\dfrac{a}{a+b}\Big/\dfrac{b}{a+b}}{\dfrac{c}{c+d}\Big/\dfrac{d}{c+d}}.$$

Also note that the counts a, b, c, and d in $\theta = ad/bc$ could just as well be replaced by the proportions a/n, b/n, c/n, and d/n.

Plackett's family of bivariate distributions (Plackett 1965) arises from an extension of this idea to bivariate distributions with continuous margins. Let X and Y be continuous random variables with a joint distribution function H, and margins F and G, respectively. Let x and y be any pair of real numbers, and let the "low" and "high" categories for the column variable correspond to the events "$X \le x$" and "$X > x$", respectively, and similarly for the row variable. Then replace the numbers a, b, c, and d in $\theta = ad/bc$ by the probabilities $H(x,y)$, $F(x) - H(x,y)$, $G(y) - H(x,y)$, and $1 - F(x) - G(y) + H(x,y)$, respectively, to obtain

$$\theta = \frac{H(x,y)[1-F(x)-G(y)+H(x,y)]}{[F(x)-H(x,y)][G(y)-H(x,y)]}. \tag{3.3.1}$$

For most joint distributions, θ will be a function of the point (x,y)—but are there joint distributions for which θ is a constant? As we shall show, the answer is yes—and these are the members of Plackett's family, which are also known as *constant global cross ratio* distributions, or *contingency-type* (or *C-type*) distributions.

Using the probability transforms $u = F(x)$, $v = G(y)$, and Sklar's theorem, we can rewrite (3.3.1) as (where C is the copula of X and Y)

$$\theta = \frac{C(u,v)\left[1 - u - v + C(u,v)\right]}{\left[u - C(u,v)\right]\left[v - C(u,v)\right]},$$

and solve for C. When $\theta = 1$, the only solution is $C = \Pi$; when $\theta \neq 1$, clearing fractions yields a quadratic in C, the roots of which are

$$C(u,v) = \frac{\left[1 + (\theta - 1)(u + v)\right] \pm \sqrt{\left[1 + (\theta - 1)(u + v)\right]^2 - 4uv\theta(\theta - 1)}}{2(\theta - 1)}. \quad (3.3.2)$$

Following (Mardia 1970), we will now show that, for $\theta > 0$ but $\theta \neq 1$, the root in (3.3.2) with the "+" sign preceding the radical is never a copula; whereas the root with the "–" sign always is.

Margins of the two roots in (3.3.2) are

$$C(u,0) = \frac{\left[1 + (\theta - 1)u\right] \pm \left[1 + (\theta - 1)u\right]}{2(\theta - 1)} \quad \text{and}$$

$$C(u,1) = \frac{\left[\theta + (\theta - 1)u\right] \pm \left[\theta - (\theta - 1)u\right]}{2(\theta - 1)}$$

and hence (for $\theta > 0$, $\theta \neq 1$) the root with the "+" sign never satisfies the boundary conditions, and the root with the "–" sign always does. Now let C_θ denote the root in (3.3.2) with the "–" sign. To show that C_θ is 2-increasing, it suffices to show that

$$\frac{\partial^2 C_\theta(u,v)}{\partial u \partial v} \geq 0 \quad \text{and} \quad C_\theta(u,v) = \int_0^u \int_0^v \frac{\partial^2 C_\theta(s,t)}{\partial s \partial t} \, dt \, ds$$

for (u,v) in $\mathbf{I}^2$. This is tedious but elementary—but also shows that each of these copulas is absolutely continuous. Thus we have the *Plackett family* of copulas: for $\theta > 0$, $\theta \neq 1$,

$$C_\theta(u,v) = \frac{\left[1 + (\theta - 1)(u + v)\right] - \sqrt{\left[1 + (\theta - 1)(u + v)\right]^2 - 4uv\theta(\theta - 1)}}{2(\theta - 1)}, \quad (3.3.3a)$$

and for $\theta = 1$,

$$C_1(u,v) = uv. \quad (3.3.3b)$$

In addition to being absolutely continuous, these copulas form a comprehensive family (like the Fréchet family in Exercise 2.4), because the limits of C_θ as θ goes to 0 and to ∞ are the bounds W and M, respectively (see Exercise 3.35). So it is not surprising that Plackett family copulas have been widely used both in modeling and as alternatives to the bivariate normal for studies of power and robustness of various statistical tests (Conway 1986; Hutchinson and Lai 1990).

In Fig. 3.18 we have scatterplots for two simulations of Plackett copulas, each using the algorithm in Exercise 3.38 with 500 observations. The one on the left is for $\theta = 20$, the one on the right is for $\theta = 0.02$.

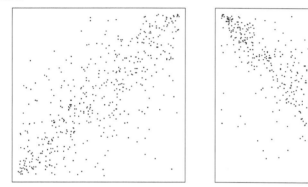

Fig. 3.18. Scatterplots for Plackett copulas with $\theta = 20$ and 0.02

To fit a Plackett copula to a data set, one must estimate the parameter θ from the data. One estimator is the maximum likelihood estimator, which must be found numerically. An attractive alternative is the observed cross-product ratio $\theta^* = ad/bc$, where $a, b, c,$ and d are the observed frequencies in the four quadrants determined by lines in $\mathbf{R}^2$ parallel to the axes through a point (p,q).

An optimum choice for (p,q) is the sample median vector, which minimizes the asymptotic variance of θ^* (Mardia 1970). In this case, $F(p) = G(q) = 1/2$, and $\theta^* = 4m^2/(1-2m)^2$, where m is the observed frequency of observations in which neither variable exceeds its median value. See (Mardia 1970) for details, and for an efficient estimator that is asymptotically equivalent to the maximum likelihood estimator.

3.3.2 Ali-Mikhail-Haq Distributions

Let X and Y be continuous random variables with joint distribution function H and marginal distribution functions F and G, respectively. When X and Y denote lifetimes of objects, such as organisms or electronic components, it is natural to talk about the "odds for survival," that is (for X, say), the ratio $P[X > x]/P[X \le x]$ of the probability of survival beyond time x to the probability of failure before time x, i.e., $\overline{F}(x)/F(x) = (1-F(x))/F(x)$. In an analogous fashion, we can define a bivariate survival odds ratio $P[X > x \text{ or } Y > y]/P[X \leq x, Y \leq y]$, or $(1-H(x,y))/H(x,y)$.

Example 3.18. Suppose X and Y have Gumbel's bivariate logistic distribution from Exercise 2.12, that is, for all x,y in $\overline{\mathbf{R}}$,

$$H(x,y) = (1+e^{-x}+e^{-y})^{-1}.$$

Then the bivariate survival odds ratio is $(1 - H(x,y))/H(x,y) = e^{-x} + e^{-y}$. But $F(x) = (1+e^{-x})^{-1}$, so that $(1 - F(x))/F(x) = e^{-x}$; and similarly for Y. It follows that

$$\frac{1-H(x,y)}{H(x,y)} = \frac{1-F(x)}{F(x)} + \frac{1-G(y)}{G(y)}. \tag{3.3.4}$$

∎

Example 3.19. Suppose X and Y are independent random variables with joint distribution function H and marginal distribution functions F and G, respectively, where $H(x,y) = F(x)G(y)$. Since $F(x) = \left(1+[(1-F(x))/F(x)]\right)^{-1}$, and similarly for G and H; we obtain

$$\frac{1-H(x,y)}{H(x,y)} = \frac{1-F(x)}{F(x)} + \frac{1-G(y)}{G(y)} + \frac{1-F(x)}{F(x)} \cdot \frac{1-G(y)}{G(y)}. \tag{3.3.5}$$

Noting the similarity between (3.3.4) and (3.3.5), Ali, Mikhail and Haq (1978) proposed searching for bivariate distributions for which the survival odds ratios satisfied

$$\frac{1-H(x,y)}{H(x,y)} = \frac{1-F(x)}{F(x)} + \frac{1-G(y)}{G(y)} + (1-\theta)\frac{1-F(x)}{F(x)} \cdot \frac{1-G(y)}{G(y)} \tag{3.3.6}$$

for some constant θ. Note that when $\theta = 1$, (3.3.6) reduces to (3.3.4); and when $\theta = 0$, (3.3.6) reduces to the independence case (3.3.5).

As with the derivation of the Plackett family in the preceding section, we use the probability transforms $u = F(x)$, $v = G(y)$ and Sklar's theorem to rewrite (3.3.6) as (where C_θ denotes the copula of X and Y)

$$\frac{1-C_\theta(u,v)}{C_\theta(u,v)} = \frac{1-u}{u} + \frac{1-v}{v} + (1-\theta)\frac{1-u}{u} \cdot \frac{1-v}{v}.$$

Solving for $C_\theta(u,v)$ yields the *Ali-Mikhail-Haq* family: for θ in $[-1,1]$,

$$C_\theta(u,v) = \frac{uv}{1-\theta(1-u)(1-v)}. \tag{3.3.7}$$

The task remains to verify that when θ is in $[-1,1]$, C_θ given by (3.3.7) is indeed a copula. It is easy to check that the boundary conditions (2.2.2a) and (2.2.2b) hold. To have $C_\theta(u,v) \geq 0$ on $\mathbf{I}^2$ requires $\theta \leq 1$, and to have $\partial^2 C_\theta(u,v)/\partial u \partial v \geq 0$ requires $\theta \geq -1$. Finally

$$C_\theta(u,v) = \int_0^u \int_0^v \frac{\partial^2 C_\theta(s,t)}{\partial s \partial t} \, dt \, ds \text{ for } (u,v) \text{ in } \mathbf{I}^2 \text{, so that the copulas in the}$$

Ali-Mikhail-Haq family are absolutely continuous. As noted in Exercise 2.32, this family is positively ordered.   ∎

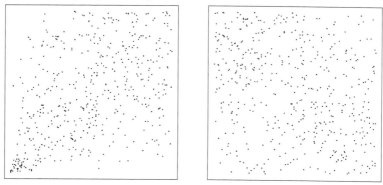

Fig. 3.19. Scatterplots for Ali-Mikhail-Haq copulas, $\theta = 1$ (left) and -1 (right)

In Fig. 3.19 we have scatterplots for simulations of the two extreme members of the Ali-Mikhail-Haq family ($\theta = 1$ and $\theta = -1$), using 500 observations and the algorithm in Exercise 3.42. These copulas are similar to the FGM copulas, in that they can only model weak dependence.

3.3.3 A Copula Transformation Method

In this section, we present a technique to transform one copula into another. It is motivated by the following problem. Let (X_1,Y_1), (X_2,Y_2), $\cdots$, (X_n,Y_n) be independent and identically distributed pairs of random variables with common joint distribution function H, copula C and marginals F (of X_i) and G (of Y_i). How are the "component-wise maxima" $X_{(n)} = \max\{X_i\}$ and $Y_{(n)} = \max\{Y_i\}$ distributed? We answer this question by finding the distribution function $H_{(n)}$ and the copula $C_{(n)}$ of $X_{(n)}$ and $Y_{(n)}$.

We first find the distribution functions $F_{(n)}$ and $G_{(n)}$ of $X_{(n)}$ and $Y_{(n)}$, respectively. Because $P[X_{(n)} \le x] = P[\text{all } X_i \le x] = (P[X_1 \le x])^n$ and similarly for $Y_{(n)}$, we have $F_{(n)}(x) = [F(x)]^n$ and $G_{(n)}(y) = [G(y)]^n$. Thus

$$H_{(n)}(x,y) = P[X_{(n)} \le x, Y_{(n)} \le y] = P[\text{all } X_i \le x, \text{all } Y_i \le y]$$
$$= [H(x,y)]^n = [C(F(x),G(y))]^n$$
$$= \left[C\big([F_{(n)}(x)]^{1/n}, [G_{(n)}(y)]^{1/n}\big) \right]^n.$$

Hence we have

$$C_{(n)}(u,v) = C^n(u^{1/n}, v^{1/n}) \text{ for } u,v \text{ in } \mathbf{I}. \tag{3.3.8}$$

Thus we have proven

Theorem 3.3.1. *If C is a copula and n a positive integer, then the function $C_{(n)}$ given by (3.3.8) is a copula. Furthermore, if (X_i,Y_i), i = 1,2,$\cdots$,n are independent and identically distributed pairs of random variables with copula C, then $C_{(n)}$ is the copula of $X_{(n)} = \max\{X_i\}$ and $Y_{(n)} = \max\{Y_i\}$.*

Example 3.20. (a) If X_i and Y_i are independent, then so are $X_{(n)}$ and $Y_{(n)}$, as $\Pi_{(n)}(u,v) = [u^{1/n} v^{1/n}]^n = uv$.

(b) If X_i and Y_i are comonotonic, then so are $X_{(n)}$ and $Y_{(n)}$, as $M_{(n)}(u,v) = [\min(u^{1/n}, v^{1/n})]^n = M(u,v)$.

(c) However, if X_i and Y_i are countermonotonic, then $X_{(n)}$ and $Y_{(n)}$ are not countermonotonic for any $n \ge 2$, as $W_{(n)}(u,v) = [\max(u^{1/n} + v^{1/n} - 1, 0)]^n$, which is not W but rather a member of the family given by (4.2.1) in the next chapter (with $\theta = -1/n$).

(d) Let C be the copula given by (2.3.4), i.e., $C(u,v) = uv/(u+v-uv)$. Then $C_{(n)}(u,v) = uv/(u^{1/n} + v^{1/n} - u^{1/n} v^{1/n})^n$, also a member of the family given by (4.2.1) (but with $\theta = 1/n$). ∎

Example 3.21. Let C be a member of the Marshall-Olkin family (3.1.3), i.e., $C(u,v) = \min(u^{1-\alpha} v, u v^{1-\beta})$. A simple computation shows that $C_{(n)} = C$ for any positive integer n. ∎

The above examples motivate the following definition:

Definition 3.3.2. A copula C is *max-stable* if for every positive real number r and all u,v in $\mathbf{I}$,

$$C(u,v) = C^r(u^{1/r}, v^{1/r}). \qquad (3.3.9)$$

Example 3.22. Let C_θ be a member of the Gumbel-Hougaard family (2.4.2), i.e.,

$$C_\theta(u,v) = \exp\left(-\left[(-\ln u)^\theta + (-\ln v)^\theta\right]^{1/\theta}\right)$$

for $\theta \geq 1$. A straightforward calculation shows that $C_\theta^r(u^{1/r}, v^{1/r}) = C_\theta(u,v)$, and hence every member of the Gumbel-Hougaard family is max-stable. ∎

The transformation in (3.3.8) above is a special case of a more general result (Genest and Rivest 2001, Klement et al. 2004), which facilitates the construction of copulas from a given copula and certain increasing functions on **I**.

Theorem 3.3.3. Let $\gamma:[0,1] \to [0,1]$ be continuous and strictly increasing with $\gamma(0) = 0$, $\gamma(1) = 1$, and let γ^{-1} denote the inverse of γ. For an arbitrary copula C, define the function C_γ by

$$C_\gamma(u,v) = \gamma^{-1}\big(C(\gamma(u),\gamma(v))\big) \text{ for } u,v \text{ in } [0,1]. \qquad (3.3.9)$$

Then C_γ is a copula if and only if γ is concave (or equivalently, γ^{-1} is convex).

Proof. (i) Assume γ^{-1} is convex. Because $\gamma(0) = 0 = \gamma^{-1}(0)$ and $\gamma(1) = 1 = \gamma^{-1}(1)$, it readily follows that C_γ satisfies the boundary conditions for a copula. Now let u_1, u_2, v_1, v_2 be in **I** such that $u_1 \leq u_2$, $v_1 \leq v_2$; and let $a = C(\gamma(u_1),\gamma(v_1))$, $b = C(\gamma(u_1),\gamma(v_2))$, $c = C(\gamma(u_2),\gamma(v_1))$, and $d = C(\gamma(u_2),\gamma(v_2))$. Because C is a copula, $a - b - c + d \geq 0$, and we must show that $\gamma^{-1}(a) - \gamma^{-1}(b) - \gamma^{-1}(c) + \gamma^{-1}(d) \geq 0$. Note that both b and c lie between a and d, so that either $a \leq b \leq c \leq d$ or $a \leq c \leq b \leq d$. If the four numbers a, b, c, d are distinct, then because γ^{-1} is convex (Roberts and Varberg 1973),

$$\frac{\gamma^{-1}(b) - \gamma^{-1}(a)}{b - a} \leq \frac{\gamma^{-1}(d) - \gamma^{-1}(c)}{d - c}.$$

But $b - a \leq d - c$, and hence $\gamma^{-1}(b) - \gamma^{-1}(a) \leq \gamma^{-1}(d) - \gamma^{-1}(c)$, as required. If two or three of the numbers a, b, c, d coincide, the proof is similar.

(ii) Assume C_γ is a copula for any copula C. For any a, d in $[0,1]$ such that $a \leq d$, let $u_1 = v_1 = \gamma^{-1}((a+1)/2)$ and $u_2 = v_2 = $

$\gamma^{-1}\big((d+1)/2\big)$ so that $\gamma(u_1) = \gamma(v_1) = (a+1)/2$ and $\gamma(u_2) = \gamma(v_2) = (d+1)/2$. Setting $C = W$ in (3.3.9), we have $W\big(\gamma(u_1),\gamma(v_1)\big) = a$, $W\big(\gamma(u_1),\gamma(v_2)\big) = W\big(\gamma(u_2),\gamma(v_1)\big) = (a+d)/2$, and $W\big(\gamma(u_2),\gamma(v_2)\big) = d$. Because W_γ is a copula, $\gamma^{-1}(a) - 2\gamma^{-1}\big((a+d)/2\big) + \gamma^{-1}(d) \geq 0$, i.e. γ^{-1} is mid-convex. But continuous mid-convex functions must be convex (Roberts and Varberg 1973), which completes the proof. □

The theorem remains true if the hypothesis $\gamma(0) = 0$ is dropped and $\gamma^{-1}(t)$ is defined on **I** as a quasi-inverse of γ. See (Durante and Sempi 2005) for details.

3.3.4 Extreme Value Copulas

Let (X_1,Y_1), (X_2,Y_2), $\cdots$, (X_n,Y_n) be independent and identically distributed pairs of random variables with a common copula C and again let $C_{(n)}$ denote the copula of the component-wise maxima $X_{(n)} = \max\{X_i\}$ and $Y_{(n)} = \max\{Y_i\}$. From Theorem 3.3.1 we know that $C_{(n)}(u,v) = C^n(u^{1/n},v^{1/n})$ for u,v in **I**. The limit of the sequence $\{C_{(n)}\}$ leads to the notion of an extreme value copula.

Definition 3.3.4. A copula C_* is an *extreme value copula* if there exists a copula C such that

$$C_*(u,v) = \lim_{n\to\infty} C^n(u^{1/n},v^{1/n}) \tag{3.3.10}$$

for u,v in **I**. Furthermore, C is said to belong to the *domain of attraction* of C_*.

Note that if the pointwise limit of a sequence of copulas exists at each point in $\mathbf{I}^2$, then the limit must be a copula (as for each rectangle in $\mathbf{I}^2$, the sequence of C-volumes will have a nonnegative limit).

Theorem 3.3.5. *A copula is max-stable if and only if it is an extreme value copula.*

Proof. Clearly every max-stable copula is an extreme value copula. Conversely, if C_* is an extreme value copula, then C_* satisfies (3.3.10) for some copula C. Hence for any positive real r,

$$C_*^r(u^{1/r},v^{1/r}) = \lim_{n\to\infty} C^{rn}(u^{1/rn},v^{1/rn}) = C_*(u,v),$$

so that C_* is max-stable. □

We now present a procedure (Pickands 1981) for constructing extreme value (or equivalently, max-stable) copulas. Let C be a max-stable copula, and let X and Y be standard exponential random variables whose survival copula is C. Thus the survival functions of X and Y are $\overline{F}(x) = e^{-x}, x > 0$, and $\overline{G}(y) = e^{-y}, y > 0$, respectively, and the joint survival function is given by

$$\overline{H}(x,y) = P(X > x, Y > y) = C(e^{-x}, e^{-y}).$$

Because C is max-stable,

$$\overline{H}(rx, ry) = C^r(e^{-x}, e^{-y}) = [\overline{H}(x,y)]^r$$

for any real $r > 0$. Define a function $A:[0,1] \to [1/2,1]$ by

$$A(t) = -\ln C(e^{-(1-t)}, e^{-t}) \tag{3.3.11}$$

or equivalently, $C(e^{-(1-t)}, e^{-t}) = \exp\{-A(t)\}$. Employing the change of variables $(x,y) = \big(r(1-t), rt\big)$ for $r > 0$, and t in $(0,1)$ [or equivalently, $(r,t) = \big(x+y, y/(x+y)\big)$], we have

$$\overline{H}(x,y) = \overline{H}\big(r(1-t), rt\big) = [\overline{H}(1-t,t)]^r$$
$$= C^r(e^{-(1-t)}, e^{-t}) = \exp\{-rA(t)\}$$
$$= \exp\big\{-(x+y)A\big(y/(x+y)\big)\big\}.$$

Because $C(u,v) = \overline{H}(-\ln u, -\ln v)$, we have proven that if C is an extreme value copula, then

$$C(u,v) = \exp\left\{\ln(uv)A\left(\frac{\ln v}{\ln(uv)}\right)\right\} \tag{3.3.12}$$

for an appropriate choice of the function A (called the *dependence function* of the extreme value copula C) in (3.3.11). For the right side of (3.3.12) to define a copula requires that $A:[0,1] \to [1/2,1]$ must satisfy the following conditions: $A(0) = A(1) = 1$, $\max\{t, 1-t\} \leq A(t) \leq 1$, and A convex. Thus the graph of A must lie in the shaded region of Fig. 3.20(a). See (Joe 1997) for details.

When $A(t) = 1$, (3.3.12) yields Π, and when $A(t) = \max\{t, 1-t\}$, (3.3.12) yields M.

Example 3.23. (a) If $A(t) = 1 - \min\big(\beta t, \alpha(1-t)\big)$ for α, β in $\mathbf{I}$, then (3.3.12) yields the Marshall-Olkin family (3.1.3) of copulas, and if (b) $A(t) = \big(t^\theta + (1-t)^\theta\big)^{1/\theta}$, $\theta \geq 1$, then (3.3.12) yields the Gumbel-Hougaard family (2.4.2). ∎

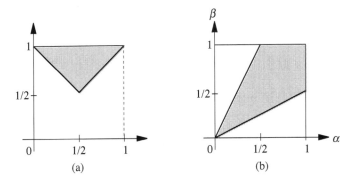

Fig. **3.20.** Regions containing (a) the graph of A in (3.3.12) and (b) (α,β) in
Example 3.24

Example 3.24. Let $A(t) = 1 - t(1-t)[\alpha t + \beta(1-t)]$. $A'(0^+) \in [-1,0]$ and
$A'(1^-) \in [0,1]$ requires α and β in **I**, and A will be convex when $\alpha \le 2\beta$
and $\beta \le 2\alpha$. So when the point (α,β) lies in the shaded region in Fig.
3.20(b), $A(t)$ will generate an extreme value copula via (3.3.12). ∎

Exercises

3.35 For the Plackett family of copulas (3.3.3), show that

(a) $C_0(u,v) = \lim_{\theta \to 0^+} C_\theta(u,v) = \dfrac{(u+v-1)+|u+v-1|}{2} = W(u,v)$,

(b) $C_\infty(u,v) = \lim_{\theta \to \infty} C_\theta(u,v) = \dfrac{(u+v)-|u-v|}{2} = M(u,v)$.

3.36 Let C_θ be a member of the Plackett family (3.3.3) of copulas,
where θ is in $(0,\infty)$.
(a) Show that $C_{1/\theta}(u,v) = u - C_\theta(u,1-v) = v - C_\theta(1-u,v)$ [see Exercise 2.6 and Theorem 2.4.4].
(b) Conclude that C_θ satisfies the functional equation $C = \hat{C}$ for
radial symmetry [see Theorem 2.7.3].

3.37 Show that the Plackett family (3.3.3) is positively ordered.

3.38 Show that the following algorithm (Johnson 1986) generates random variates (u,v) from a Plackett distribution with parameter θ:

1. Generate two independent uniform $(0,1)$ variates u, t;
2. Set $a = t(1-t)$; $b = \theta + a(\theta-1)^2$; $c = 2a(u\theta^2 + 1 - u) + \theta(1-2a)$; and $d = \sqrt{\theta} \cdot \sqrt{\theta + 4au(1-u)(1-\theta)^2}$;
3. Set $v = [c - (1-2t)d]/2b$;
4. The desired pair is (u,v).

3.39 (a) Show that the Farlie-Gumbel-Morgenstern family (3.2.10) is a first-order approximation to the Plackett family, i.e., if C_θ in (3.3.3), with θ in $[0,2]$, is expanded in a Taylor series in powers of $(\theta-1)$, then the first two terms are

$$uv + (\theta - 1)uv(1-u)(1-v).$$

(b) Similarly, show that a second-order approximation to the Plackett family consists of the copulas with cubic sections given by (3.2.20) with $a = (\theta-1) - (\theta-1)^2/2$ and $b = (\theta-1)^2/2$ for θ in $[0,3]$.

3.40 Let C_θ denote a member of the Ali-Mikhail-Haq family (3.3.7). Show that

$$C_\theta(u,v) = uv \sum_{k=0}^{\infty} [\theta(1-u)(1-v)]^k$$

and hence the FGM family (3.2.10) is a first-order approximation to the Ali-Mikhail-Haq family, and the iterated FGM family (3.2.17) of Lin (1987) with $\varphi = \theta$ is a second-order approximation.

3.41 (a) Show that the harmonic mean of two Ali-Mikhail-Haq copulas is again an Ali-Mikhail-Haq copula, i.e., if C_α and C_β are given by (3.3.7), then the harmonic mean of C_α and C_β is $C_{(\alpha+\beta)/2}$.
(b) Show that each Ali-Mikhail-Haq copula is a weighted harmonic mean of the two extreme members of the family, i.e., for all θ in $[-1,1]$,

$$C_\theta(u,v) = \frac{1}{\dfrac{1-\theta}{2} \cdot \dfrac{1}{C_{-1}(u,v)} + \dfrac{1+\theta}{2} \cdot \dfrac{1}{C_{+1}(u,v)}}.$$

[Cf. Exercise 3.21.]

3.42 Show that the following algorithm (Johnson 1986) generates random variates (u,v) from an Ali-Mikhail-Haq distribution with parameter θ:

    1. Generate two independent uniform $(0,1)$ variates u, t;

    2. Set $a = 1 - u$; $b = -\theta(2at+1)+2\theta^2 a^2 t+1$; and

    $c = \theta^2(4a^2 t - 4at+1)-\theta(4at-4t+2)+1$;

    3. Set $v = 2t(a\theta-1)^2\big/\big(b+\sqrt{c}\big)$;

    4. The desired pair is (u,v).

3.4 Copulas with Specified Properties

In this short section, we investigate copulas with certain well-known analytical or functional properties.

3.4.1 Harmonic Copulas

Let C be a copula with continuous second-order partial derivatives on $(0,1)^2$. Then C is *harmonic* in $\mathbf{I}^2$ if C satisfies *Laplace's equation* in $(0,1)^2$:

$$\nabla^2 C(u,v) = \frac{\partial^2}{\partial u^2} C(u,v) + \frac{\partial^2}{\partial v^2} C(u,v) = 0.$$

Clearly Π is harmonic. It is the only harmonic copula, because for any other harmonic copula C, $C-\Pi$ would also be harmonic and equal to 0 on the boundary of $\mathbf{I}^2$ and hence equal to 0 on all of $\mathbf{I}^2$.

Closely related notions are subharmonic and superharmonic copulas. A copula C is *subharmonic* if $\nabla^2 C(u,v) \geq 0$ and *superharmonic* if $\nabla^2 C(u,v) \leq 0$. For example, it is an elementary calculus exercise to show that if C_θ is a FGM copula given by (3.2.10), then C_θ is subharmonic for $\theta \in [-1,0]$ and superharmonic for $\theta \in [0,1]$.

3.4.2 Homogeneous Copulas

Definition 3.4.1. A copula C is *homogeneous of degree* k if for some real number k and all u,v,λ in $\mathbf{I}$,

$$C(\lambda u, \lambda v) = \lambda^k C(u,v). \tag{3.4.1}$$

Example 3.25. (a) Since $(\lambda u)(\lambda v) = \lambda^2 uv$, Π is homogeneous of degree 2, and since $\min(\lambda u, \lambda v) = \lambda \min(u,v)$, M is homogeneous of degree 1.

(b) Let C_θ be a member of the Cuadras-Augé family (2.2.10), $\theta \in [0,1]$. Then

$$C_\theta(\lambda u, \lambda v) = [M(\lambda u, \lambda v)]^\theta [\Pi(\lambda u, \lambda v)]^{1-\theta}$$
$$= \lambda^\theta [M(u,v)]^\theta \cdot (\lambda^2)^{1-\theta} [\Pi(u,v)]^{1-\theta}$$
$$= \lambda^{2-\theta} C_\theta(u,v).$$

Thus C_θ is homogeneous of degree $2-\theta$.   ∎

There are no other homogeneous copulas, as the following theorem demonstrates.

Theorem 3.4.2. *Suppose C is homogeneous of degree k. Then* (i) $1 \leq k \leq 2$, *and* (ii) *C is a member of the Cuadras-Augé family* (2.2.10) *with* $\theta = 2 - k$.

Proof. Setting $u = v = 1$ in (2.9.1) yields $C(\lambda, \lambda) = \lambda^k$, hence the diagonal section δ_C of C is given by $\delta_C(t) = t^k$. Invoking Exercise 2.8(a) yields $2t - 1 \leq t^k \leq t$ for t in $\mathbf{I}$, so that $1 \leq k \leq 2$. Setting $v = 1$ in (2.9.1) yields $C(\lambda u, \lambda) = \lambda^k u = (\lambda u)\lambda^{k-1}$. Hence $C(u,v) = uv^{k-1}$ for $u \leq v$, and similarly $C(u,v) = u^{k-1}v$ for $v \leq u$. Thus C is a Cuadras-Augé copula with $\theta = 2 - k$ as claimed.   □

3.4.3 Concave and Convex Copulas

Definition 3.4.3. A copula C is *concave* (*convex*) if for all (a,b), (c,d) in $\mathbf{I}^2$ and all λ in $\mathbf{I}$,

$$C\big(\lambda a + (1-\lambda)c, \lambda b + (1-\lambda)d\big) \geq (\leq) \lambda C(a,b) + (1-\lambda)C(c,d). \quad (3.4.2)$$

Equivalently, C is concave if the set of points in the unit cube $\mathbf{I}^3$ below the graph of $C(u,v)$ is a convex set, and C is convex if the set of points in the unit cube above the graph of $C(u,v)$ is a convex set.

Example 3.26. (a) It is easily verified that M is concave. It is the only concave copula, because if C were concave, then setting $(a,b) = (1,1)$ and $(c,d) = (0,0)$ in (3.4.2) yields $C(\lambda, \lambda) \geq \lambda$. But this implies that $\delta_C(t) = t$ on $\mathbf{I}$, and hence (as a result of Exercise 2.8(b)) C must be M.

(b) It is also easily verified that W is convex. It is the only convex copula, because if C were convex, then setting $(a,b) = (1,0)$ and $(c,d) =$

(0,1) in (3.4.2) yields $C(\lambda,1-\lambda) \leq 0$. Hence $C(t,1-t) = 0$ on **I**, and as a result of Exercise 2.9, C must be W. ∎

Thus convexity and concavity are conditions too strong to be of much interest for copulas. Hence we consider weaker versions of these properties.

Suppose that only the vertical or the horizontal sections of a copula C are concave. As we shall see in Sect. 5.2.3, many copulas have this property, and this geometric property of the graph of C corresponds to a statistical positive dependence property known as stochastic monotonicity.

We now weaken the notions in Definition 3.4.3 by replacing the weighted average of $C(a,b)$ and $C(c,d)$ on the right in (3.4.2) by the minimum or the maximum of $C(a,b)$ and $C(c,d)$:

Definition 3.4.4. A copula C is *quasi-concave* if for all (a,b), (c,d) in **I**^{2} and all λ in **I**,

$$C(\lambda a + (1-\lambda)c, \lambda b + (1-\lambda)d) \geq \min\{C(a,b),C(c,d)\}, \qquad (3.4.3)$$

and C is *quasi-convex* if for all (a,b), (c,d) in **I**^{2} and all λ in **I**,

$$C(\lambda a + (1-\lambda)c, \lambda b + (1-\lambda)d) \leq \max\{C(a,b),C(c,d)\}. \qquad (3.4.4)$$

In the next theorem, we show that the quasi-concavity of a copula C is equivalent to a property of the level sets of C.

Theorem 3.4.5 (Alsina et al. 2005). *Let C be a copula, and let L_t be the function whose graph is the upper boundary of the level set* $\{(u,v) \in \mathbf{I}^2 | C(u,v) = t\}$, *i.e.,* $L_t(u) = \sup\{v \in \mathbf{I} | C(u,v) = t\}$ *for all u in **I**. Then C is quasi-concave if and only if the function L_t is convex for all t in $[0,1)$.*

Proof. Suppose that L_t is convex for each t in $[0,1)$, so that each of the sets $L(t) = \{(u,v) \in \mathbf{I}^2 | C(u,v) \geq t\}$ is convex. Choose points $P = (u_1,v_1)$, $Q = (u_2,v_2)$ in **I**^{2} and let $a = \min\{C(u_1,v_1),C(u_2,v_2)\}$. Because $C(u_1,v_1) \geq a$ and $C(u_2,v_2) \geq a$, both P and Q are in $L(a)$, hence the entire line segment joining P and Q lies in $L(a)$, and thus C is quasi-concave.

In the other direction, assume L_a is not convex for some a in $[0,1)$, so that the set $L(a)$ is not convex. Hence there exist points P and Q in $L(a)$ and a point (u,v) on the segment joining P and Q such that $C(u,v) < a$. Hence C is not quasi-concave. □

The next example shows that W is the only quasi-convex copula.

Example 3.27. Suppose C is quasi-convex. Then setting $(a,b) = (1,0)$ and $(c,d) = (0,1)$ in (3.4.4) yields $C(\lambda,1-\lambda) \leq 0$, so that as in Example 3.23(b), C must be W.   ∎

Closely related to quasi-concavity and -convexity are the notions of Schur-concavity and -convexity:

Definition 3.4.6. A copula C is *Schur-concave* if for all a,b,λ in $\mathbf{I}$,

$$C(a,b) \leq C\big(\lambda a + (1-\lambda)b, \lambda b + (1-\lambda)a\big), \qquad (3.4.5)$$

and *Schur-convex* when the inequality in (3.4.5) is reversed.

Note that W is the only Schur-convex copula, because setting $(a,b) = (1,0)$ yields $0 \geq C(\lambda,1-\lambda)$ (see Examples 3.26(b) and 3.27).

If C is Schur-concave, then setting $\lambda = 0$ in (3.4.5) yields $C(a,b) \leq C(b,a)$ for all (a,b) in $\mathbf{I}^2$, hence C must be symmetric. Thus the Schur-concavity of a copula can be interpreted geometrically as follows: the graph of a section formed by intersecting the surface $z = C(u,v)$ with the plane $u + v = t$ (t in $[0,2]$) is symmetric in the plane $u + v = t$ with respect to the vertical line through $(t/2,t/2)$ and descends in both directions from a maximum at $(t/2,t/2,C(t/2,t/2))$.

It is easy to show that M, W, and Π are Schur-concave and that any convex linear combination of Schur-concave copulas is a Schur-concave copula. Thus every member of the Fréchet and Mardia families in Exercise 2.4 is Schur-concave.

The next example shows that Schur-concavity neither implies nor is implied by quasi-concavity:

Example 3.28. (a) Let $C = (M+W)/2$. Because C is a member of both the Fréchet and Mardia families, it is Schur-concave. Some of the contours of C are illustrated in Fig. 3.21(a). These contours are the graphs of the functions L_t in Theorem 3.4.5. Because L_t fails to be convex for t in $(0,1/4)$, C is not quasi-concave.

(b) Let $C_{\alpha,\beta}$ be any copula from the family in Exercise 3.8 with $\alpha \neq \beta$. The contours of $C_{\alpha,\beta}$ can be readily seen (see Figs. 3.8(a) and 3.21(b)) to be convex, so that $C_{\alpha,\beta}$ is quasi-concave. But $C_{\alpha,\beta}$ is not symmetric, hence it is not Schur-concave.   ∎

Finally, we note that if a copula is both quasi-concave and symmetric, then it is Schur-concave: set $(c,d) = (b,a)$ in (3.4.3). For further properties of Schur-concave copulas and additional examples, see (Durante and Sempi 2003).

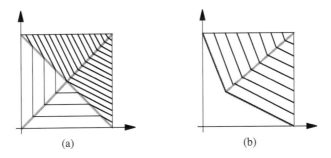

Fig. 3.21. Some contours of the copulas in Example 3.28

3.5 Constructing Multivariate Copulas

First, a word of caution: Constructing n-copulas is difficult. Few of the procedures discussed earlier in this chapter have n-dimensional analogs. In this section, we will outline some of the problems associated with the construction of n-copulas and provide references to some techniques. Most of our illustrations will be for 3-copulas; however, when appropriate, we will provide the n-dimensional version of the relevant theorems.

Recall that 2-copulas join or "couple" one-dimensional distribution functions to form bivariate distribution functions. The "naive" approach to constructing multidimensional distributions via copulas would be to use 2-copulas to join or couple other 2-copulas, as the following example illustrates:

Example 3.29. (a) Define a 3-place function C via $C(u,v,w) = \Pi(M(u,v),w) = w \cdot \min(u,v)$. Then, as shown in Example 2.23(a), C is a 3-copula.

(b) Define a 3-place function C via $C(u,v,w) = W(M(u,v),w)$. Then $W(M(u,v),w) = \min(u,v) - \min(u,v,1-w)$, and hence C is the 3-copula in Example 2.23(b). ∎

Unfortunately, this procedure can fail:

Example 3.30. Define a 3-place function C via $C(u,v,w) = W(W(u,v),w) = \max(u+v+w-2,0)$. Thus $C = W^3$, which is not a 3-copula (see Exercise 2.35). Note that each of the 2-margins of W^3 is W, and it is impossible in set of three random variables X, Y, and Z for each random variable to be almost surely a decreasing function of each of the remaining two. ∎

In fact, this procedure—replacing one of the arguments in a 2-copula with another 2-copula—often fails. If C_1 and C_2 are 2-copulas such that $C_2(C_1(u,v),w)$ is a 3-copula, we say that C_1 is *directly compatible* with C_2 (Quesada Molina and Rodríguez Lallena 1994). The following theorem provides criteria for direct compatibility when one of C_1 or C_2 is M, W, or Π. Its proof can be found in (Quesada Molina and Rodríguez Lallena 1994).

Theorem 3.5.1. 1. *Every 2-copula C is directly compatible with Π;*
2. *The only 2-copula directly compatible with M is M;*
3. *The only 2-copula directly compatible with W is M;*
4. *M is directly compatible with every 2-copula C;*
5. *W is directly compatible only with Π; and*

6. *Π is directly compatible with a 2-copula C if and only if for all v_1, v_2, w_1, w_2 in $\mathbf{I}$ such that $v_1 \le v_2$ and $w_1 \le w_2$, the function*

$$u \mapsto V_C\big([uv_1, uv_2] \times [w_1, w_2]\big)$$

is nondecreasing on $\mathbf{I}$.

An important class of copulas for which this procedure—endowing a 2-copula with a multivariate margin—often succeeds is the class of Archimedean copulas. Archimedean n-copulas are discussed in Sect. 4.6.

From Sklar's theorem, we know that if C is a 2-copula, and F and G are univariate distribution functions, then $C(F(x),G(y))$ is always a two dimensional distribution function. Can we extend this procedure to higher dimensions by replacing F and G by multivariate distributions functions? That is, given $m+n \ge 3$, for what 2-copulas C is it true that if $F(\mathbf{x})$ is an m-dimensional distribution function and $G(\mathbf{y})$ is an n-dimensional distribution function, then $C(F(\mathbf{x}),G(\mathbf{y}))$ is an $(m+n)$-dimensional distribution function? The answer is provided in the following "impossibility" theorem (Genest et al. 1995):

Theorem 3.5.2. *Let m and n be positive integers such that $m+n \ge 3$, and suppose that C is a 2-copula such that $H(\mathbf{x},\mathbf{y}) = C(F(\mathbf{x}),G(\mathbf{y}))$ is an $(m+n)$-dimensional distribution function with margins $H(\mathbf{x},\infty) = F(\mathbf{x})$ and $H(\infty,\mathbf{y}) = G(\mathbf{y})$ for all m-dimensional distribution functions $F(\mathbf{x})$ and n-dimensional distribution functions $G(\mathbf{y})$. Then $C = \Pi$.*

The following theorem (Schweizer and Sklar 1983) presents related results for the cases when the 2-copula C in the preceding theorem is Π or M, and the multidimensional distribution functions F and G are copulas (or, if the dimension is 1, the identity function):

Theorem 3.5.3. *Let m and n be integers ≥ 2. Let C_1 be an m-copula, and C_2 an n-copula.*

1. *Let C be the function from $\mathbf{I}^{m+n}$ to $\mathbf{I}$ given by*

$$C(x_1, x_2, \cdots, x_{m+n}) = M\big(C_1(x_1, x_2, \cdots, x_m), C_2(x_{m+1}, x_{m+2}, \cdots, x_{m+n})\big).$$

Then C is an $(m+n)$-copula if and only if $C_1 = M^m$ and $C_2 = M^n$.
  2. Let C', C'', and C''' be the functions defined by

$$C'(x_1,x_2,\cdots,x_{m+1}) = \Pi\big(C_1(x_1,x_2,\cdots,x_m),x_{m+1}\big),$$

$$C''(x_1,x_2,\cdots,x_{n+1}) = \Pi\big(x_1,C_2(x_2,x_3,\cdots,x_{n+1})\big),$$

$$C'''(x_1,x_2,\cdots,x_{m+n}) = \Pi\big(C_1(x_1,x_2,\cdots,x_m),C_2(x_{m+1},x_{m+2},\cdots,x_{m+n})\big).$$

Then C' is always an $(m+1)$-copula, C'' is always an $(n+1)$-copula, and C''' is always an $(m+n)$-copula.

The results in the preceding theorems illustrate some aspects of what has become known as *the compatibility problem*. Recall from Sect. 2.10 that if C is an n-copula, and we set $n - k$ (for $1 \leq k < n$) of the arguments of C equal to 1, then the result is one of the $\binom{n}{k}$ k-margins of C. In the opposite direction, however, a given set of $\binom{n}{k}$ k-copulas rarely are the k-margins of any n-copula. If they are, then these $\binom{n}{k}$ k-copulas are said to be *compatible*.

The compatibility problem has a long history. To facilitate our discussion, let $\mathbf{C}^3(C_{12})$ denote the class of 3-copulas of continuous random variables X, Y, and Z such that the 2-copula of X and Y is C_{12} (i.e., $C_{XY} = C_{12}$); $\mathbf{C}^3(C_{12},C_{13})$ the class of 3-copulas for which $C_{XY} = C_{12}$ and $C_{XZ} = C_{13}$; and similarly in higher dimensions. Note that parts 4, 5, and 6 of Theorem 3.5.1 deal with the class $\mathbf{C}^3(C_{12})$ when C_{12} is M, Π, or W. For the more general problem of constructing a trivariate joint distribution function given the three univariate margins and one bivariate margin, see (Joe 1997).

Necessary and sufficient conditions for a 3-copula C to have specified 2-margins C_{12} and C_{13} (i.e., to be a member of $\mathbf{C}^3(C_{12},C_{13})$) were first discussed (in terms of Fréchet-Hoeffding classes—i.e., joint distribution functions with given margins—rather than copulas) in (Dall'Aglio 1960, 1972). Compatibility questions and construction procedures associated with the classes $\mathbf{C}^3(C_{12},C_{13},C_{23})$, $\mathbf{C}^4(C_{123},C_{124},C_{134},C_{234})$, and $\mathbf{C}^n(C_{ij},1 \leq i < j \leq n)$ are discussed (again in terms of joint distributions functions rather than copulas) in (Joe 1996, 1997). In 1983, Schweizer and Sklar noted that "the problem of extending [these results] to higher dimensions is perhaps the most important open question concerning copulas;" and that remains true today.

The classes of n-copulas in the preceding paragraph all have over-lapping margins, that is, given 2- or 3-margins that share a common one-dimensional margin. For construction procedures (many of which are based on conditional distributions) in the case for which the given margins are nonoverlapping, see (Kellerer 1964; Rüschendorf 1985; Cuadras 1992; Marco and Ruiz 1992; Chakak and Koehler 1995; Li et al. 1996b) and the references therein.

We conclude this section with an n-dimensional extension of one of the families discussed earlier in this chapter.

Example 3.31. *Farlie-Gumbel-Morgenstern n-copulas.* The FGM family (3.2.10) has the following extension to a $(2^n - n - 1)$-parameter family of n-copulas, $n \geq 3$ (Johnson and Kotz 1975):

$$C(\mathbf{u}) = u_1 u_2 \cdots u_n \left[1 + \sum_{k=2}^{n} \sum_{1 \leq j_1 < \cdots < j_k \leq n} \theta_{j_1 j_2 \cdots j_k} \bar{u}_{j_1} \bar{u}_{j_2} \cdots \bar{u}_{j_k} \right]$$

(where $\bar{u} = 1 - u$). Each copula in this family is absolutely continuous with density

$$\frac{\partial^n C(\mathbf{u})}{\partial u_1 \cdots \partial u_n} = 1 + \sum_{k=2}^{n} \sum_{1 \leq j_1 < \cdots < j_k \leq n} \theta_{j_1 j_2 \cdots j_k} (1 - 2u_{j_1})(1 - 2u_{j_2}) \cdots (1 - 2u_{j_k}).$$

Because $C(\mathbf{u})$ is quadratic in each variable, the density $\partial^n C(\mathbf{u})/\partial u_1 \cdots \partial u_n$ is linear in each variable. Hence the density will be nonnegative on $\mathbf{I}^n$ if and only if it is nonnegative at each of the 2^n vertices of $\mathbf{I}^n$, which leads to the following 2^n constraints for the parameters (Cambanis 1977):

$$1 + \sum_{k=2}^{n} \sum_{1 \leq j_1 < \cdots < j_k \leq n} \varepsilon_{j_1} \varepsilon_{j_2} \cdots \varepsilon_{j_k} \theta_{j_1 j_2 \cdots j_k} \geq 0, \quad \varepsilon_{j_1}, \varepsilon_{j_2}, \cdots, \varepsilon_{j_n} \in \{-1, +1\},$$

(as a consequence, each parameter must satisfy $|\theta| \leq 1$).

Note that each k-margin, $2 \leq k < n$, of an FGM n-copula is an FGM k-copula. See (Conway 1983) for applications and additional references. ∎

4 Archimedean Copulas

In this chapter, we discuss an important class of copulas known as Archimedean copulas. These copulas find a wide range of applications for a number of reasons: (1) the ease with which they can be constructed; (2) the great variety of families of copulas which belong to this class; and (3) the many nice properties possessed by the members of this class. As mentioned in the Introduction, Archimedean copulas originally appeared not in statistics, but rather in the study of probabilistic metric spaces, where they were studied as part of the development of a probabilistic version of the triangle inequality. For an account of this history, see (Schweizer 1991) and the references cited therein.

4.1 Definitions

Let X and Y be continuous random variables with joint distribution function H and marginal distribution function F and G, respectively. When X and Y are independent, $H(x,y) = F(x)G(y)$ for all x,y in $\overline{\mathbf{R}}$, and this is the only instance in which the joint distribution function factors into a product of a function of F and a function of G. But in the past chapter we saw cases in which *a function of H* does indeed factor into a product of a function of F and a function of G. In (3.3.6) of Sect. 3.3, we observed that the joint and marginal distribution functions for members of the Ali-Mikhail-Haq family satisfy the relationship

$$\frac{1-H(x,y)}{H(x,y)} = \frac{1-F(x)}{F(x)} + \frac{1-G(y)}{G(y)} + (1-\theta)\frac{1-F(x)}{F(x)}\cdot\frac{1-G(y)}{G(y)}.$$

With a little algebra, this can be rewritten as

$$1+(1-\theta)\frac{1-H(x,y)}{H(x,y)} = \left[1+(1-\theta)\frac{1-F(x)}{F(x)}\right]\cdot\left[1+(1-\theta)\frac{1-G(y)}{G(y)}\right],$$

that is, $\lambda(H(x,y)) = \lambda(F(x))\lambda(G(y))$, where $\lambda(t) = 1+(1-\theta)(1-t)/t$. Equivalently, whenever we can write $\lambda(H(x,y)) = \lambda(F(x))\lambda(G(y))$ for a function λ (which must be positive on the interval $(0,1)$) then on setting

$\varphi(t) = -\ln\lambda(t)$, we can also write H as a sum of functions of the marginals F and G, i.e., $\varphi(H(x,y)) = \varphi(F(x)) + \varphi(G(y))$, or for copulas,

$$\varphi(C(u,v)) = \varphi(u) + \varphi(v). \qquad (4.1.1)$$

Example 4.1. The copula $\hat{C}_\theta$ given by (2.6.3) in Example 2.14 satisfies (4.1.1) with $\varphi(t) = t^{-1/\theta} - 1$. The copula C_θ from (2.4.2) in Exercise 2.13 satisfies (4.1.1) with $\varphi(t) = (-\ln t)^\theta$. ∎

Because we are interested in expressions that we can use for the construction of copulas, we want to solve the relation $\varphi(C(u,v)) = \varphi(u) + \varphi(v)$ for $C(u,v)$, that is, $C(u,v) = \varphi^{[-1]}(\varphi(u) + \varphi(v))$ for an appropriately defined "inverse" $\varphi^{[-1]}$. This can be done as follows:

Definition 4.1.1. Let φ be a continuous, strictly decreasing function from $\mathbf{I}$ to $[0,\infty]$ such that $\varphi(1) = 0$. The *pseudo-inverse* of φ is the function $\varphi^{[-1]}$ with $\mathrm{Dom}\,\varphi^{[-1]} = [0,\infty]$ and $\mathrm{Ran}\,\varphi^{[-1]} = \mathbf{I}$ given by

$$\varphi^{[-1]}(t) = \begin{cases} \varphi^{-1}(t), & 0 \le t \le \varphi(0), \\ 0, & \varphi(0) \le t \le \infty. \end{cases} \qquad (4.1.2)$$

Note that $\varphi^{[-1]}$ is continuous and nonincreasing on $[0,\infty]$, and strictly decreasing on $[0,\varphi(0)]$. Furthermore, $\varphi^{[-1]}(\varphi(u)) = u$ on $\mathbf{I}$, and

$$\varphi\big(\varphi^{[-1]}(t)\big) = \begin{cases} t, & 0 \le t \le \varphi(0), \\ \varphi(0), & \varphi(0) \le t \le \infty, \end{cases}$$
$$= \min(t,\varphi(0)).$$

Finally, if $\varphi(0) = \infty$, then $\varphi^{[-1]} = \varphi^{-1}$.

Lemma 4.1.2. *Let φ be a continuous, strictly decreasing function from $\mathbf{I}$ to $[0,\infty]$ such that $\varphi(1) = 0$, and let $\varphi^{[-1]}$ be the pseudo-inverse of φ defined by (4.1.2). Let C be the function from $\mathbf{I}^2$ to $\mathbf{I}$ given by*

$$C(u,v) = \varphi^{[-1]}(\varphi(u) + \varphi(v)). \qquad (4.1.3)$$

Then C satisfies the boundary conditions (2.2.2a) and (2.2.2b) for a copula.

Proof. $C(u,0) = \varphi^{[-1]}(\varphi(u)+\varphi(0)) = 0$, and $C(u,1) = \varphi^{[-1]}(\varphi(u)+\varphi(1)) = \varphi^{[-1]}(\varphi(u)) = u$. By symmetry, $C(0,v) = 0$ and $C(1,v) = v$. □

In the following lemma, we obtain a necessary and sufficient condition for the function C in (4.1.3) to be 2-increasing.

Lemma 4.1.3. *Let* φ, $\varphi^{[-1]}$ *and* C *satisfy the hypotheses of Lemma 4.1.2. Then* C *is 2-increasing if and only if for all* v *in* **I**, *whenever* $u_1 \le u_2$,

$$C(u_2,v) - C(u_1,v) \le u_2 - u_1. \tag{4.1.4}$$

Proof. Because (4.1.4) is equivalent to $V_C([u_1,u_2] \times [v,1]) \ge 0$, it holds whenever C is 2-increasing. Hence assume that C satisfies (4.1.4). Choose v_1, v_2 in **I** such that $v_1 \le v_2$, and note that $C(0,v_2) = 0 \le v_1 \le v_2 = C(1,v_2)$. But C is continuous (because φ and $\varphi^{[-1]}$ are), and thus there is a t in **I** such that $C(t, v_2) = v_1$, or $\varphi(v_2) + \varphi(t) = \varphi(v_1)$. Hence

$$C(u_2,v_1) - C(u_1,v_1) = \varphi^{[-1]}(\varphi(u_2)+\varphi(v_1)) - \varphi^{[-1]}(\varphi(u_1)+\varphi(v_1)),$$
$$= \varphi^{[-1]}(\varphi(u_2)+\varphi(v_2)+\varphi(t)) - \varphi^{[-1]}(\varphi(u_1)+\varphi(v_2)+\varphi(t)),$$
$$= C(C(u_2,v_2),t) - C(C(u_1,v_2),t),$$
$$\le C(u_2,v_2) - C(u_1,v_2),$$

so that C is 2-increasing. □

We are now ready to state and prove the main result of this section.

Theorem 4.1.4. *Let* φ *be a continuous, strictly decreasing function from* **I** *to* $[0,\infty]$ *such that* $\varphi(1) = 0$, *and let* $\varphi^{[-1]}$ *be the pseudo-inverse of* φ *defined by* (4.1.2). *Then the function* C *from* $\mathbf{I}^2$ *to* **I** *given by* (4.1.3) *is a copula if and only if* φ *is convex.*

Proof (Alsina et al. 2005). We have already shown that C satisfies the boundary conditions for a copula, and as a consequence of the preceding lemma, we need only prove that (4.1.4) holds if and only if φ is convex [note that φ is convex if and only if $\varphi^{[-1]}$ is convex]. Observe that (4.1.4) is equivalent to

$$u_1 + \varphi^{[-1]}(\varphi(u_2)+\varphi(v)) \le u_2 + \varphi^{[-1]}(\varphi(u_1)+\varphi(v))$$

for $u_1 \le u_2$, so that if we set $a = \varphi(u_1)$, $b = \varphi(u_2)$, and $c = \varphi(v)$, then (4.1.4) is equivalent to

$$\varphi^{[-1]}(a) + \varphi^{[-1]}(b+c) \le \varphi^{[-1]}(b) + \varphi^{[-1]}(a+c), \tag{4.1.5}$$

where $a \geq b$ and $c \geq 0$. Now suppose (4.1.4) holds, i.e., suppose that $\varphi^{[-1]}$ satisfies (4.1.5). Choose any s, t in $[0,\infty]$ such that $0 \leq s < t$. If we set $a = (s+t)/2$, $b = s$, and $c = (t-s)/2$ in (4.1.5), we have

$$\varphi^{[-1]}\left(\frac{s+t}{2}\right) \leq \frac{\varphi^{[-1]}(s) + \varphi^{[-1]}(t)}{2}. \qquad (4.1.6)$$

Thus $\varphi^{[-1]}$ is midconvex, and because $\varphi^{[-1]}$ is continuous it follows that $\varphi^{[-1]}$ is convex.

In the other direction, assume $\varphi^{[-1]}$ is convex. Fix a, b, and c in $\mathbf{I}$ such that $a \geq b$ and $c \geq 0$; and let $\gamma = (a-b)/(a-b+c)$. Now $a = (1-\gamma)b + \gamma(a+c)$ and $b+c = \gamma b + (1-\gamma)(a+c)$, and hence

$$\varphi^{[-1]}(a) \leq (1-\gamma)\varphi^{[-1]}(b) + \gamma\varphi^{[-1]}(a+c)$$

and

$$\varphi^{[-1]}(b+c) \leq \gamma\varphi^{[-1]}(b) + (1-\gamma)\varphi^{[-1]}(a+c).$$

Adding these inequalities yields (4.1.5), which completes the proof. □

Copulas of the form (4.1.3) are called *Archimedean* copulas (the meaning of the term "Archimedean" will be explained in Sect. 4.3). The function φ is called a *generator* of the copula. If $\varphi(0) = \infty$, we say that φ is a *strict* generator. In this case, $\varphi^{[-1]} = \varphi^{-1}$ and $C(u,v) = \varphi^{-1}(\varphi(u) + \varphi(v))$ is said to be a *strict* Archimedean copula. Figure 4.1 illustrates generators and their quasi-inverses in the strict and non-strict cases. To be precise, the function φ is an *additive generator* of C. If we set $\lambda(t) = \exp(-\varphi(t))$ and $\lambda^{[-1]}(t) = \varphi^{[-1]}(-\ln t)$, then $C(u,v) = \lambda^{[-1]}[\lambda(u)\lambda(v)]$, so that λ is a *multiplicative generator*. In the sequel, we will deal primarily with additive generators.

Example 4.2. (a) Let $\varphi(t) = -\ln t$ for t in $[0,1]$. Because $\varphi(0) = \infty$, φ is strict. Thus $\varphi^{[-1]}(t) = \varphi^{-1}(t) = \exp(-t)$, and generating C via (4.1.3) yields $C(u,v) = \exp(-[(-\ln u) + (-\ln v)]) = uv = \Pi(u,v)$. Thus Π is a strict Archimedean copula.

(b) Let $\varphi(t) = 1-t$ for t in $[0,1]$. Then $\varphi^{[-1]}(t) = 1-t$ for t in $[0,1]$ and 0 for $t > 1$; i.e., $\varphi^{[-1]}(t) = \max(1-t,0)$. Again using (4.1.3), $C(u,v) = \max(u + v - 1,0) = W(u,v)$. Hence W is also Archimedean.

(c) M is not Archimedean—see Exercise 4.2. ∎

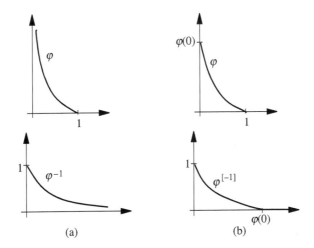

Fig. 4.1. Strict (a) and non-strict (b) generators and inverses

Example 4.3. Let $\varphi_\theta(t) = \ln(1 - \theta \ln t)$ for θ in $(0,1]$. Because $\varphi_\theta(0) = \infty$, φ_θ is strict, and $\varphi_\theta^{[-1]}(t) = \varphi_\theta^{-1}(t) = \exp[(1 - e^t)/\theta]$. If C_θ denotes the Archimedean copula generated by φ_θ, then $C_\theta(u,v) = uv\exp(-\theta \ln u \ln v)$, the survival copula for Gumbel's bivariate exponential distribution (see Example 2.13).   ∎

In the next section, we will present a number of one-parameter families of Archimedean copulas and in Sect. 4.3 will study the fundamental properties of Archimedean copulas, using many of those one-parameter families as examples. We conclude this section with two theorems concerning some algebraic properties of Archimedean copulas.

Theorem 4.1.5. *Let C be an Archimedean copula with generator φ. Then:*

1. *C is symmetric; i.e., $C(u,v) = C(u,v)$ for all u,v in $\mathbf{I}$;*
2. *C is associative, i.e., $C(C(u,v),w) = C(u,C(v,w))$ for all u,v,w in $\mathbf{I}$;*

3. *If $c > 0$ is any constant, then $c\varphi$ is also a generator of C.*

The proof of Theorem 4.1.5 is left as an exercise. Furthermore, it is also easy to show (see Exercise 4.2) that the diagonal section δ_C of an Archimedean copula C satisfies $\delta_C(u) < u$ for all u in $(0,1)$. The following theorem states that this property and associativity characterize Archimedean copulas. Its proof can be found in (Ling 1965).

Theorem 4.1.6. *Let C be an associative copula such that $\delta_C(u) < u$ for all u in $(0,1)$. Then C is Archimedean.*

Although there does not seem to be a statistical interpretation for random variables with an associative copula, associativity will be a useful property when we construct multivariate Archimedean copulas in Sect. 4.6.

4.2 One-parameter Families

As seen in Examples 4.2 and 4.3, Archimedean copulas can be constructed at will using Theorem 4.1.4—we need only find functions that will serve as generators, that is, continuous decreasing convex functions φ from $\mathbf{I}$ to $[0,\infty]$ with $\varphi(1) = 0$—and define the corresponding copulas via (4.1.4). For another example let $\varphi(t) = (1/t) - 1$, we then obtain the copula $C(u,v) = uv/(u+v-uv)$, which we first encountered in Example 2.8 and Exercises 2.12 and 2.21. Because this copula is a member of several of the families which we present below, we henceforth designate it as "$\Pi/(\Sigma-\Pi)$."

In Table 4.1 [adapted from (Alsina et al. 2005)], we list some important one-parameter families of Archimedean copulas, along with their generators, the range of the parameter, and some special and limiting cases. The limiting cases are computed by standard methods, including l'Hôpital's rule, and by Theorems 4.4.7 and 4.4.8 in Sect. 4.4.

As noted earlier, one reason for the usefulness of Archimedean copulas in statistical modeling is the variety of dependence structures present in the various families. Following Table 4.1, we present a sampling of scatterplots from simulations for eight of the families (Figs. 4.2-4.9). In each case, we used 500 sample points and the algorithm in Exercise 4.15 or 4.16. Additional scatterplots for members of the families in Table 4.1 may be found in (Armstrong 2003).

In the following example, we show how one family of Archimedean copulas arises in a statistical setting.

Example 4.4 (Schmitz 2004). Let $\{X_1,X_2,\cdots,X_n\}$ be a set of independent and identically distributed continuous random variables with distribution function F, and let $X_{(1)} = \min\{X_1,X_2,\cdots,X_n\}$ and $X_{(n)} = \max\{X_1,X_2,\cdots,X_n\}$. We now find the copula $C_{1,n}$ of $X_{(1)}$ and $X_{(n)}$. The distribution functions F_n of $X_{(n)}$ and F_1 of $X_{(1)}$ are given by $F_n(x) = [F(x)]^n$ and $F_1(x) = 1-[1-F(x)]^n$. For convenience, we will first find the joint distribution function H^Δ and copula C^Δ of $-X_{(1)}$ and $X_{(n)}$, rather than $X_{(1)}$ and $X_{(n)}$:

$$H^{\Delta}(s,t) = P\left[-X_{(1)} \le s, X_{(n)} \le t\right]$$
$$= P\left[-s \le X_{(1)}, X_{(n)} \le t\right]$$
$$= P\left[\text{all } X_i \text{ in } [-s,t]\right]$$
$$= \begin{cases} [F(t) - F(-s)]^n, & -s \le t, \\ 0, & -s > t, \end{cases}$$
$$= \left[\max(F(t) - F(-s), 0)\right]^n.$$

From Corollary 2.3.7 we have $C^{\Delta}(u,v) = H^{\Delta}(G^{(-1)}(u), F_n^{(-1)}(v))$, where G now denotes the distribution function of $-X_{(1)}$, $G(x) = [1 - F(-x)]^n$. Let $u = [1 - F(-s)]^n$ and $v = [F(t)]^n$, so that $F(-s) = 1 - u^{1/n}$ and $F(t) = v^{1/n}$. Thus $C^{\Delta}(u,v) = \left[\max\left(u^{1/n} + v^{1/n} - 1, 0\right)\right]^n$, a member of the Clayton family (4.2.1) in Table 4.1 (with $\theta = -1/n$).

Invoking part 2 of Theorem 2.4.4 now yields
$$C_{1,n}(u,v) = v - C^{\Delta}(1 - u, v)$$
$$= v - \left[\max\left((1 - u)^{1/n} + v^{1/n} - 1, 0\right)\right]^n.$$

Although $X_{(1)}$ and $X_{(n)}$ are clearly not independent ($C_{1,n} \ne \Pi$), they are asymptotically independent because (using the fact that the Clayton copula with $\theta = 0$ is Π) $\lim_{n \to \infty} C_{1,n}(u,v) = v - \Pi(1 - u, v) = uv$. ∎

4.3 Fundamental Properties

In this section, we will investigate some of the basic properties of Archimedean copulas. For convenience, let Ω denote the set of continuous strictly decreasing convex functions φ from $\mathbf{I}$ to $[0,\infty]$ with $\varphi(1) = 0$.

By now the reader is surely wondering about the meaning of the term "Archimedean" for these copulas. Recall the Archimedean axiom for the positive real numbers: If a,b are positive real numbers, then there exists an integer n such that $na > b$. An Archimedean copula behaves like a binary operation on the interval $\mathbf{I}$, in that the copula C assigns to each pair u,v in $\mathbf{I}$ a number $C(u,v)$ in $\mathbf{I}$. From Theorem 4.1.5, we see that the "operation" C is commutative and associative, and preserves order as a consequence of (2.2.5), i.e., $u_1 \le u_2$ and $v_1 \le v_2$ implies $C(u_1, v_1) \le C(u_2, v_2)$ (algebraists call $(\mathbf{I}, C)$ an ordered Abelian semigroup). [Discussion continues after Fig. 4.9.]

Table 4.1. One-parameter

(4.2.#)	$C_\theta(u,v)$	$\varphi_\theta(t)$
1	$\left[\max\left(u^{-\theta}+v^{-\theta}-1,0\right)\right]^{-1/\theta}$	$\frac{1}{\theta}\left(t^{-\theta}-1\right)$
2	$\max\left(1-\left[(1-u)^\theta+(1-v)^\theta\right]^{1/\theta},0\right)$	$(1-t)^\theta$
3	$\dfrac{uv}{1-\theta(1-u)(1-v)}$	$\ln\dfrac{1-\theta(1-t)}{t}$
4	$\exp\left(-\left[(-\ln u)^\theta+(-\ln v)^\theta\right]^{1/\theta}\right)$	$(-\ln t)^\theta$
5	$-\dfrac{1}{\theta}\ln\left(1+\dfrac{(e^{-\theta u}-1)(e^{-\theta v}-1)}{e^{-\theta}-1}\right)$	$-\ln\dfrac{e^{-\theta t}-1}{e^{-\theta}-1}$
6	$1-\left[(1-u)^\theta+(1-v)^\theta-(1-u)^\theta(1-v)^\theta\right]^{1/\theta}$	$-\ln\left[1-(1-t)^\theta\right]$
7	$\max\left(\theta uv+(1-\theta)(u+v-1),0\right)$	$-\ln\left[\theta t+(1-\theta)\right]$
8	$\max\left(\dfrac{\theta^2 uv-(1-u)(1-v)}{\theta^2-(\theta-1)^2(1-u)(1-v)},0\right)$	$\dfrac{1-t}{1+(\theta-1)t}$
9	$uv\exp(-\theta\ln u\ln v)$	$\ln(1-\theta\ln t)$
10	$uv\Big/\left[1+(1-u^\theta)(1-v^\theta)\right]^{1/\theta}$	$\ln\left(2t^{-\theta}-1\right)$
11	$\left[\max\left(u^\theta v^\theta-2(1-u^\theta)(1-v^\theta),0\right)\right]^{1/\theta}$	$\ln\left(2-t^\theta\right)$
12	$\left(1+\left[(u^{-1}-1)^\theta+(v^{-1}-1)^\theta\right]^{1/\theta}\right)^{-1}$	$\left(\dfrac{1}{t}-1\right)^\theta$
13	$\exp\left(1-\left[(1-\ln u)^\theta+(1-\ln v)^\theta-1\right]^{1/\theta}\right)$	$(1-\ln t)^\theta-1$
14	$\left(1+\left[(u^{-1/\theta}-1)^\theta+(v^{-1/\theta}-1)^\theta\right]^{1/\theta}\right)^{-\theta}$	$(t^{-1/\theta}-1)^\theta$

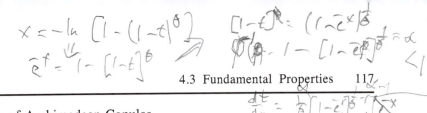

Families of Archimedean Copulas

$\theta \in$	Strict	Limiting and Special Cases	(4.2.#)
$[-1,\infty)\setminus\{0\}$	$\theta \geq 0$	$C_{-1} = W,\ C_0 = \Pi,\ C_1 = \dfrac{\Pi}{\Sigma - \Pi},\ C_\infty = M$	1
$[1,\infty)$	no	$C_1 = W,\ C_\infty = M$	2
$[-1,1)$	yes	$C_0 = \Pi,\ C_1 = \dfrac{\Pi}{\Sigma - \Pi}$	3
$[1,\infty)$	yes	$C_1 = \Pi,\ C_\infty = M$	4
$(-\infty,\infty)\setminus\{0\}$	yes	$C_{-\infty} = W,\ C_0 = \Pi,\ C_\infty = M$	5
$[1,\infty)$	yes	$C_1 = \Pi,\ C_\infty = M$	6
$(0,1]$	no	$C_0 = W,\ C_1 = \Pi$	7
$[1,\infty)$	no	$C_1 = W,\ C_\infty = \dfrac{\Pi}{\Sigma - \Pi}$	8
$(0,1]$	yes	$C_0 = \Pi$	9
$(0,1]$	yes	$C_0 = \Pi$	10
$(0,1/2]$	no	$C_0 = \Pi$	11
$[1,\infty)$	yes	$C_1 = \dfrac{\Pi}{\Sigma - \Pi},\ C_\infty = M$	12
$(0,\infty)$	yes	$C_1 = \Pi,\ C_\infty = M$	13
$[1,\infty)$	yes	$C_1 = \dfrac{\Pi}{\Sigma - \Pi},\ C_\infty = M$	14

Table 4.1. One-parameter

(4.2.#)	$C_\theta(u,v)$	$\varphi_\theta(t)$
15	$\left\{\max\left(1-\left[(1-u^{1/\theta})^\theta+(1-v^{1/\theta})^\theta\right]^{1/\theta},0\right)\right\}^\theta$	$\left(1-t^{1/\theta}\right)^\theta$
16	$\frac{1}{2}\left(S+\sqrt{S^2+4\theta}\right),\ S=u+v-1-\theta\left(\frac{1}{u}+\frac{1}{v}-1\right)$	$\left(\frac{\theta}{t}+1\right)(1-t)$
17	$\left(1+\frac{[(1+u)^{-\theta}-1][(1+v)^{-\theta}-1]}{2^{-\theta}-1}\right)^{-1/\theta}-1$	$-\ln\frac{(1+t)^{-\theta}-1}{2^{-\theta}-1}$
18	$\max\left(1+\theta/\ln\left[e^{\theta/(u-1)}+e^{\theta/(v-1)}\right],0\right)$	$e^{\theta/(t-1)}$
19	$\theta/\ln\left(e^{\theta/u}+e^{\theta/v}-e^\theta\right)$	$e^{\theta/t}-e^\theta$
20	$\left[\ln\left(\exp(u^{-\theta})+\exp(v^{-\theta})-e\right)\right]^{-1/\theta}$	$\exp\left(t^{-\theta}\right)-e$
21	$1-(1-\{\max([1-(1-u)^\theta]^{1/\theta}+$ $[1-(1-v)^\theta]^{1/\theta}-1,0)\}^\theta)^{1/\theta}$	$1-\left[1-(1-t)^\theta\right]^{1/\theta}$
22	$\max\left(\left[1-(1-u^\theta)\sqrt{1-(1-v^\theta)^2}\right.\right.$ $\left.\left.-(1-v^\theta)\sqrt{1-(1-u^\theta)^2}\right]^{1/\theta},0\right)$	$\arcsin\left(1-t^\theta\right)$

Notes on some of the families in Table 4.1:

(4.2.1) This family of copulas was discussed by Clayton (1978), Oakes (1982, 1986), Cox and Oakes (1984), and Cook and Johnson (1981, 1986). Genest and MacKay (1986) call this the generalized *Cook and Johnson* family; Hutchinson and Lai (1990) call it the *Pareto* family of copulas—see Example 2.14; while Genest and Rivest (1993) call it the *Clayton* family, as shall we. It is one of only two families (the other is (4.2.5)) in the table that are comprehensive.

(4.2.3) This is the Ali-Mikhail-Haq family, which we derived algebraically in Sect. 3.3.2. Also see Example 4.8 in the next section.

(4.2.4) This family of copulas was first discussed by Gumbel (1960b), hence many authors refer to it as the *Gumbel* family. However, because Gumbel's name is attached to another Archimedean family (4.2.9) and this family also appears in Hougaard (1986), Hutchinson and Lai (1990) refer to it as the *Gumbel-Hougaard* family. We encountered this family in Exercise 2.13 in conjunction with type B bivariate extreme value distributions. Also see (Genest and Rivest 1989).

Families of Archimedean Copulas

$\theta \in$	Strict	Limiting and Special Cases	(4.2.#)
$[1,\infty)$	no	$C_1 = W,\ C_\infty = M$	15
$[0,\infty)$	$\theta > 0$	$C_0 = W,\ C_\infty = \dfrac{\Pi}{\Sigma - \Pi}$	16
$(-\infty,\infty)\backslash\{0\}$	yes	$C_{-1} = \Pi,\ C_\infty = M$	17
$[2,\infty)$	no	$C_\infty = M$	18
$(0,\infty)$	yes	$C_0 = \dfrac{\Pi}{\Sigma - \Pi},\ C_\infty = M$	19
$(0,\infty)$	yes	$C_0 = \Pi,\ C_\infty = M$	20
$[1,\infty)$	no	$C_1 = W,\ C_\infty = M$	21
$(0,1]$	no	$C_0 = \Pi$	22

Notes on some of the families in Table 4.1 (continued):

(4.2.5) This is the *Frank* family, which first appeared in Frank (1979) in a non-statistical context. Some of the statistical properties of this family were discussed in (Nelsen 1986; Genest 1987). These are the *only* Archimedean copulas which satisfy the functional equation $C(u,v) = \hat{C}(u,v)$ in Theorem 2.7.3 for radial symmetry—see (Frank 1979) for a proof of this remarkable result. As noted above, this is one of two comprehensive families in the table.

(4.2.6) This family is discussed in (Joe 1993, 1997), and the co-copulas for members of this family appear in (Frank 1981).

(4.2.9) The copulas in this family are the survival copulas associated with Gumbel's bivariate exponential distribution (Gumbel 1960a)—see Examples 2.9 and 2.13. Although many authors refer to these copulas as another *Gumbel* family, Hutchinson and Lai (1990) call it the *Gumbel-Barnett* family, as Barnett (1980) first discussed it as a family of copulas, i.e., after the margins of the bivariate exponential were translated to uniform (0,1) margins.

(4.2.15) This family is discussed in (Genest and Ghoudi 1994).

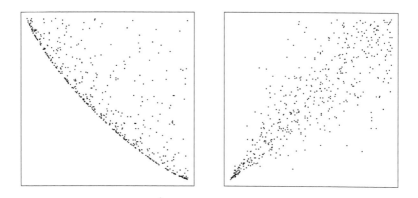

Fig. 4.2. Scatterplots for copulas (4.2.1), $\theta = -0.8$ (left) and $\theta = 4$ (right)

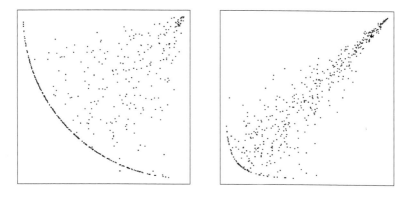

Fig. 4.3. Scatterplots for copulas (4.2.2), $\theta = 2$ (left) and $\theta = 8$ (right)

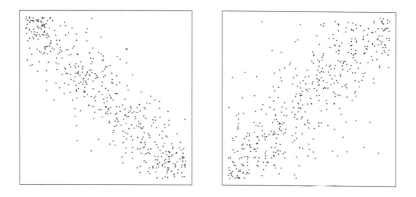

Fig. 4.4. Scatterplots for copulas (4.2.5), $\theta = -12$ (left) and $\theta = 8$ (right)

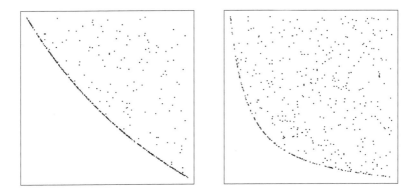

Fig. 4.5. Scatterplots for copulas (4.2.7), $\theta = 0.4$ (left) and $\theta = 0.9$ (right)

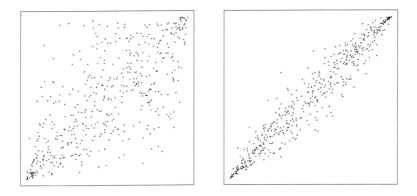

Fig. 4.6. Scatterplots for copulas (4.2.12), $\theta = 1.5$ (left) and $\theta = 4$ (right)

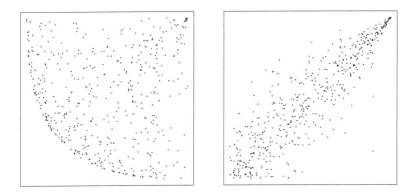

Fig. 4.7. Scatterplots for copulas (4.2.15), $\theta = 1.5$ (left) and $\theta = 4$ (right)

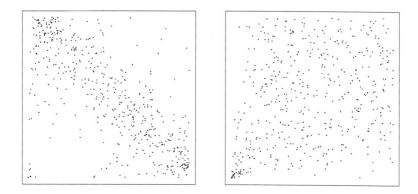

Fig. 4.8. Scatterplots for copulas (4.2.16), $\theta = 0.01$ (left) and $\theta = 1$ (right)

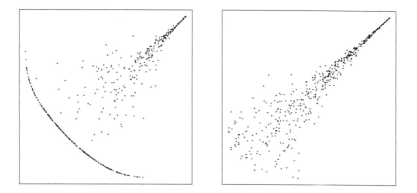

Fig. 4.9. Scatterplots for copulas (4.2.18), $\theta = 2$ (left) and $\theta = 6$ (right)

For any u in $\mathbf{I}$, we can define the *C-powers* u_C^n of u recursively: $u_C^1 = u$, and $u_C^{n+1} = C(u, u_C^n)$ [note that u_C^2 belongs to the diagonal section $\delta_C(u)$ of C]. The version of the Archimedean axiom for $(\mathbf{I}, C)$ is, For any two numbers u, v in $(0,1)$, there exists a positive integer n such that $u_C^n < v$. The next theorem shows that Archimedean copulas satisfy this version of the Archimedean axiom and hence merit their name. The term "Archimedean" for these copulas was introduced in (Ling 1965).

Theorem 4.3.1. *Let C be an Archimedean copula generated by φ in Ω. Then for any u, v in $\mathbf{I}$, there exists a positive integer n such that $u_C^n < v$.*

Proof. Let u, v be any elements of $(0,1)$. The nth C-power u_C^n of u is readily seen to be $\varphi^{[-1]}(n\varphi(u))$. Since $\varphi(u)$ and $\varphi(v)$ are positive real numbers, the Archimedean axiom applies, and thus there is an integer n such that $n\varphi(u) > \varphi(v)$. But because $v > 0$, $\varphi(v) < \varphi(0)$, and hence $v = \varphi^{[-1]}(\varphi(v)) > \varphi^{[-1]}(n\varphi(u)) = u_C^n$. [Note that the convexity of φ is not required in the proof of this theorem.] □

For an account of the history of the representation of associative functions, which dates back to the early work of Niels Abel, see (Schweizer and Sklar 1983; Alsina et al. 2005).

In the next theorem, we set the groundwork for determining which Archimedean copulas are absolutely continuous, and which ones have singular components. Recall (Sects. 2.2 and 3.4.3) that the level sets of a copula C are given by $\left\{(u,v) \in \mathbf{I}^2 \middle| C(u,v) = t\right\}$. For an Archimedean copula and for $t > 0$, this level set consists of the points on the *level curve* $\varphi(u) + \varphi(v) = \varphi(t)$ in $\mathbf{I}^2$ that connects the points $(1,t)$ and $(t,1)$. We will often write the level curve as $v = L_t(u)$, as solving for v as a function of u yields

$$v = L_t(u) = \varphi^{[-1]}\big(\varphi(t) - \varphi(u)\big) = \varphi^{-1}\big(\varphi(t) - \varphi(u)\big), \qquad (4.3.1)$$

where the last step (replacing $\varphi^{[-1]}$ by φ^{-1}) is justified because $\varphi(t) - \varphi(u)$ is in the interval $[0, \varphi(0))$. For $t = 0$, we call $\left\{(u,v) \in \mathbf{I}^2 \middle| C(u,v) = 0\right\}$ the *zero set* of C, and denote it $Z(C)$. For many Archimedean copulas, $Z(C)$ is simply the two line segments $\{0\} \times \mathbf{I}$ and $\mathbf{I} \times \{0\}$. For others, $Z(C)$ has positive area, and for such a zero set the boundary curve $\varphi(u) + \varphi(v) = \varphi(0)$, i.e., $v = L_0(u)$, of $Z(C)$ is called the *zero curve* of C. See Fig. 4.2 for an illustration of the latter case—the member of family (4.2.2) in Table 4.1 with $\theta = 2$, in which the level curves and zero curve are quarter circles. Indeed, the graph of this copula is one-quarter of a circular cone whose vertex is one unit above $(1,1)$.

In Fig. 4.10, the level curves are convex. This must be the case for all Archimedean copulas (but not all copulas—see Example 3.28 and Exercise 4.4), as the following theorem shows.

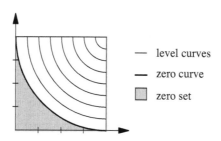

Fig. 4.10. Graphs of some level curves, the zero set, and the zero curve of the Archimedean copula in (4.2.2) with $\theta = 2$

Theorem 4.3.2. *The level curves of an Archimedean copula are convex.*

Proof. Let C be an Archimedean copula with generator φ. For t in $[0,1)$, the level curves of C are given by (4.3.1), and we need only show that L_t is midconvex because it is continuous. Now φ is convex, so

$$\varphi(t) - \varphi\left(\frac{u_1 + u_2}{2}\right) \geq \varphi(t) - \frac{\varphi(u_1) + \varphi(u_2)}{2}$$

$$= \frac{[\varphi(t) - \varphi(u_1)] + [\varphi(t) - \varphi(u_2)]}{2};$$

and because φ^{-1} is decreasing and convex, we have

$$L_t\left(\frac{u_1 + u_2}{2}\right) = \varphi^{-1}\left[\varphi(t) - \varphi\left(\frac{u_1 + u_2}{2}\right)\right]$$

$$\leq \varphi^{-1}\left[\frac{[\varphi(t) - \varphi(u_1)] + [\varphi(t) - \varphi(u_2)]}{2}\right]$$

$$\leq \frac{1}{2}\left[\varphi^{-1}\left(\varphi(t) - \varphi(u_1)\right) + \varphi^{-1}\left(\varphi(t) - \varphi(u_2)\right)\right]$$

$$= \frac{L_t(u_1) + L_t(u_2)}{2}. \qquad \square$$

The C-measure carried by each of the level curves of an Archimedean copula C is given in the following theorem (Alsina et al. 2005).

Theorem 4.3.3. *Let C be an Archimedean copula generated by φ in Ω.*

1. For t in $(0,1)$, the C-measure of the level curve $\varphi(u) + \varphi(v) = \varphi(t)$ is given by

$$\varphi(t)\left[\frac{1}{\varphi'(t^-)}-\frac{1}{\varphi'(t^+)}\right], \qquad (4.3.2)$$

where $\varphi'(t^-)$ and $\varphi'(t^+)$ denote the one-sided derivatives of φ at t. In particular, if $\varphi'(t)$ exists—which is the case for all but at most a countably infinite set of points—then this C-measure is 0.

2. If C is not strict, then the C-measure of the zero curve $\varphi(u) + \varphi(v) = \varphi(0)$ is equal to

$$-\frac{\varphi(0)}{\varphi'(0^+)}, \qquad (4.3.3)$$

and thus equal to 0 whenever $\varphi'(0^+) = -\infty$.

Proof. We first note that because φ is convex, the one-sided derivatives $\varphi'(t^-)$ and $\varphi'(t^+)$ exist in $(0,1]$ and $[0,1)$, respectively (Roberts and Varberg 1973). Let t be in $(0,1)$, and set $w = \varphi(t)$. Let n be a fixed positive integer, and consider the partition of the interval $[t,1]$ induced by the regular partition $\{0,w/n,\cdots,kw/n,\cdots,w\}$ of $[0,w]$, i.e., the partition $\{t=t_0,t_1,\cdots,t_n=1\}$ where $t_{n-k} = \varphi^{[-1]}(kw/n)$, $k = 0,1,\cdots,n$. Because $w < \varphi(0)$, it follows from (4.1.2) that

$$C(t_j,t_k)=\varphi^{[-1]}\big(\varphi(t_j)+\varphi(t_k)\big)$$
$$=\varphi^{[-1]}\left(\frac{n-j}{n}w+\frac{n-k}{n}w\right)=\varphi^{[-1]}\left(w+\frac{n-j-k}{n}w\right).$$

In particular, $C(t_j,t_{n-j})=\varphi^{[-1]}(w)=t$.

Denote the rectangle $[t_{k-1},t_k]\times[t_{n-k},t_{n-k+1}]$ by R_k, and let $S_n = \cup_{k=1}^n R_k$ [see Fig. 4.11(a)]. From the convexity of $\varphi^{[-1]}$ it follows that

$$0\le t_1-t_0\le t_2-t_1\le\cdots\le t_n-t_{n-1}=1-t_{n-1},$$

and clearly $\lim_{n\to\infty}(1-t_{n-1}) = 1-\varphi^{[-1]}(0) = 0$. Hence the C-measure of the level curve $\varphi(u)+\varphi(v)=\varphi(t)$ is given by $\lim_{n\to\infty}V_C(S_n)$. For each k we have

$$V_C(R_k)=C(t_{k-1},t_{n-k})-t-t+C(t_k,t_{n-k+1})$$
$$=\left[\varphi^{[-1]}(w+w/n)-\varphi^{[-1]}(w)\right]-\left[\varphi^{[-1]}(w)-\varphi^{[-1]}(w-w/n)\right].$$

Thus

$$V_C(S_n) = \sum_{k=1}^{n} V_C(R_k)$$

$$= w\left[\frac{\varphi^{[-1]}(w+w/n) - \varphi^{[-1]}(w)}{w/n} - \frac{\varphi^{[-1]}(w) - \varphi^{[-1]}(w-w/n)}{w/n}\right]$$

from which (4.3.2) follows by taking the limit as $n\to\infty$.

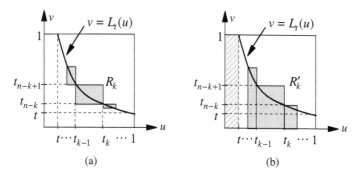

Fig. 4.11. Rectangles R_k and R_k' in the proofs of Theorems 4.3.3 and 4.3.4

For a non-strict C and $t = 0$, $\varphi(0)$ is finite and $C(u,v) = 0$ in $Z(C)$, i.e., on and below the level curve $\varphi(u)+\varphi(v)=\varphi(0)$. Thus for each k, $V_C(R_k) = C(t_k,t_{n-k+1})$, from which, using the above argument, (4.3.3) follows. □

Example 4.5. Let θ be in $(0,1]$, and let φ_θ be the piecewise linear function in Ω whose graph connects $(0,2-\theta)$ to $(\theta/2,1-(\theta/2))$ to $(1,0)$, as illustrated in part (a) of Fig. 4.12. The slopes of the two line segments in the graph are $-(2-\theta)/\theta$ and -1. If C_θ is the Archimedean copula generated by φ_θ, then it follows from (4.3.2) that the C_θ-measure of the level curve $\varphi_\theta(u)+\varphi_\theta(v) = \varphi_\theta(\theta/2)$ is

$$\left(1-\frac{\theta}{2}\right)\left[-\frac{\theta}{2-\theta}+1\right]=1-\theta;$$

and from (4.3.3) that the C_θ-measure of the zero curve $\varphi_\theta(u)+\varphi_\theta(v)=\varphi_\theta(0)$ is θ. Because these measures sum to one, the Archimedean copulas in this family are singular, and the support of C_θ consists of the level curve $\varphi_\theta(u)+\varphi_\theta(v)=\varphi_\theta(\theta/2)$ and the zero curve, as illustrated in part (b) of Fig. 4.12. [Note that both $\lim_{\theta\to 0^+} C_\theta$ and C_1 are

W.] Indeed, if the generator φ of an Archimedean copula C is a piece-wise linear function in Ω, then C must be singular. ∎

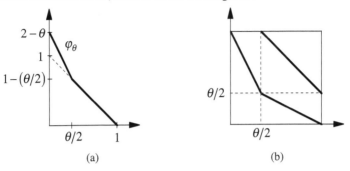

(a)   (b)

Fig. 4.12. The generator and support of the copula in Example 4.5

Using the same methods as in the proof of Theorem 4.3.3, we can find the C-measure of the region in $\mathbf{I}^2$ lying on, or below and to the left of, each level curve.

Theorem 4.3.4. *Let C be an Archimedean copula generated by φ in Ω. Let $K_C(t)$ denote the C-measure of the set $\{(u,v) \in \mathbf{I}^2 | C(u,v) \le t\}$, or equivalently, of the set $\{(u,v) \in \mathbf{I}^2 | \varphi(u) + \varphi(v) \ge \varphi(t)\}$. Then for any t in* $\mathbf{I}$,

$$K_C(t) = t - \frac{\varphi(t)}{\varphi'(t^+)}. \qquad (4.3.4)$$

Proof. Let t be in $(0,1)$, and set $w = \varphi(t)$. Let n be a fixed positive integer, and consider the same partitions of $[t,1]$ and $[0,w]$ as appear in the proof of Theorem 4.3.3. Let R'_k denote the rectangle $[t_{k-1}, t_k] \times [0, t_{n-k+1}]$, and set $S'_n = \cup_{k=1}^{n} R'_k$ [see Fig. 4.11(b)]. Proceeding as before, $K_C(t)$ is given by the sum of the C-measure of $[0,t] \times \mathbf{I}$ and $\lim_{n \to \infty} V_C(S'_n)$, i.e., $K_C(t) = t + \lim_{n \to \infty} V_C(S'_n)$. For each k we have

$$V_C(R'_k) = C(t_k, t_{n-k+1}) - t = \varphi^{[-1]}(w - w/n) - \varphi^{[-1]}(w),$$

and hence

$$V_C(S'_n) = \sum_{k=1}^{n} V_C(R'_k)$$

$$= -w \left[\frac{\varphi^{[-1]}(w) - \varphi^{[-1]}(w - w/n)}{w/n} \right]$$

from which (4.3.4) follows by taking the limit as $n \to \infty$.   □

The following corollary, which is a generalization of Theorem 4.3.4, is required for the proof of Theorem 4.4.7 in the next section.

Corollary 4.3.5. *Let C be an Archimedean copula generated by φ in Ω. Let $K'_C(s,t)$ denote the C-measure of the set $\{(u,v) \in \mathbf{I}^2 | u \le s, C(u,v) \le t\}$. Then for any s,t in $\mathbf{I}$,*

$$K'_C(s,t) = \begin{cases} s, & s \le t \\ t - \dfrac{\varphi(t) - \varphi(s)}{\varphi'(t^+)}, & s > t. \end{cases} \qquad (4.3.5)$$

Proof. When $s \le t$, $K'_C(s,t) = s$, as $\{(u,v) \in \mathbf{I}^2 | u \le s, C(u,v) \le t\} = \{(u,v) \in \mathbf{I}^2 | u \le s\}$. Assume $s > t$. Proceeding as in Theorems 4.3.3 and 4.3.4, let $z = \varphi(s)$ and consider the partition of the interval $[t,s]$ (rather than $[t,1]$) induced by the regular partition of the interval $[z,w]$ (rather than the interval $[0,w]$). Here $t_{n-k} = \varphi^{[-1]}(z + [k(w-z)/n])$, $k = 0,1,\cdots,n$, and hence $C(t_k, L(t_{k-1})) = \varphi^{[-1]}(w - (w-z)/n)$. Thus $V_C(R'_k) = \varphi^{[-1]}(w - (w-z)/n) - \varphi^{[-1]}(w)$, and the rest of the proof is analogous to that of Theorem 4.3.4. Note that (4.3.5) reduces to (4.3.4) when $s = 1$.   □

A special subclass of Archimedean copulas consists of those for which the generator is twice differentiable, i.e., when the copula C has a generator φ in Ω such that $\varphi'(t) < 0$ and $\varphi''(t) > 0$ for all t in $(0,1)$. For such copulas, Genest and MacKay (1986a,b) proved part 2 of Theorem 4.3.3 by using (2.4.1) to find the C-measure $A_C(1,1)$ of the absolutely continuous component of C. For copulas in this subclass, the support of the singular component (if any) consists of the zero curve; and moreover when C is absolutely continuous its density is given by

$$-\frac{\varphi''(C(u,v))\varphi'(u)\varphi'(v)}{[\varphi'(C(u,v))]^3}. \qquad (4.3.6)$$

Example 4.6. (a) Each of the copulas in the Clayton family (4.2.1) in Table 4.1 is non-strict for θ in the interval $[-1,0)$. However, $\varphi'_\theta(0^+) = -\infty$ and $\varphi''_\theta(t) > 0$ for any such θ, and hence every member of the Clayton family is absolutely continuous.

(b) The copulas in family (4.2.2) are all non-strict, and an elementary computation shows that the C_θ-measure of the zero curve is

$-\varphi_\theta(0)\big/\varphi'_\theta(0^+) = 1/\theta$. Thus the members of this family (aside from W) have both a singular and an absolutely continuous component. ∎

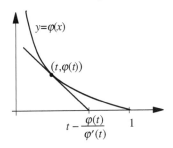

Fig. 4.13. A geometric interpretation of $K_C(t) = t - \big(\varphi(t)/\varphi'(t)\big)$

When the generator φ of an Archimedean copula is continuously differentiable, the C-measure of the set $\left\{(u,v) \in \mathbf{I}^2 \big| C(u,v) \le t\right\}$ given by (4.3.4) is $K_C(t) = t - \big(\varphi(t)/\varphi'(t)\big)$. As noted in (Genest and MacKay 1986b), there is a geometric interpretation of this result — $K_C(t)$ is the x-intercept of the line tangent to the graph of $y = \varphi(x)$ at the point $(t,\varphi(t))$, as shown in Fig. 4.13. Furthermore, when φ is non-strict, the x-intercept $K_C(0) = -\varphi(0)/\varphi'(0)$ of the line tangent to $y = \varphi(x)$ at its y-intercept $(0,\varphi(0))$ is the C-measure of the zero curve (which is positive when $\varphi'(0) > -\infty$).

The following corollary presents a probabilistic interpretation of Theorem 4.3.4 and Corollary 4.3.5 that will be useful in Chapter 5 when we consider the population version of the nonparametric measure of association known as Kendall's tau for Archimedean copulas.

Corollary 4.3.6. *Let U and V be uniform $(0,1)$ random variables whose joint distribution function is the Archimedean copula C generated by φ in Ω. Then the function K_C given by (4.3.4) is the distribution function of the random variable $C(U,V)$. Furthermore, the function K'_C given by (4.3.5) is the joint distribution function of U and $C(U,V)$.*

The next theorem (Genest and Rivest 1993) is an extension of Corollary 4.3.6. An application of this theorem is the algorithm for generating random variates from distributions with Archimedean copulas given in Exercise 4.15.

Theorem 4.3.7. *Under the hypotheses of Corollary 4.3.6, the joint distribution function $H(s,t)$ of the random variables $S = \varphi(U)/\big[\varphi(U)+\varphi(V)\big]$ and $T = C(U,V)$ is given by $H(s,t) = s \cdot K_C(t)$ for* *all (s,t) in $\mathbf{I}^2$. Hence S and T are independent, and S is uniformly distributed on $(0,1)$.*

Proof. We present a proof for the case when C is absolutely continuous. For a proof in the general case, see (Genest and Rivest 1993). The joint density $h(s,t)$ of S and T is given by

$$h(s,t) = \frac{\partial^2}{\partial u \partial v} C(u,v) \cdot \left| \frac{\partial(u,v)}{\partial(s,t)} \right|$$

in terms of s and t, where $\partial^2 C(u,v)/\partial u \partial v$ is given by (4.3.6) and $\partial(u,v)/\partial(s,t)$ denotes the Jacobian of the transformation $\varphi(u) = s\varphi(t)$, $\varphi(v) = (1-s)\varphi(t)$. But

$$\frac{\partial(u,v)}{\partial(s,t)} = \frac{\varphi(t)\varphi'(t)}{\varphi'(u)\varphi'(v)},$$

and hence

$$h(s,t) = \left(-\frac{\varphi''(t)\varphi'(u)\varphi'(v)}{[\varphi'(t)]^3} \right) \cdot \left(-\frac{\varphi(t)\varphi'(t)}{\varphi'(u)\varphi'(v)} \right) = \frac{\varphi''(t)\varphi(t)}{[\varphi'(t)]^2}.$$

Thus

$$H(s,t) = \int_0^s \int_0^t \frac{\varphi''(y)\varphi(y)}{[\varphi'(y)]^2} \, dy \, dx = s \cdot \left[y - \frac{\varphi(y)}{\varphi'(y)} \right]_0^t = s \cdot K_C(t),$$

and the conclusion follows.   □

If one has a copula C that is associative and for which $\delta_C(u) < u$ on $(0,1)$, then it must be Archimedean by Theorem 4.1.6. The next theorem yields a technique for finding generators of such copulas.

Theorem 4.3.8. *Let C be an Archimedean copula with generator φ in Ω. Then for almost all u,v in $\mathbf{I}$,*

$$\varphi'(u) \frac{\partial C(u,v)}{\partial v} = \varphi'(v) \frac{\partial C(u,v)}{\partial u}. \tag{4.3.7}$$

Proof. Because φ is convex, φ' exists almost everywhere in $(0,1)$. From Theorem 2.2.7, the partials $\partial C(u,v)/\partial u$ and $\partial C(u,v)/\partial v$ exist for almost all u,v in $\mathbf{I}$. Hence, applying the chain rule to $\varphi(C(u,v)) = \varphi(u) + \varphi(v)$, we have

$$\varphi'(C(u,v)) \frac{\partial C(u,v)}{\partial u} = \varphi'(u), \text{ and } \varphi'(C(u,v)) \frac{\partial C(u,v)}{\partial v} = \varphi'(v).$$

But because φ is strictly decreasing, $\varphi'(t) \neq 0$ wherever it exists, from which (4.3.7) now follows.   □

The next two examples illustrate the use of this theorem, in conjunction with Theorems 4.1.5 and 4.1.6, to determine if a particular copula is Archimedean, and when it is, what a generator might be.

Example 4.7. The Farlie-Gumbel-Morgenstern family of copulas was introduced in Example 3.12 in Sect. 3.2.5. Are any members of this family Archimedean? If so, they must be associative. But it is easy to show that if C_θ is given by (3.2.10), then

$$C_\theta\left(\frac{1}{4}, C_\theta\left(\frac{1}{2}, \frac{1}{3}\right)\right) \neq C_\theta\left(C_\theta\left(\frac{1}{4}, \frac{1}{2}\right), \frac{1}{3}\right)$$

for all θ in $[-1,1]$ except 0. Hence, except for Π, Farlie-Gumbel-Morgenstern copulas are not Archimedean.   ∎

Example 4.8. The Ali-Mikhail-Haq family of copulas was derived by algebraic methods in Sect. 3.3.2. It is easy but tedious to show that when C_θ is given by (3.3.7), then $C_\theta\left(u, C_\theta(v,w)\right) = C_\theta\left(C_\theta(u,v), w\right)$ for u,v,w in $\mathbf{I}$ and all θ in $[-1,1]$, and that $C_\theta(u,u) < u$ for all u in $(0,1)$. Hence, by Theorem 4.1.6, each Ali-Mikhail-Haq copula C_θ is Archimedean. To find a generator, we evaluate the partial derivatives of C_θ and invoke (4.3.7) to obtain

$$\frac{\varphi'_\theta(u)}{\varphi'_\theta(v)} = \frac{\partial C_\theta(u,v)/\partial u}{\partial C_\theta(u,v)/\partial v} = \frac{v\left[1-\theta(1-v)\right]}{u\left[1-\theta(1-u)\right]}.$$

Hence $\varphi'_\theta(t) = -c_\theta/\left(t[1-\theta(1-t)]\right)$ (where $c_\theta > 0$ since $\varphi'_\theta(t) < 0$), from which it follows that a generator is given by

$$\varphi_\theta(t) = \frac{c_\theta}{1-\theta}\ln\frac{1-\theta(1-t)}{t} \text{ for } \theta \text{ in } [-1,1), \text{ and } \varphi_1(t) = c_1\left(\frac{1}{t}-1\right).$$

Upon setting $c_1 = 1$ and $c_\theta = 1-\theta$ for θ in $[-1,1)$, we obtain the expression for φ_θ given in (4.2.3).   ∎

As a consequence of Example 3.11, a source of generators of Archimedean copulas consists of inverses of Laplace transforms of distribution functions. That is, if $\Lambda(\theta)$ is a distribution function with $\Lambda(0) = 0$ and

$$\psi(t) = \int_0^\infty e^{-\theta t} d\Lambda(\theta),$$

then $\varphi = \psi^{-1}$ generates a strict Archimedean copula—see (3.2.7).

Example 4.9. If Λ is a gamma distribution function with parameters $\alpha = 1/\theta$ and $\beta = 1$ for $\theta > 0$, then the Laplace transform of Λ is $\psi(t) =$

$(1+t)^{-1/\theta}$. Hence $\varphi(t) = \psi^{-1}(t) = t^{-\theta} - 1$, which generates the strict subfamily of (4.2.1). For further examples, see (Joe 1993). ∎

We close this section by noting that there are estimation procedures for selecting the Archimedean copula which best fits a given random sample (Genest and Rivest 1993), and for estimating the parameter θ in a given Archimedean family (Shih and Louis 1995).

Exercises

4.1   Prove Theorem 4.1.5.

4.2   The diagonal section of an Archimedean copula C with generator φ in Ω is given by $\delta_C(u) = \varphi^{[-1]}[2\varphi(u)]$. Prove that if C is Archimedean, then for u in $(0,1)$, $\delta_C(u) < u$. Conclude that M is not an Archimedean copula.

4.3   Show that $\varphi : I \to [0,\infty]$ is in Ω if and only if $1 - \varphi^{[-1]}(t)$ is a unimodal distribution function on $[0,\infty]$ with mode at zero.

4.4   The converse of Theorem 4.3.2 is false. Using the copulas in Example 3.3 and Exercise 3.8, show that non-Archimedean copulas can have (a) non-convex level curves, and (b) convex level curves.

4.5   Let C be an Archimedean copula. Prove that C is strict if and only if $C(u,v) > 0$ for (u,v) in $(0,1]^2$.

4.6   This exercise shows that different Archimedean copulas can have the same zero set. Let

$$\varphi_1(t) = \arctan\frac{1-t}{1+t} \quad \text{and} \quad \varphi_2(t) = \ln\frac{\sqrt{2}+1-t}{\sqrt{2}-1+t}.$$

(a) Show that φ_1 and φ_2 are in Ω, and hence generate Archimedean copulas C_1 and C_2, respectively.
(b) Show that

$$C_1(u,v) = \max\left(\frac{uv+u+v-1}{1+u+v-uv},0\right)$$

and

$$C_2(u,v) = \max\left(\frac{uv+u+v-1}{3-u-v+uv},0\right).$$

(c) Show that C_1 and C_2 have the same zero curve $v = (1-u)/(1+u)$, from which it follows that $Z(C_1) = Z(C_2)$.

4.7 Let C_1 and C_2 be non-strict Archimedean copulas with generators φ_1 and φ_2, respectively, normalized (via Theorem 4.1.5, part 3) so that $\varphi_1(0) = 1 = \varphi_2(0)$. Let $\psi(t) = \varphi_1 \circ \varphi_2^{-1}(t)$ for t in $\mathbf{I}$. Prove that $Z(C_1) = Z(C_2)$ if and only if $\psi(t) + \psi(1-t) = 1$ for all t in $\mathbf{I}$, i.e., if and only if the graph of ψ is symmetric with respect to the point $(1/2, 1/2)$. (Alsina et al. 2005).

4.8 Let C_θ be a member of the Frank family (4.2.5) of copulas for θ in $\mathbf{R}$.
(a) Show that $C_{-\theta}(u,v) = u - C_\theta(u, 1-v) = v - C_\theta(1-u, v)$ [see Exercise 2.6 and Theorem 2.4.4].
(b) Conclude that C_θ satisfies the functional equation $C = \hat{C}$ for radial symmetry [see Theorem 2.7.3].
[Cf. Exercise 3.36.]

4.9 (a) Show that the Farlie-Gumbel-Morgenstern family (3.2.10) is a first-order approximation to the Frank family, i.e., if C_θ in (4.2.5), with θ in $[-2,2]$, is expanded in a Taylor series in powers of θ, then the first two terms are

$$uv + \frac{\theta}{2} uv(1-u)(1-v).$$

(b) Similarly, show that a second-order approximation to the Frank family consists of the copulas with cubic sections given by (3.2.20) with $a = \theta/2$ and $b = \theta^2/12$ for θ in $[-2,2]$.

4.10 (a) Show that the geometric mean of two Gumbel-Barnett copulas is again a Gumbel-Barnett copula, i.e., if C_α and C_β are given by (4.2.9), then the geometric mean of C_α and C_β is $C_{(\alpha+\beta)/2}$.
(b) Show that each Gumbel-Barnett copula is a weighted geometric mean of the two extreme members of the family, i.e., for all θ in $[0,1]$,

$$C_\theta(u,v) = \left[C_0(u,v)\right]^{1-\theta} \cdot \left[C_1(u,v)\right]^{\theta}.$$

[Cf. Exercises 2.5, 3.21 and 3.41.]

4.11 Prove that, as in Example 4.6(a), $\varphi'_\theta(0^+) = -\infty$ and $\varphi''_\theta(t) > 0$ holds for copulas in families (4.2.11), (4.2.15), (4.2.21), and (4.2.22) in Table 4.1, and hence that members of these families are absolutely continuous.

4.12 Prove that, as in Example 4.6(b), the C_θ-measure of the zero curve is $1/\theta$ for the members of families (4.2.8) and (4.2.18) in Table 4.1; and is $-\theta\ln\theta/(1-\theta)$ for family (4.2.7).

4.13 Prove that every Archimedean copula is Schur-concave (Durante and Sempi 2003). [Hint: use Theorems 3.4.5 and 4.3.2, and note that every Archimedean copula is symmetric.]

4.14 Let C be a copula, and define $C_\gamma(u,v) = \gamma^{-1}\big(C(\gamma(u),\gamma(v))\big)$, where γ and γ^{-1} satisfy the properties in Theorem 3.3.3.
(a) Show that Π_γ is a strict Archimedean copula. [Hint: is $-\ln\gamma(t)$ a generator?]
(b) Show that W_γ is a non-strict Archimedean copula. [Hint: is $1-\gamma(t)$ a generator?]
(c) More generally, show that if C is Archimedean, so is C_γ.

4.15 Use Theorem 4.3.7 to show that the following algorithm generates random variates (u,v) whose joint distribution function is an Archimedean copula C with generator φ in Ω:
    1. Generate two independent uniform $(0,1)$ variates s and t;
    2. Set $w = K_C^{(-1)}(t)$, where K_C is given by (4.3.4);
    3. Set $u = \varphi^{[-1]}(s\varphi(w))$ and $v = \varphi^{[-1]}((1-s)\varphi(w))$;
    4. The desired pair is (u,v).

4.16 Show that the following algorithm (Genest and MacKay 1986a) generates random variates (u,v) whose joint distribution function is an Archimedean copula C with generator φ in Ω:
    1. Generate two independent uniform $(0,1)$ variates u and t;
    2. Set $w = \varphi'^{(-1)}(\varphi'(u)/t)$;
    3. Set $v = \varphi^{[-1]}(\varphi(w)-\varphi(u))$;
    4. The desired pair is (u,v).

4.17 Show that the following algorithm (Devroye 1986) generates random variates (u,v) whose joint distribution function is the Clayton copula (4.2.1) with parameter $\theta > 0$:

    1. Generate two independent exponential (mean $\mu = 1$) variates x and y; and a gamma ($\alpha = \theta, \beta = 1$) variate z, independent of x and y;

    2. Set $u = \left[1+(x/z)\right]^{-\theta}$ and $v = \left[1+(y/z)\right]^{-\theta}$;

    3. The desired pair is (u,v).

4.4 Order and Limiting Cases

Recall from Definition 2.8.1 the concordance ordering of copulas — C_1 is smaller than C_2 ($C_1 \prec C_2$) if $C_1(u,v) \leq C_2(u,v)$ for all u,v in $\mathbf{I}$. Also recall that a family $\{C_\theta\}$ of copulas is positively ordered if $C_\alpha \prec C_\beta$ whenever $\alpha \leq \beta$; and negatively ordered if $C_\alpha \succ C_\beta$ whenever $\alpha \leq \beta$. In Exercise 2.32, we saw that the Ali-Mikhail-Haq family (4.2.3) of Archimedean copulas is positively ordered.

Example 4.10. Let C_1 and C_2 be the members of the Gumbel-Barnett family (4.2.9) with parameters θ_1 and θ_2, respectively. If $\theta_1 \leq \theta_2$, then $-\theta_1 \ln u \ln v \geq -\theta_2 \ln u \ln v$ for u,v in $(0,1)$, from which it follows that $C_1 \succ C_2$. Hence the Gumbel-Barnett family of copulas is negatively ordered. ∎

Example 4.11. Let C_1 and C_2 be the members of family (4.2.19) with parameters θ_1 and θ_2, respectively. Using Definition 2.8.1 requires determining the sense of the inequality (if one exists) between

$$\frac{\theta_1}{\ln\!\left(e^{\theta_1/u} + e^{\theta_1/v} - e^{\theta_1}\right)} \quad \text{and} \quad \frac{\theta_2}{\ln\!\left(e^{\theta_2/u} + e^{\theta_2/v} - e^{\theta_2}\right)}$$

when $\theta_1 \leq \theta_2$. ∎

As the preceding example shows, it is often not easy to verify directly via Definition 2.8.1 that a pair of copulas are ordered. For Archimedean copulas, the situation is often simpler in that the concordance order is determined by properties of the generators. For the first of these results we need the notion of a subadditive function:

Definition 4.4.1. A function f defined on $[0,\infty)$ is *subadditive* if for all x,y in $[0,\infty)$,

$$f(x+y) \leq f(x) + f(y). \tag{4.4.1}$$

The next theorem (Schweizer and Sklar 1983) characterizes the concordance ordering of Archimedean copulas in terms of the subadditivity of composites of generators and their inverses.

Theorem 4.4.2. *Let C_1 and C_2 be Archimedean copulas generated, respectively, by φ_1 and φ_2 in Ω. Then $C_1 \prec C_2$ if and only if $\varphi_1 \circ \varphi_2^{[-1]}$ is subadditive.*

Proof. Let $f = \varphi_1 \circ \varphi_2^{[-1]}$. Note that f is continuous, nondecreasing, and $f(0) = 0$. From (4.1.3), $C_1 \prec C_2$ if and only if for all u,v in I,

$$\varphi_1^{[-1]}\big(\varphi_1(u)+\varphi_1(v)\big) \le \varphi_2^{[-1]}\big(\varphi_2(u)+\varphi_2(v)\big). \tag{4.4.2}$$

Let $x = \varphi_2(u)$ and $y = \varphi_2(v)$, then (4.4.2) is equivalent to

$$\varphi_1^{[-1]}\big(f(x)+f(y)\big) \le \varphi_2^{[-1]}(x+y) \tag{4.4.3}$$

for all x,y in $[0, \varphi_2(0)]$. Moreover if $x > \varphi_2(0)$ or $y > \varphi_2(0)$, then each side of (4.4.3) is equal to 0.

Now suppose that $C_1 \prec C_2$. Applying φ_1 to both sides of (4.4.3) and noting that $\varphi_1 \circ \varphi_1^{[-1]}(w) \le w$ for all $w \ge 0$ yields (4.4.1) for all x,y in $[0,\infty)$, hence f is subadditive. Conversely, if f satisfies (4.4.1), then applying $\varphi_1^{[-1]}$ to both sides and noting that $\varphi_1^{[-1]} \circ f = \varphi_2^{[-1]}$ yields (4.4.2), completing the proof. $\square$

Verifying the subadditivity of a function such as $f = \varphi_1 \circ \varphi_2^{[-1]}$ may still be as difficult as verifying directly that a pair of copulas satisfies Definition 2.8.1. So we now present several corollaries that give sufficient conditions for the subadditivity of $\varphi_1 \circ \varphi_2^{[-1]}$, and hence for the copula C_1 to be smaller than C_2. The first requires the following lemma from (Schweizer and Sklar 1983), which relates subadditivity to concavity.

Lemma 4.4.3. *Let f be defined on $[0,\infty)$. If f is concave and $f(0) = 0$, then f is subadditive.*

Proof. Let x,y be in $[0,\infty)$. If $x + y = 0$, then $x = y = 0$, so that with $f(0) = 0$, (4.4.1) is trivial. So assume $x + y > 0$, so that

$$x = \frac{x}{x+y}(x+y) + \frac{y}{x+y}(0) \text{ and } y = \frac{x}{x+y}(0) + \frac{y}{x+y}(x+y).$$

If f is concave and $f(0) = 0$, then

$$f(x) \ge \frac{x}{x+y}f(x+y) + \frac{y}{x+y}f(0) = \frac{x}{x+y}f(x+y)$$

and

$$f(y) \geq \frac{x}{x+y}f(0) + \frac{y}{x+y}f(x+y) = \frac{y}{x+y}f(x+y),$$

from which (4.4.1) follows and f is subadditive. $\qquad\Box$
  Combining Lemma 4.4.3 and Theorem 4.4.2 yields Corollary 4.4.4.

Corollary 4.4.4. *Under the hypotheses of Theorem 4.4.2, if $\varphi_1 \circ \varphi_2^{[-1]}$ is concave, then $C_1 \prec C_2$.*

Example 4.12. Let C_1 and C_2 be members of the Gumbel-Hougaard family (4.2.4) with parameters θ_1 and θ_2, so that the generators of C_1 and C_2 are φ_1 and φ_2, respectively, where $\varphi_k(t) = (-\ln t)^{\theta_k}$ for $k = 1,2$. Then $\varphi_1 \circ \varphi_2^{[-1]}(t) = t^{\theta_1/\theta_2}$. So if $\theta_1 \leq \theta_2$, then $\varphi_1 \circ \varphi_2^{[-1]}$ is concave and $C_1 \prec C_2$. Hence the Gumbel-Hougaard family is positively ordered. $\qquad\blacksquare$
  Another useful test for the concordance ordering of Archimedean copulas is the following result from (Genest and MacKay 1986a).

Corollary 4.4.5. *Under the hypotheses of Theorem 4.4.2, if φ_1/φ_2 is nondecreasing on $(0,1)$, then $C_1 \prec C_2$.*

Proof. Let g be the function from $(0,\infty)$ to $(0,\infty)$ defined by $g(t) = f(t)/t$, where again $f = \varphi_1 \circ \varphi_2^{[-1]}$. Assume φ_1/φ_2 is nondecreasing on $(0,\infty)$. Because $g \circ \varphi_2 = \varphi_1/\varphi_2$ and φ_2 is decreasing, it follows that g is nonincreasing on $(0,\varphi_2(0))$, and hence on $(0,\infty)$. Thus for all $x,y \geq 0$, $x[g(x+y) - g(x)] + y[g(x+y) - g(y)] \leq 0$, or $(x+y)g(x+y) \leq xg(x) + yg(y)$. Hence f is subadditive, which completes the proof. $\qquad\Box$

Example 4.13. Let C_1 and C_2 be members of family (4.2.2) with parameters θ_1 and θ_2, that is, the generators of C_1 and C_2 are φ_1 and φ_2, respectively, where $\varphi_k(t) = (1-t)^{\theta_k}$ for $k = 1,2$. Then $\varphi_1(t)/\varphi_2(t) = (1-t)^{\theta_1-\theta_2}$. So if $\theta_1 \leq \theta_2$, then φ_1/φ_2 is nondecreasing on $(0,1)$ and $C_1 \prec C_2$. Hence this family is also positively ordered. $\qquad\blacksquare$
  Yet another test—often the easiest to use—is the following, an extension of a result in (Genest and MacKay 1986a). The proof is from (Alsina et al. 2005).

Corollary 4.4.6. *Under the hypotheses of Theorem 4.4.2, if φ_1 and φ_2 are continuously differentiable on $(0,1)$, and if φ_1'/φ_2' is nondecreasing on $(0,1)$, then $C_1 \prec C_2$.*

Proof. Because both φ_1 and φ_2 are decreasing on $(0,1)$, both φ_1' and φ_2' are negative on $(0,1)$. Let $g = \varphi_1/\varphi_2$ and $f = \varphi_1'/\varphi_2'$, and assume f is nondecreasing. Because f is also continuous, $\lim_{t \to 1^-} f(t)$ exists (finite or infinite). But because $\lim_{t \to 1^-} \varphi_1(t) = 0 = \lim_{t \to 1^-} \varphi_2(t)$, l'Hôpital's rule applies, and $\lim_{t \to 1^-} f(t) = \lim_{t \to 1^-} g(t)$. Now

$$g' = \frac{\varphi_2 \varphi_1' - \varphi_1 \varphi_2'}{\varphi_2^2} = \left(\frac{\varphi_1'}{\varphi_2'} - \frac{\varphi_1}{\varphi_2} \right) \frac{\varphi_2'}{\varphi_2} = (f - g) \frac{\varphi_2'}{\varphi_2}. \tag{4.4.4}$$

By Corollary 4.4.5, we need only show that g' is nonnegative, or equivalently, because φ_2'/φ_2 is negative, that $f(t) - g(t) \leq 0$ on $(0,1)$. Suppose not, that is, suppose there is a t_0 in $(0,1)$ such that $f(t_0) - g(t_0) > 0$. Then

$$g(t_0) < f(t_0) \leq \lim_{t \to 1^-} f(t) = \lim_{t \to 1^-} g(t).$$

But by (4.4.4), $g'(t_0) < 0$, and hence there is a t_1 in $(t_0,1)$ such that $g(t_1) < g(t_0)$ and $g'(t_1) = 0$. But then $g(t_1) < g(t_0) < f(t_0) \leq f(t_1)$, so that by (4.4.4), $g'(t_1) < 0$, a contradiction. $\qquad \square$

Example 4.14. Let C_1 and C_2 be members of the Clayton family (4.2.1) with parameters θ_1 and θ_2, and generators φ_1 and φ_2, respectively, where $\varphi_k(t) = (t^{-\theta_k} - 1)/\theta_k$ for $k = 1,2$. Then $\varphi_1'(t)/\varphi_2'(t) = t^{\theta_2 - \theta_1}$. So if $\theta_1 \leq \theta_2$, then φ_1'/φ_2' is nondecreasing on $(0,1)$ and $C_1 \prec C_2$. Hence the Clayton family is also positively ordered. $\qquad \blacksquare$

Example 4.15. Let C_1 and C_2 be members of family (4.2.19) with parameters θ_1 and θ_2, and generators φ_1 and φ_2, respectively, where $\varphi_k(t) = e^{\theta/t} - e^{\theta}$ for $k = 1, 2$. Then

$$\varphi_1'(t)/\varphi_2'(t) = \frac{\theta_1}{\theta_2} \exp\left(\frac{\theta_1 - \theta_2}{t} \right).$$

So if $\theta_1 \leq \theta_2$, then φ_1'/φ_2' is nondecreasing on $(0,1)$ and $C_1 \prec C_2$. Hence the family (4.2.19) is positively ordered [cf. Example 4.11]. $\qquad \blacksquare$

Whether a totally ordered family of copulas is positively or negatively ordered is a matter of taste or convenience. The direction of the order can be easily changed by reparameterization. For example if the parameter space is $(-\infty,\infty)$ or $(0,\infty)$, then replacing θ by $-\theta$ or $1/\theta$, respectively, will suffice.

In the preceding four examples, we have seen that four of the families of Archimedean copulas from Table 4.1 are ordered. However, there are families of Archimedean copulas that are not ordered, as the next example demonstrates.

Example 4.16. The family of Archimedean copulas (4.2.10) is neither positively nor negatively ordered. A simple calculation shows that for θ in $(0,1)$, $C_{\theta/2}(u,v) \le C_\theta(u,v)$ for u,v in $\mathbf{I}$ if and only if $u^{\theta/2} + v^{\theta/2} \le 1$. ∎

We conclude this section with two theorems that can often be used to determine whether or not M, Π, or W are limiting members of an Archimedean family. The first applies to Archimedean limits such as W or Π, and is from (Genest and MacKay 1986a). Because M is not Archimedean, we treat it separately in the second theorem, which is also from (Genest and MacKay 1986a)—the proof is from (Alsina et al. 2005).

Theorem 4.4.7. Let $\{C_\theta | \theta \in \Theta\}$ be a family of Archimedean copulas with differentiable generators φ_θ in Ω. Then $C = \lim C_\theta$ is an Archimedean copula if and only if there exists a function φ in Ω such that for all s,t in $(0,1)$,

$$\lim \frac{\varphi_\theta(s)}{\varphi_\theta'(t)} = \frac{\varphi(s)}{\varphi'(t)}, \tag{4.4.5}$$

where "$\lim$" denotes the appropriate one-sided limit as θ approaches an end point of the parameter interval Θ.

Proof. Let (U_θ, V_θ) be uniform $(0,1)$ random variables with joint distribution function C_θ, and let K_θ' denote the joint distribution function of the random variables U_θ and $C_\theta(U_\theta, V_\theta)$. Then from Corollaries 4.3.5 and 4.3.6; we have

$$K_\theta'(s,t) = P[U_\theta \le s, C(U_\theta, V_\theta) \le t] = t - \frac{\varphi_\theta(t)}{\varphi_\theta'(t)} + \frac{\varphi_\theta(s)}{\varphi_\theta'(t)} \tag{4.4.6}$$

whenever $0 < t < s < 1$. Now let U and V be uniform $(0,1)$ random variables with joint distribution function C, let K' denote the joint distribution function of the random variables U and $C(U,V)$. Assume that $C = \lim C_\theta$ is an Archimedean copula with generator φ in Ω. It now follows that

$$\lim K_\theta'(s,t) = K'(s,t) = t - \frac{\varphi(t)}{\varphi'(t)} + \frac{\varphi(s)}{\varphi'(t)} \tag{4.4.7}$$

for $0 < t < s < 1$, thus equation (4.4.5) is a consequence of (4.4.6) and (4.4.7).

In the other direction, assume that (4.4.5) holds. Hence there is a set of positive constants c_θ such that for all t in $(0,1]$, $\lim c_\theta \varphi_\theta(t) = \varphi(t)$. It follows that the limit of $\varphi_\theta^{[-1]} / c_\theta$ is $\varphi^{[-1]}$, and thus for fixed u, v in $\mathbf{I}$,

$$\lim \varphi_\theta^{[-1]}[\varphi_\theta(u) + \varphi_\theta(v)] = \varphi^{[-1]}[\varphi(u) + \varphi(v)],$$

which completes the proof.   □

Because the generator of W is $\varphi(t) = 1 - t$, W will be the limit of a family $\{C_\theta | \theta \in \Theta\}$ if $\lim \varphi_\theta(s)/\varphi'_\theta(t) = s - 1$; and because the generator of Π is $\varphi(t) = -\ln t$, Π will be the limit of a family $\{C_\theta | \theta \in \Theta\}$ if $\lim \varphi_\theta(s)/\varphi'_\theta(t) = t\ln s$.

Example 4.17. (a) For the family of Archimedean copulas given by (4.2.7) in Table 4.1, $\varphi_\theta(t) = -\ln[\theta t + (1-\theta)]$ for θ in $(0,1]$. Hence, using l'Hôpital's rule,

$$\lim_{\theta \to 0^+} \frac{\varphi_\theta(s)}{\varphi'_\theta(t)} = \lim_{\theta \to 0^+} \frac{\ln[\theta s + (1-\theta)]}{\theta/[\theta t + (1-\theta)]} = \lim_{\theta \to 0^+} \frac{[\theta t + (1-\theta)]^2(s-1)}{\theta s + (1-\theta)} = s - 1$$

for s,t in $(0,1)$. Thus $C_0 = W$.

(b) For the same family, we have

$$\lim_{\theta \to 1^-} \frac{\varphi_\theta(s)}{\varphi'_\theta(t)} = \lim_{\theta \to 1^-} \frac{\ln[\theta s + (1-\theta)]}{\theta/[\theta t + (1-\theta)]} = t\ln s$$

for s,t in $(0,1)$. Thus $C_1 = \Pi$.   ■

Theorem 4.4.8. Let $\{C_\theta | \theta \in \Theta\}$ *be a family of Archimedean copulas with differentiable generators* φ_θ *in* Ω. *Then* $\lim C_\theta(u,v) = M(u,v)$ *if and only if*

$$\lim \frac{\varphi_\theta(t)}{\varphi'_\theta(t)} = 0 \text{ for } t \text{ in } (0,1),$$

where "lim" denotes the appropriate one-sided limit as θ *approaches an end point of the parameter interval* Θ.

Proof. Let γ denote the end point of Θ and assume $\lim \varphi_\theta(t)/\varphi'_\theta(t) = 0$. Fix an arbitrary t in $(0,1)$ and choose ε in $(0,t)$. Then $0 \le -\varphi_\theta(t)/\varphi'_\theta(t) \le \varepsilon$ for θ sufficiently close to γ (when γ is finite) or for $|\theta|$ sufficiently large (when γ is infinite). Because $t - \varphi_\theta(t)/\varphi'_\theta(t)$ is the t-intercept of the tangent line to the graph of $y = \varphi_\theta(x)$ at the point $(t,\varphi_\theta(t))$ (see Fig. 4.13), invoking the convexity of φ_θ to compare the y-coordinates of $y = \varphi_\theta(x)$ and the above tangent line when $x = t + \varphi_\theta(t)/\varphi'_\theta(t)$ yields, for these θ,

$$\varphi_\theta\left(t+\frac{\varphi_\theta(t)}{\varphi_\theta'(t)}\right) > 2\varphi_\theta(t),$$

thus

$$C_\theta(t,t) = \varphi_\theta^{[-1]}(2\varphi_\theta(t)) > t+\frac{\varphi_\theta(t)}{\varphi_\theta'(t)} > t-\varepsilon.$$

Hence $\lim C_\theta(t,t) = t$, so that $\lim C_\theta(u,v) = M(u,v)$. The converse follows by reversing the argument. $\qquad\square$

Example 4.18. For the family of Archimedean copulas given by (4.2.12) in Table 4.1, $\varphi_\theta(t) = ((1/t)-1)^\theta$ for θ in $[1,\infty)$ so that $\varphi_\theta(t)/\varphi_\theta'(t) = (t^2-t)/\theta$, and hence $\lim_{\theta\to\infty} \varphi_\theta(t)/\varphi_\theta'(t) = 0$ for all t in $(0,1)$. Thus $C_\infty = M$. $\qquad\blacksquare$

4.5 Two-parameter Families

In this section, we will consider some two-parameter families of Archimedean copulas. The first subsection deals with parametric families generated by composing a generator φ in Ω with the power function $t \mapsto t^\theta$, $\theta > 0$. In the second subsection, we consider a two-parameter family that contains every Archimedean copula that is a rational function on the complement of its zero set.

4.5.1 Families of Generators

In this section, we first examine methods of constructing families of generators of Archimedean copulas from a single generator φ in Ω. Assume that φ is a generator in Ω, for example, $\varphi(t) = (1/t) - 1$, or $\varphi(t) = -\ln t$. From such a φ, we can create parametric families of generators, which can then, in turn, be used to create families of Archimedean copulas.

Theorem 4.5.1. *Let φ be in Ω, let α and β be positive real numbers, and define*

$$\varphi_{\alpha,1}(t) = \varphi(t^\alpha) \text{ and } \varphi_{1,\beta}(t) = [\varphi(t)]^\beta. \qquad (4.5.1)$$

1. *If $\beta \geq 1$, then $\varphi_{1,\beta}$ is an element of Ω.*
2. *If α is in $(0,1]$, then $\varphi_{\alpha,1}$ is an element of Ω.*

3. *If φ is twice differentiable and $t\varphi'(t)$ is nondecreasing on $(0,1)$,*
then $\varphi_{\alpha,1}$ is an element of Ω for all $\alpha > 0$.

The proof is elementary and consists of a straightforward verification
that the two compositions of φ with the power function are decreasing
and convex for the specified values of the parameters α and β. Following (Oakes 1994), we will refer to a family of generators
$\left\{\varphi_{\alpha,1} \in \Omega \middle| \varphi_{\alpha,1}(t) = \varphi(t^{\alpha})\right\}$ as the *interior power family associated with φ*
and a family $\left\{\varphi_{1,\beta} \in \Omega \middle| \varphi_{1,\beta}(t) = [\varphi(t)]^{\beta}\right\}$ as the *exterior power family*
associated with φ. We let $C_{\alpha,1}$ and $C_{1,\beta}$ denote the copulas generated
by $\varphi_{\alpha,1}$ and $\varphi_{1,\beta}$, respectively.

Example 4.18. The interior power family associated with $\varphi(t) = (1/t) -$
1 for $\alpha > 0$ generates a subfamily of the Clayton family (4.2.1) in Table
4.1; and the exterior power family associated with $\varphi(t) = -\ln t$ generates
the Gumbel-Hougaard family (4.2.4) in Table 4.1. Other interior power
families include (4.2.9), (4.2.10), (4.2.20), and (4.2.22); and other exterior power families include (4.2.2) and (4.2.12).   ∎

The next example illustrates the ease with which two-parameter families of Archimedean copulas can be constructed by using Theorem
4.5.1 to add a parameter to one of the one-parameter families in Table
4.1.

Example 4.19. (Fang et al. 2000) For θ in $[-1,1]$, $\varphi_{\theta}(t) =$
$\ln\big([1-\theta(1-t)]/t\big)$ [with $\varphi_1(t) = (1/t) - 1$] generates an Ali-Mikhail-Haq
copula [(4.2.3) in Table 4.1]. Because $t\varphi'_{\theta}(t)$ is nondecreasing for θ in
$[0,1]$, the interior power family generated by φ_{θ} is the two-parameter
family given by

$$C_{\theta\alpha,1}(u,v) = \frac{uv}{[1 - \theta(1 - u^{1/\alpha})(1 - v^{1/\alpha})]^{\alpha}}$$

for u,v in $\mathbf{I}$, $\alpha > 0$, $0 \le \theta \le 1$. This family also appears in (Genest and
Rivest 2001). Note that $C_{0\alpha,1} = \Pi$ and that $C_{1\alpha,1}$ is a member of the
Clayton family (4.2.1).   ∎

Example 4.20. Let (X_1,Y_1), (X_2,Y_2), $\cdots$, (X_n,Y_n) be independent and
identically distributed pairs of random variables with a common Archimedean copula C with generator φ. Let $C_{(n)}$ denote the copula of
the component-wise maxima $X_{(n)} = \max\{X_i\}$ and $Y_{(n)} = \max\{Y_i\}$.
From Theorem 3.3.1 we have

$$C_{(n)}(u,v) = \left[\varphi^{[-1]}\left(\varphi(u^{1/n}) + \varphi(v^{1/n})\right)\right]^n$$

for u,v in $\mathbf{I}$. The generator of $C_{(n)}$ is $\varphi_{1/n,1}(t) = \varphi(t^{1/n})$, and thus the copula of the component-wise maxima is a member of the interior power family generated by φ.   ∎

In Example 3.22, we observed that each Gumbel-Hougaard copula is max-stable and hence an extreme value copula. Are there other Archimedean extreme value copulas? The answer is no (Genest and Rivest 1989):

Theorem 4.5.2. *Gumbel-Hougaard copulas* (4.2.4) *are the only Archimedean extreme value copulas.*

Proof. Assume φ generates an Archimedean extreme value copula C. From part 3 of Theorem 4.1.5, we may assume φ is scaled so that $\varphi(1/e)$ = 1. Because C is max-stable, we have $\varphi(t^s) = c_s\varphi(t)$ for $s > 0$, t in $(0,1]$ (we have replaced $1/r$ in Definition 3.3.2 by s for convenience). Now let $x = -\ln t$, then $\varphi(e^{-sx}) = c_s\varphi(e^{-x})$, so that if we set $g(x) = \varphi(e^{-x})$, then $g(sx) = c_s g(x)$ for $s,x > 0$. Because $g(1) = 1$, $c_s = g(s)$, and we have $g(sx) = g(s)g(x)$ for $s,x > 0$. This is a variant of Cauchy's equation, the solution to which (Aczél 1966) is $g(x) = x^\theta$. Hence $\varphi(t) = g(-\ln t) = (-\ln t)^\theta$, which generates the Gumbel-Hougaard family (4.2.4).   □

As the examples in Sect. 4.4 illustrate, many of the interior and exterior power families of Archimedean copulas are ordered.

Theorem 4.5.3. *Let φ be in Ω, and let $\varphi_{\alpha,1}$ and $\varphi_{1,\beta}$ be given by* (4.5.1). *Further assume that $\varphi_{\alpha,1}$ and $\varphi_{1,\beta}$ generate copulas $C_{\alpha,1}$ and $C_{1,\beta}$, respectively. [It follows that $\beta \geq 1$ and that α is an element of a subset A of $(0,\infty)$, which includes $(0,1]$.]*

1. *If $1 \leq \beta_1 \leq \beta_2$, then $C_{1,\beta_1} \prec C_{1,\beta_2}$.*

2. *If $\varphi\left([\varphi^{[-1]}(t)]^\theta\right)$ is subadditive for all θ in $(0,1)$, and if α_1, α_2 are in A, then $\alpha_1 \leq \alpha_2$ implies $C_{\alpha_1,1} \prec C_{\alpha_2,1}$.*

Proof. Part 1 follows from Corollary 4.4.5, because when $\beta_1 \leq \beta_2$, $\varphi_{1,\beta_1}(t)/\varphi_{1,\beta_2}(t) = [\varphi(t)]^{\beta_1-\beta_2}$ is nondecreasing. Part 2 follows from Theorem 4.4.2, because $\varphi_{\alpha_1,1} \circ \varphi_{\alpha_2,1}^{[-1]}(t) = \varphi\left([\varphi^{[-1]}(t)]^{\alpha_1/\alpha_2}\right)$.   □

Example 4.21. Let $\varphi(t) = (1/t) - 1$, and consider the copulas $C_{\alpha,1}$ generated by $\varphi_{\alpha,1}$ for $\alpha > 0$ [this is family (4.2.1) from Table 4.1]. Here

$\varphi\left([\varphi^{[-1]}(t)]^\theta\right) = (t+1)^\theta - 1$, which is concave on $(0,1)$, and

$\varphi\left([\varphi^{[-1]}(0)]^\theta\right) = 0$. Hence by Lemma 4.4.3 and part 2 of the above theorem, this family is positively ordered (which of course has been shown earlier, in Example 4.14). ∎

Note that every exterior power family of Archimedean copulas is positively ordered. This is definitely not the case for interior power families—recall Example 4.16, where it was shown that the interior power family (4.2.10) is not ordered.

Corollary 4.5.4. *Under the hypotheses of Theorem 4.5.3, if $\varphi_\alpha'/\varphi_\alpha$ is nonincreasing in α, then $\alpha_1 \le \alpha_2$ implies $C_{\alpha_1,1} \prec C_{\alpha_2,1}$.*

An examination of Table 4.1 shows that all the interior power families in the table include Π as a limiting case, while all the exterior power families include M as a limiting case.

Theorem 4.5.5. *Let φ be in Ω, and let $\varphi_{\alpha,1}$ and $\varphi_{1,\beta}$ be given by (4.5.1). Further assume that $\varphi_{\alpha,1}$ and $\varphi_{1,\beta}$ generate copulas $C_{\alpha,1}$ and $C_{1,\beta}$, respectively, where $\beta \ge 1$ and α is an element of a subset of $(0,\infty)$ which includes $(0,1]$.*

1. If φ is continuously differentiable and $\varphi'(1) \ne 0$, then

$$C_{0,1}(u,v) = \lim_{\alpha \to 0^+} C_{\alpha,1}(u,v) = \Pi(u,v).$$

2.

$$C_{1,\infty}(u,v) = \lim_{\beta \to \infty} C_{1,\beta}(u,v) = M(u,v).$$

Proof. Appealing to Theorems 4.4.7 and 4.4.8, we have

$$\lim_{\alpha \to 0^+} \frac{\varphi_{\alpha,1}(s)}{\varphi_{\alpha,1}'(t)} = \lim_{\alpha \to 0^+} \frac{\varphi(s^\alpha)}{\varphi'(t^\alpha)\alpha t^{\alpha-1}} = \frac{t}{\varphi'(1)} \lim_{\alpha \to 0^+} \varphi'(s^\alpha) \cdot s^\alpha \ln s = t \ln s$$

and $\quad \lim_{\beta \to \infty} \dfrac{\varphi_{1,\beta}(t)}{\varphi_{1,\beta}'(t)} = \lim_{\beta \to \infty} \dfrac{[\varphi(t)]^\beta}{\beta[\varphi(t)]^{\beta-1}\varphi'(t)} = \lim_{\beta \to \infty} \dfrac{\varphi(t)}{\beta\varphi'(t)} = 0.$ □

We are now in a position to create two-parameter families of Archimedean copulas by using generators which are the composites given by

$$\varphi_{\alpha,\beta}(t) = \left[\varphi(t^\alpha)\right]^\beta. \tag{4.5.2}$$

We illustrate the procedure via two examples.

Example 4.22. The function $\varphi(t) = (1/t) - 1$ generates the copula $C(u,v) = uv/(u+v-uv)$, which we denoted " $\Pi/(\Sigma-\Pi)$ " in Table 4.1. Using

(4.5.2), we now let $\varphi_{\alpha,\beta}(t) = (t^{-\alpha} - 1)^{\beta}$ for $\alpha > 0, \beta \geq 1$. This generates the two-parameter family of Archimedean copulas

$$C_{\alpha,\beta}(u,v) = \left\{ \left[(u^{-\alpha} - 1)^{\beta} + (v^{-\alpha} - 1)^{\beta} \right]^{1/\beta} + 1 \right\}^{-1/\alpha}. \quad (4.5.3)$$

Using Theorem 4.5.5, we can extend the parameter range to include $\alpha = 0$ and $\beta = \infty$ because $C_{0,1} = \Pi$, $C_{0,\beta}$ is the Gumbel family (4.2.4) in Table 4.1, and $C_{\alpha,\infty} = M$. Furthermore, the subfamily $C_{\alpha,1}$ is the "$\theta \geq 1$" portion of the Clayton family (4.2.1) in Table 4.1; and for $\alpha = 1/\beta, \beta \geq 1$, we get family (4.2.14). From Theorem 4.5.3, it is easy to verify that this family is positively ordered by both parameters, that is, if $\alpha_1 \leq \alpha_2$ and $\beta_1 \leq \beta_2$, then $C_{\alpha_1,\beta_1} \prec C_{\alpha_2,\beta_2}$. This family has been used as a family of survival copulas for a bivariate Weibull model, see (Lu and Bhattacharyya 1990) for details. ∎

Example 4.23. Let $\varphi(t) = 1 - t$, the generator of W. Using (4.5.2), let $\varphi_{\alpha,\beta}(t) = (1 - t^{\alpha})^{\beta}$ for α in $(0,1], \beta \geq 1$. This generates the two-parameter family of Archimedean copulas

$$C_{\alpha,\beta}(u,v) = \max\left(\left\{ 1 - \left[(1-u^{\alpha})^{\beta} + (1-v^{\alpha})^{\beta} \right]^{1/\beta} \right\}^{1/\alpha}, 0 \right). \quad (4.5.4)$$

Note that $C_{1,1} = W$, $C_{0,1} = \Pi$, and $C_{\alpha,\infty} = M$. Four subfamilies of (4.5.4) appear in Table 4.1: for $\beta = 1$, we get the "$\theta \in (-1,0]$" portion of (4.2.1); for $\alpha = 1$, we have family (4.2.2); for $\alpha\beta = 1$, we have family (4.2.15); and in the limit as α goes to zero, we have family (4.2.4). As with the preceding example, this family is also positively ordered by both parameters. ∎

One-parameter families of Archimedean copulas *not* in Table 4.1 can be readily constructed from two-parameter families such as those in the above examples. For example, set $\beta = \alpha + 1, \alpha \geq 0$ in (4.5.3); or $\beta = 1/(1-\alpha), 0 \leq \alpha < 1$ in either (4.5.3) or (4.5.4); in each instance we obtain a one-parameter family that is positively ordered and includes both Π and M as limiting cases.

Other choices for φ in (4.5.2) lead to other two-parameter families—for example, $\varphi(t) = (1-t)/(1+t)$, $\varphi(t) = \ln(1 - \ln t)$, $\varphi(t) = (1/t) - t$, $\varphi(t) = \exp[(1/t) - 1] - 1$, and so on.

4.5.2 Rational Archimedean Copulas

In Table 4.1, it is easy to find families of Archimedean copulas that are rational functions on $\mathbf{I}^2 \setminus Z(C)$, i.e., copulas $C(u,v)$ such that if $C(u,v) > 0$, then $C(u,v) = P(u,v)/Q(u,v)$ where P and Q are polynomials—for example, families (4.2.3), (4.2.7), and (4.2.8). Are there others? We call such copulas *rational*, and answer the question affirmatively by constructing a two-parameter family of all rational Archimedean copulas.

Because Archimedean copulas must be symmetric and associative (recall Theorem 4.1.5), our starting point is the following theorem (Alsina et al. 2005), which we state without proof:

Theorem 4.5.6. *Let R be a rational 2-place real function reduced to lowest terms, i.e., let*

$$R(u,v) = \frac{P(u,v)}{Q(u,v)}$$

where P and Q are relatively prime polynomials, neither of which is identically zero. Then R is symmetric and associative if and only if

$$R(u,v) = \frac{a_1 uv + b_1(u+v) + c_1}{a_2 + b_2(u+v) + c_2 uv} \qquad (4.5.5)$$

where

$$\begin{aligned} b_1 b_2 &= c_1 c_2, \\ b_1^2 + b_2 c_1 &= a_1 c_1 + a_2 b_1, \\ b_2^2 + b_1 c_2 &= a_2 c_2 + a_1 b_2. \end{aligned} \qquad (4.5.6)$$

Now let C be a function with domain $\mathbf{I}^2$ given by (4.5.5) and (4.5.6) on the complement of its zero set. In order for C to be a copula, we must impose further restrictions on the six coefficients in (4.5.5). The boundary condition $C(u,1) = u$ requires $R(u,1) = u$, or equivalently,

$$(b_2 + c_2)u^2 + (a_2 + b_2 - a_1 - b_1)u + (b_1 + c_1) = 0$$

for all u in $\mathbf{I}$. Hence $c_1 = -b_1$, $c_2 = -b_2$, and $a_1 + b_1 = a_2 + b_2$, and thus

$$R(u,v) = \frac{(a_1 + b_1)uv - b_1(1-u)(1-v)}{(a_2 + b_2) - b_2(1-u)(1-v)}.$$

Because R is not constant, we have $a_1 + b_1 = a_2 + b_2 \neq 0$, and, upon setting $\alpha = b_2/(a_2 + b_2)$ and $\beta = b_1/(a_1 + b_1)$, it follows that on the complement of its zero set, a rational Archimedean copula must have the form

$$C_{\alpha,\beta}(u,v) = \frac{uv - \beta(1-u)(1-v)}{1 - \alpha(1-u)(1-v)} \qquad (4.5.7)$$

for appropriate values of α and β.

In order to find the values of α and β so that $C_{\alpha,\beta}$ in (4.5.7) will be a copula, we will first find a function $\varphi_{\alpha,\beta}$ that generates $C_{\alpha,\beta}$, and then determine α and β so that $\varphi_{\alpha,\beta}$ is continuous, strictly decreasing and convex on (0,1). To find a candidate for $\varphi_{\alpha,\beta}$, we appeal to Theorem 4.3.8: If $C_{\alpha,\beta}$ is an Archimedean copula, then its generator $\varphi_{\alpha,\beta}$ must satisfy

$$\frac{\varphi'_{\alpha,\beta}(u)}{\varphi'_{\alpha,\beta}(v)} = \frac{\alpha v^2 + (1-\alpha-\beta)v + \beta}{\alpha u^2 + (1-\alpha-\beta)u + \beta},$$

so that

$$\varphi'_{\alpha,\beta}(t) = \frac{-c_{\alpha,\beta}}{\alpha t^2 + (1-\alpha-\beta)t + \beta} \qquad (4.5.8)$$

and

$$\varphi''_{\alpha,\beta}(t) = \frac{c_{\alpha,\beta}[2\alpha t + (1-\alpha-\beta)]}{\left[\alpha t^2 + (1-\alpha-\beta)t + \beta\right]^2},$$

where $c_{\alpha,\beta}$ is a constant. Assume that $\varphi'_{\alpha,\beta}(t) < 0$ and $\varphi''_{\alpha,\beta}(t) > 0$ on (0,1). Because $\varphi'_{\alpha,\beta}(0^+) = -c_{\alpha,\beta}/\beta$ and $\varphi'_{\alpha,\beta}(1^-) = -c_{\alpha,\beta}$, we have $c_{\alpha,\beta} > 0$ and $\beta \geq 0$. Then $\varphi''_{\alpha,\beta}(t) > 0$ if $2\alpha t + (1-\alpha-\beta) > 0$ for t in (0,1), which requires that $\alpha + \beta \leq 1$ and $\beta - \alpha \leq 1$. Conversely, the conditions $\beta \geq 0, \alpha + \beta \leq 1$, and $\beta - \alpha \leq 1$ (or equivalently, $0 \leq \beta \leq 1-|\alpha|$) are sufficient to insure that $2\alpha t + (1-\alpha-\beta) > 0$, which in turn implies that the denominator of (4.5.8) is strictly positive on (0,1), and hence, with $c_{\alpha,\beta} > 0$, to give $\varphi'_{\alpha,\beta}(t) < 0$ and $\varphi''_{\alpha,\beta}(t) > 0$ on (0,1). Thus (4.5.8) has a solution $\varphi_{\alpha,\beta}$ that is continuous, strictly decreasing, and convex on $\mathbf{I}$, and which generates $C_{\alpha,\beta}$ in (4.5.7). Hence we have

Theorem 4.5.7. *The function $C_{\alpha,\beta}$ defined on $\mathbf{I}^2$ by*

$$C_{\alpha,\beta}(u,v) = \max\left(\frac{uv - \beta(1-u)(1-v)}{1-\alpha(1-u)(1-v)}, 0\right) \qquad (4.5.9)$$

is a (rational Archimedean) copula if and only if $0 \leq \beta \leq 1-|\alpha|$.

Note that $C_{0,0} = \Pi$, $C_{0,1} = W$, and $C_{1,0} = \Pi/(\Sigma - \Pi)$.

The parameter space for $C_{\alpha,\beta}$ consists of the points in the α,β-plane that are on and inside the triangle with vertices $(-1,0)$, $(0,1)$, and $(1,0)$, as illustrated in Fig. 4.14. The curve in the first quadrant will play a role when we discuss the generators of $C_{\alpha,\beta}$.

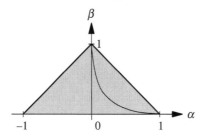

Fig. 4.14. The parameter space for $C_{\alpha,\beta}$ given by (4.5.9)

When $C_{\alpha,\beta}$ is a rational Archimedean copula, Exercise 4.5 tells us that $C_{\alpha,\beta}$ is strict if and only if $C_{\alpha,\beta}(u,v) > 0$ for u,v in $(0,1]$, which is equivalent to $\beta = 0$ in (4.5.9). Thus we have

Corollary 4.5.8. *A rational Archimedean copula is strict if and only if $\beta = 0$ in (4.5.9), i.e., if and only if it is a member of the Ali-Mikhail-Haq family (4.2.3).*

When $\beta = 1$, $\alpha = 0$ and $C_{0,1} = W$. When β is in $(0,1)$, the zero curve of $C_{\alpha,\beta}$ is a portion of the graph of $uv - \beta(1-u)(1-v) = 0$, a rectangular hyperbola with asymptotes $u = -\beta/(1-\beta)$ and $v = -\beta/(1-\beta)$, one branch of which passes through $(0,1)$ and $(1,0)$. Indeed, all the level curves of $C_{\alpha,\beta}$ for β in $[0,1)$ are portions of hyperbolas—see Exercise 4.22.

In order to obtain an explicit expression for the generator of $C_{\alpha,\beta}$, we need only integrate both sides of equation (4.5.8) to find $\varphi_{\alpha,\beta}$. There are three cases to consider, depending on the whether the discriminant $\Delta = (1-\alpha-\beta)^2 - 4\alpha\beta$ of the quadratic in the denominator of $\varphi'_{\alpha,\beta}(t)$ in (4.5.8) is positive, zero, or negative. But within the parameter space for α and β illustrated in Fig. 4.14, $\Delta = 0$ if and only if $\sqrt{\alpha} + \sqrt{\beta} = 1$, i.e., if and only if the point (α,β) is on the curve in the first quadrant of Fig. 4.14, a portion of a parabola whose axis is $\beta = \alpha$. Furthermore, $\Delta > 0$ if and only if $\sqrt{\alpha} + \sqrt{\beta} < 1$, i.e., for (α,β) below and to the left of the curve; and $\Delta < 0$ if and only if $\sqrt{\alpha} + \sqrt{\beta} > 1$, i.e., for (α,β) above and to the right of the curve.

It is now a simple matter to exhibit the generators $\varphi_{\alpha,\beta}$ explicitly:

$$\varphi_{\alpha,\beta}(t) = \begin{cases} \dfrac{1-t}{1+t\sqrt{\alpha/\beta}}, & \sqrt{\alpha}+\sqrt{\beta}=1, \\[3mm] \ln\dfrac{2-\left(1+\alpha-\beta-\sqrt{\Delta}\right)(1-t)}{2-\left(1+\alpha-\beta+\sqrt{\Delta}\right)(1-t)}, & \sqrt{\alpha}+\sqrt{\beta}<1, \\[3mm] \arctan\dfrac{(1-t)\sqrt{-\Delta}}{2-(1+\alpha-\beta)(1-t)}, & \sqrt{\alpha}+\sqrt{\beta}>1. \end{cases}$$

We conclude this section by displaying some of the one-parameter subfamilies of rational Archimedean copulas.

1. When $\beta = 0$, we obtain the Ali-Mikhail-Haq family (4.2.3) with $\theta = \alpha$.

2. When $\alpha = 0$, we obtain family (4.2.7) with $\theta = 1 - \beta$.

3. When $\sqrt{\alpha} + \sqrt{\beta} = 1$, we obtain family (4.2.8) with $\theta = 1/\sqrt{\beta}$.

4. When $\beta - \alpha = 1$, we set $\theta = -\alpha = 1 - \beta$ with θ in [0,1] to obtain the family

$$C_\theta(u.v) = \max\left(\frac{uv - (1-\theta)(1-u)(1-v)}{1+\theta(1-u)(1-v)}, 0 \right)$$

with generators

$$\varphi_\theta(t) = \ln\frac{1+\sqrt{\theta}(1-t)}{1-\sqrt{\theta}(1-t)}.$$

Note that $C_0 = W$.

5. When $\alpha + \beta = 1$, we set $\theta = \alpha = 1 - \beta$ with θ in [0,1] to obtain the family

$$C_\theta(u.v) = \max\left(\frac{uv - (1-\theta)(1-u)(1-v)}{1-\theta(1-u)(1-v)}, 0 \right)$$

with generators

$$\varphi_\theta(t) = \arctan\frac{(1-t)\sqrt{\theta(1-\theta)}}{1-\theta(1-t)}.$$

Note that $C_0 = W$ and $C_1 = \Pi/(\Sigma - \Pi)$.

Exercises

4.18 (a) Show that the following families of Archimedean copulas in Table 4.1 are positively ordered: (4.2.3), (4.2.5)-(4.2.8), (4.2.12)-(4.2.18), (4.2.20), and (4.2.21).
(b) Show that families (4.2.11) and (4.2.22) from Table 4.1 are negatively ordered.

4.19 Let C be a strict Archimedean copula generated by φ in Ω. Prove that if $-\ln\varphi^{-1}$ is concave on $(0,\infty)$, then $C \succ \Pi$.

4.20 Prove that the only absolutely continuous rational Archimedean copulas are the members of the Ali-Mikhail-Haq family (4.2.3).

4.21 Let $C_{\alpha,\beta}$ be a rational Archimedean copula, as given by (4.5.9), with $\beta > 0$. Show that the probability mass on the zero curve is given by

$$\sqrt{\beta}, \qquad\qquad\qquad\qquad \text{if } \sqrt{\alpha}+\sqrt{\beta}=1,$$

$$\frac{\beta}{\sqrt{\Delta}}\ln\frac{1-\alpha+\beta+\sqrt{\Delta}}{1-\alpha+\beta-\sqrt{\Delta}}, \quad \text{if } \sqrt{\alpha}+\sqrt{\beta}<1, \text{ and}$$

$$\frac{2\beta}{\sqrt{-\Delta}}\arctan\frac{\sqrt{-\Delta}}{1-\alpha+\beta}, \quad \text{if } \sqrt{\alpha}+\sqrt{\beta}>1,$$

where $\Delta = (1-\alpha-\beta)^2 - 4\alpha\beta$.

4.22 Let $C_{\alpha,\beta}$ be a rational Archimedean copula, as given by (4.5.9). Show that if β is in $[0,1)$, then the level curve $C_{\alpha,\beta}(u,v) = t$ for t in $\mathbf{I}$ is a portion of one branch of the rectangular hyperbola whose asymptotes are $u = (\alpha t - \beta)/(1+\alpha t - \beta)$ and $v = (\alpha t - \beta)/(1+\alpha t - \beta)$.

4.23 Consider the family $\{C_{\alpha,\beta}\}$ of rational Archimedean copulas, where $C_{\alpha,\beta}$ is given by (4.5.9) with $0 \le \beta \le 1-|\alpha|$. Show that this family is positively ordered by α and negatively ordered by β —that is, if $\alpha_1 \le \alpha_2$ and $\beta_1 \ge \beta_2$, then $C_{\alpha_1,\beta_1} \prec C_{\alpha_2,\beta_2}$.

4.6 Multivariate Archimedean Copulas

We now turn our attention to the construction of Archimedean n-copulas. Recall Example 4.2(a), in which we wrote the product copula Π in the form $\Pi(u,v) = uv = \exp(-[(-\ln u)+(-\ln v)])$. The extension of this idea to n dimensions, with $\mathbf{u} = (u_1,u_2,\cdots,u_n)$, results in writing the n-dimensional product copula Π^n in the form

$$\Pi^n(\mathbf{u}) = (u_1,u_2,\cdots,u_n) = \exp(-[(-\ln u_1)+(-\ln u_2)+\cdots+(-\ln u_n)]).$$

This leads naturally to the following generalization of (4.1.3):

$$C^n(\mathbf{u}) = \varphi^{[-1]}(\varphi(u_1)+\varphi(u_2)+\cdots+\varphi(u_n)), \tag{4.6.1}$$

(where the superscript on C denotes dimension).

The functions C^n in (4.6.1) are the *serial iterates* (Schweizer and Sklar 1983) of the Archimedean 2-copula generated by φ, that is, if we set $C^2(u_1,u_2) = C(u_1,u_2) = \varphi^{[-1]}(\varphi(u_1)+\varphi(u_2))$, then for $n \geq 3$, $C^n(u_1,u_2,\cdots,u_n) = C(C^{n-1}(u_1,u_2,\cdots,u_{n-1}),u_n)$ [recall from Theorem 4.1.5 that Archimedean copulas are symmetric and associative]. But note that this technique of composing copulas generally fails, as was illustrated in Sect. 3.4.

Using $\varphi(t) = 1-t$ in (4.6.1) generates W^n, and W^n fails to be a copula for any $n > 2$ (Exercise 2.35). Theorem 4.1.4 gives the properties of φ (continuous, strictly decreasing and convex, with $\varphi(1) = 0$) needed for C^n in (4.6.1) to be a copula for $n = 2$. What additional properties of φ (and $\varphi^{[-1]}$) will insure that C^n in (4.6.1) is a copula for $n \geq 3$?

One answer involves the derivatives of $\varphi^{[-1]}$, and requires that those derivatives alternate in sign.

Definition 4.6.1 (Widder 1941). A function $g(t)$ is *completely monotonic* on an interval J if it is continuous there and has derivatives of all orders that alternate in sign, i.e., if it satisfies

$$(-1)^k \frac{d^k}{dt^k} g(t) \geq 0 \tag{4.6.2}$$

for all t in the interior of J and $k = 0,1,2,\cdots$.

As a consequence, if $g(t)$ is completely monotonic on $[0,\infty)$ and $g(c) = 0$ for some (finite) $c > 0$, then g must be identically zero on $[0,\infty)$

(Widder 1941). So if the pseudo-inverse $\varphi^{[-1]}$ of an Archimedean generator φ is completely monotonic, it must be positive on $[0,\infty)$, i.e., φ is strict and $\varphi^{[-1]} = \varphi^{-1}$.

The following theorem (Kimberling 1974) gives necessary and sufficient conditions for a strict generator φ to generate Archimedean n-copulas for all $n \geq 2$. See also (Schweizer and Sklar 1983, Alsina et al. 2005).

Theorem 4.6.2. *Let φ be a continuous strictly decreasing function from $\mathbf{I}$ to $[0,\infty]$ such that $\varphi(0) = \infty$ and $\varphi(1) = 0$, and let φ^{-1} denote the inverse of φ. If C^n is the function from $\mathbf{I}^n$ to $\mathbf{I}$ given by (4.6.1), then C^n is an n-copula for all $n \geq 2$ if and only if φ^{-1} is completely monotonic on $[0,\infty)$.*

Example 4.23. Let $\varphi_\theta(t) = t^{-\theta} - 1$ for $\theta > 0$, which generates a subfamily of the bivariate Clayton family (4.2.1), the subfamily whose generators are strict. Here $\varphi_\theta^{-1}(t) = (1+t)^{-1/\theta}$, which is easily shown to be complete monotonic on $[0,\infty)$. Thus we can generalize the Clayton family of 2-copulas to a family of n-copulas for $\theta > 0$ and any $n \geq 2$:

$$C_\theta^n(\mathbf{u}) = \left(u_1^{-\theta} + u_2^{-\theta} + \cdots + u_n^{-\theta} - n + 1\right)^{-1/\theta}. \qquad \blacksquare$$

Note that the subfamily of the Clayton family (4.2.1) of copulas considered in the preceding example contains only copulas that are larger than Π. The following corollary guarantees that this must occur when φ^{-1} is completely monotonic.

Corollary 4.6.3. *If the inverse φ^{-1} of a strict generator φ of an Archimedean copula C is completely monotonic, then $C \succ \Pi$.*

Proof. As a consequence of Exercise 4.17, we need only show that $-\ln\varphi^{-1}$ is concave on $(0,\infty)$. This is equivalent to requiring that (letting g denote φ^{-1} for simplicity) $g \cdot g'' - (g')^2 \geq 0$ on $(0,\infty)$. But this inequality holds for completely monotonic functions (Widder 1941). $\qquad \square$

Three additional useful results are the following—the first is from (Widder 1941), the next two are from (Feller 1971):

1. If g is completely monotonic and f is *absolutely monotonic*, i.e., $d^k f(t)/dt^k \geq 0$ for $k = 0,1,2,\cdots$, then the composite $f \circ g$ is completely monotonic;

2. If f and g are completely monotonic, so is their product fg;

3. If f is completely monotonic and g is a positive function with a completely monotone derivative, then $f \circ g$ is completely monotonic. In particular, e^{-g} is completely monotonic.

Example 4.24. Let $\varphi_\theta(t) = -\ln((e^{-\theta t} - 1)/(e^{-\theta} - 1))$, which generates the bivariate Frank family (4.2.5). Although all the generators of this family are strict, we must, as a consequence of Corollary 4.6.3, restrict θ to $(0, \infty)$, the values of θ for which $C_\theta \succ \Pi$. For the Frank family, $\varphi_\theta^{-1}(t)$ is given by

$$\varphi_\theta^{-1}(t) = -\frac{1}{\theta} \ln\left[1 - (1 - e^{-\theta})e^{-t}\right].$$

But for $\theta > 0$, the function $f(x) = -\ln(1-x)/\theta$ is absolutely monotonic for x in $(0,1)$ and $g(t) = (1 - e^{-\theta})e^{-t}$ is completely monotonic for t in $[0, \infty)$, from which it follows that φ_θ^{-1} is completely monotonic on $[0, \infty)$. Thus for $\theta > 0$, we can generalize the Frank family of 2-copulas to a family of n-copulas for any $n \geq 2$:

$$C_\theta^n(\mathbf{u}) = -\frac{1}{\theta} \ln\left(1 + \frac{(e^{-\theta u_1} - 1)(e^{-\theta u_2} - 1) \cdots (e^{-\theta u_n} - 1)}{(e^{-\theta} - 1)^{n-1}}\right).$$

When $\theta < 0$, φ_θ^{-1} fails to be completely monotonic. ∎

Example 4.25. Let $\varphi_\theta(t) = (-\ln t)^\theta$, $\theta \geq 1$, which generates the bivariate Gumbel-Hougaard family (4.2.4). Here $\varphi_\theta^{-1}(t) = \exp(-t^{1/\theta})$. But because e^{-x} is completely monotonic and $t^{1/\theta}$ is a positive function with a completely monotonic derivative, φ_θ^{-1} is completely monotonic. Thus we can generalize the Gumbel-Hougaard family of 2-copulas to a family of n-copulas for $\theta \geq 1$ and any $n \geq 2$:

$$C_\theta^n(\mathbf{u}) = \exp\left(-\left[(-\ln u_1)^\theta + (-\ln u_2)^\theta + \cdots + (-\ln u_n)^\theta\right]^{1/\theta}\right). \quad ∎$$

Other families in Table 4.1 can be extended to n-copulas (for values of the parameter θ for which C_θ is larger than Π). See Exercise 4.24.

The procedure in the preceding example can be generalized to any exterior power family of generators associated with a strict generator φ whose inverse is completely monotonic.

Lemma 4.6.4. *Let φ be a strict generator whose inverse is completely monotonic on $[0,\infty)$, and set $\varphi_{1,\beta}(t) = [\varphi(t)]^{\beta}$ for $\beta \geq 1$. Then $\varphi_{1,\beta}^{-1}$ is completely monotonic on $[0,\infty)$.*

Example 4.26. The two-parameter family of copulas presented in Example 4.23 can be extended to a two-parameter family of n-copulas. Let $\varphi_{\alpha,\beta}(t) = (t^{-\alpha} - 1)^{\beta}$ for $\alpha > 0$, $\beta \geq 1$. Because the inverse of $\varphi_{\alpha,1}(t) = t^{-\alpha} - 1$ is completely monotonic on $[0,\infty)$ (see Example 4.23), Lemma 4.6.4 insures that $\varphi_{\alpha,\beta}^{-1}$ is completely monotonic. Hence

$$C_{\alpha,\beta}^{n}(\mathbf{u}) = \left\{ \left[(u_1^{-\alpha} - 1)^{\beta} + (u_2^{-\alpha} - 1)^{\beta} + \cdots + (u_n^{-\alpha} - 1)^{\beta} \right]^{1/\beta} + 1 \right\}^{-1/\alpha}$$

is an n-copula for $\alpha > 0$, $\beta \geq 1$, and each $n \geq 2$. ∎

Another source of generators for Archimedean n-copulas consists of the inverses of Laplace transforms of distribution functions (see Examples 3.11 and 4.9), as the following lemma (Feller 1971) shows:

Lemma 4.6.5. *A function ψ on $[0,\infty)$ is the Laplace transform of a distribution function Λ if and only if ψ is completely monotonic and $\psi(0) = 1$.*

The arguments in (Alsina et al. 2005) for the proof of Theorem 4.6.2 can be used to partially extend the theorem to the case when $\varphi^{[-1]}$ is m-monotonic on $[0,\infty)$ for some $m \geq 2$, that is, the derivatives of $\varphi^{[-1]}$ up to and including the mth are defined and alternate in sign, i.e., (4.6.2) holds for $k = 0,1,2,\cdots,m$, on $(0,\infty)$. In such cases, if $\varphi^{[-1]}$ is m-monotonic on $[0,\infty)$, then the function C^n given by (4.6.1) is an n-copula for $2 \leq n \leq m$.

Example 4.27. Let $\varphi_\theta(t) = t^{-\theta} - 1$ for $\theta \in [-1,0)$, which generates the non-strict subfamily of the bivariate Clayton family (4.2.1). Here $\varphi_\theta^{-1}(t) = (1+t)^{-1/\theta}$, which is readily shown to be m-monotonic on $[0,\infty)$ when $\theta > -1/(m-1)$. Thus we can generalize the Clayton family of 2-copulas with a given $\theta \in [-1,0)$, to a family of n-copulas for $n < 1 - (1/\theta)$. ∎

Although it is fairly simple to generate Archimedean n-copulas, they do have their limitations. First of all, in general all the k-margins of an Archimedean n-copula are identical. Secondly, the fact that there are usually only one or two parameters limits the nature of the dependence structure in these families.

Exercises

4.24 Show that the inverse of the generator of each of the following families in Table 4.1 is completely monotone for the values of the parameter θ for which $C_\theta \succ \Pi$: (4.2.3), (4.2.6), (4.2.12), (4.2.13), (4.2.14), and (4.2.19) [these are in addition to families (4.2.1), (4.2.4), and (4.2.5), which were examined in Examples 4.23, 4.24, and 4.25].

4.25 Let $\varphi(t) = 1/t - t$ [this is the $\theta = 1$ member of familiy (4.2.16)], and let C be the (strict) Archimedean copula generated by φ.

(a) Show that $C \succ \Pi$. [Hint: Corollary 4.4.6.]

(b) Show that φ^{-1} is 3-monotonic but not 4-monotonic.
Conclude that the converse of Corollary 4.6.3 does not hold.

5 Dependence

In this chapter, we explore ways in which copulas can be used in the study of dependence or association between random variables. As Jogdeo (1982) notes,

> Dependence relations between random variables is one of the most widely studied subjects in probability and statistics. The nature of the dependence can take a variety of forms and unless some specific assumptions are made about the dependence, no meaningful statistical model can be contemplated.

There are a variety of ways to discuss and to measure dependence. As we shall see, many of these properties and measures are, in the words of Hoeffding (1940, 1941), "scale-invariant," that is, they remain unchanged under strictly increasing transformations of the random variables. As we noted in the Introduction, "...it is precisely the copula which captures those properties of the joint distribution which are invariant under almost surely strictly increasing transformations" (Schweizer and Wolff 1981). As a consequence of Theorem 2.4.3, "scale-invariant" properties and measures are expressible in terms of the copula of the random variables. The focus of this chapter is an exploration of the role that copulas play in the study of dependence.

Dependence properties and measures of association are interrelated, and so there are many places where we could begin this study. Because the most widely known scale-invariant measures of association are the population versions of Kendall's tau and Spearman's rho, both of which "measure" a form of dependence known as concordance, we will begin there.

A note on terminology: we shall reserve the term "correlation coefficient" for a measure of the linear dependence between random variables (e.g., Pearson's product-moment correlation coefficient) and use the more modern term "measure of association" for measures such as Kendall's tau and Spearman's rho.

5.1 Concordance

Informally, a pair of random variables are concordant if "large" values of one tend to be associated with "large" values of the other and "small" values of one with "small" values of the other. To be more precise, let (x_i, y_i) and (x_j, y_j) denote two observations from a vector (X, Y) of continuous random variables. We say that (x_i, y_i) and (x_j, y_j) are *concordant* if $x_i < x_j$ and $y_i < y_j$, or if $x_i > x_j$ and $y_i > y_j$. Similarly, we say that (x_i, y_i) and (x_j, y_j) are *discordant* if $x_i < x_j$ and $y_i > y_j$ or if $x_i > x_j$ and $y_i < y_j$. Note the alternate formulation: (x_i, y_i) and (x_j, y_j) are concordant if $(x_i - x_j)(y_i - y_j) > 0$ and discordant if $(x_i - x_j)(y_i - y_j) < 0$.

5.1.1 Kendall's tau

The sample version of the measure of association known as Kendall's tau is defined in terms of concordance as follows (Kruskal 1958; Hollander and Wolfe 1973; Lehmann 1975): Let $\{(x_1, y_1), (x_2, y_2), \cdots, (x_n, y_n)\}$ denote a random sample of n observations from a vector (X, Y) of continuous random variables. There are $\binom{n}{2}$ distinct pairs (x_i, y_i) and (x_j, y_j) of observations in the sample, and each pair is either concordant or discordant—let c denote the number of concordant pairs and d the number of discordant pairs. Then Kendall's tau for the sample is defined as

$$t = \frac{c-d}{c+d} = (c-d) \Big/ \binom{n}{2}. \tag{5.1.1}$$

Equivalently, t is the probability of concordance minus the probability of discordance for a pair of observations (x_i, y_i) and (x_j, y_j) that is chosen randomly from the sample. The population version of Kendall's tau for a vector (X, Y) of continuous random variables with joint distribution function H is defined similarly. Let (X_1, Y_1) and (X_2, Y_2) be independent and identically distributed random vectors, each with joint distribution function H. Then the population version of Kendall's tau is defined as the probability of concordance minus the probability of discordance:

$$\tau = \tau_{X,Y} = P[(X_1 - X_2)(Y_1 - Y_2) > 0] - P[(X_1 - X_2)(Y_1 - Y_2) < 0] \tag{5.1.2}$$

(we shall use Latin letters for sample statistics and Greek letters for the corresponding population parameters).

In order to demonstrate the role that copulas play in concordance and measures of association such as Kendall's tau, we first define a "concordance function" Q, which is the difference of the probabilities of concordance and discordance between two vectors (X_1, Y_1) and

(X_2,Y_2) of continuous random variables with (possibly) different joint distributions H_1 and H_2, but with common margins F and G. We then show that this function depends on the distributions of (X_1,Y_1) and (X_2,Y_2) only through their copulas.

Theorem 5.1.1. *Let (X_1,Y_1) and (X_2,Y_2) be independent vectors of continuous random variables with joint distribution functions H_1 and H_2, respectively, with common margins F (of X_1 and X_2) and G (of Y_1 and Y_2). Let C_1 and C_2 denote the copulas of (X_1,Y_1) and (X_2,Y_2), respectively, so that $H_1(x,y) = C_1(F(x),G(y))$ and $H_2(x,y) = C_2(F(x),G(y))$. Let Q denote the difference between the probabilities of concordance and discordance of (X_1,Y_1) and (X_2,Y_2), i.e., let*

$$Q = P[(X_1 - X_2)(Y_1 - Y_2) > 0] - P[(X_1 - X_2)(Y_1 - Y_2) < 0]. \quad (5.1.3)$$

Then

$$Q = Q(C_1,C_2) = 4\iint_{\mathbf{I}^2} C_2(u,v)\,dC_1(u,v) - 1. \quad (5.1.4)$$

Proof. Because the random variables are continuous, $P[(X_1 - X_2)(Y_1 - Y_2) < 0] = 1 - P[(X_1 - X_2)(Y_1 - Y_2) > 0]$ and hence

$$Q = 2P[(X_1 - X_2)(Y_1 - Y_2) > 0] - 1. \quad (5.1.5)$$

But $P[(X_1 - X_2)(Y_1 - Y_2) > 0] = P[X_1 > X_2, Y_1 > Y_2] + P[X_1 < X_2, Y_1 < Y_2]$, and these probabilities can be evaluated by integrating over the distribution of one of the vectors (X_1,Y_1) or (X_2,Y_2), say (X_1,Y_1). First we have

$$P[X_1 > X_2, Y_1 > Y_2] = P[X_2 < X_1, Y_2 < Y_1],$$

$$= \iint_{\mathbf{R}^2} P[X_2 \le x, Y_2 \le y]\,dC_1(F(x),G(y)),$$

$$= \iint_{\mathbf{R}^2} C_2(F(x),G(y))\,dC_1(F(x),G(y)),$$

so that employing the probability transforms $u = F(x)$ and $v = G(y)$ yields

$$P[X_1 > X_2, Y_1 > Y_2] = \iint_{\mathbf{I}^2} C_2(u,v)\,dC_1(u,v).$$

Similarly,

$$P[X_1 < X_2, Y_1 < Y_2]$$

$$= \iint_{\mathbf{R}^2} P[X_2 > x, Y_2 > y]\,dC_1(F(x),G(y)),$$

$$= \iint_{\mathbf{R}^2} [1 - F(x) - G(y) + C_2(F(x),G(y))]\,dC_1(F(x),G(y)),$$

$$= \iint_{\mathbf{I}^2} [1 - u - v + C_2(u,v)]\,dC_1(u,v).$$

But because C_1 is the joint distribution function of a pair (U,V) of uniform $(0,1)$ random variables, $E(U) = E(V) = 1/2$, and hence

$$P[X_1 < X_2, Y_1 < Y_2] = 1 - \frac{1}{2} - \frac{1}{2} + \iint_{\mathbf{I}^2} C_2(u,v)\, dC_1(u,v),$$

$$= \iint_{\mathbf{I}^2} C_2(u,v)\, dC_1(u,v).$$

Thus

$$P[(X_1 - X_2)(Y_1 - Y_2) > 0] = 2\iint_{\mathbf{I}^2} C_2(u,v)\, dC_1(u,v),$$

and the conclusion follows upon substitution in (5.1.5).   □

Because the concordance function Q in Theorem 5.1.1 plays an important role throughout this section, we summarize some of its useful properties in the following corollary, whose proof is left as an exercise.

Corollary 5.1.2. *Let C_1, C_2, and Q be as given in Theorem 5.1.1. Then*

1. *Q is symmetric in its arguments: $Q(C_1, C_2) = Q(C_2, C_1)$.*
2. *Q is nondecreasing in each argument: if $C_1 \prec C_1'$ and $C_2 \prec C_2'$ for all (u,v) in $\mathbf{I}^2$, then $Q(C_1, C_2) \leq Q(C_1', C_2')$.*
3. *Copulas can be replaced by survival copulas in Q, i.e., $Q(C_1, C_2) = Q(\hat{C}_1, \hat{C}_2)$.*

Example 5.1. The function Q is easily evaluated for pairs of the basic copulas M, W and Π. First, recall that the support of M is the diagonal $v = u$ in $\mathbf{I}^2$ (see Example 2.11). Because M has uniform $(0,1)$ margins, it follows that if g is an integrable function whose domain is $\mathbf{I}^2$, then

$$\iint_{\mathbf{I}^2} g(u,v)\, dM(u,v) = \int_0^1 g(u,u)\, du.$$

Hence we have

$$Q(M,M) = 4\iint_{\mathbf{I}^2} \min(u,v)\, dM(u,v) - 1 = 4\int_0^1 u\, du - 1 = 1;$$

$$Q(M,\Pi) = 4\iint_{\mathbf{I}^2} uv\, dM(u,v) - 1 = 4\int_0^1 u^2\, du - 1 = 1/3;\ \text{and}$$

$$Q(M,W) = 4\iint_{\mathbf{I}^2} \max(u+v-1,0)\, dM(u,v) - 1 = 4\int_{1/2}^1 (2u-1)\, du - 1 = 0.$$

Similarly, because the support of W is the secondary diagonal $v = 1 - u$, we have

$$\iint_{\mathbf{I}^2} g(u,v)\, dW(u,v) = \int_0^1 g(u,1-u)\, du,$$

and thus

$$Q(W,\Pi) = 4\iint_{\mathbf{I}^2} uv \; dW(u,v) - 1 = 4\int_0^1 u(1-u)\,du - 1 = -1/3; \text{ and}$$

$$Q(W,W) = 4\iint_{\mathbf{I}^2} \max(u+v-1,0)\,dW(u,v) - 1 = 4\int_0^1 0\,du - 1 = -1.$$

Finally, because $d\Pi(u,v) = du\,dv,$

$$Q(\Pi,\Pi) = 4\iint_{\mathbf{I}^2} uv \; d\Pi(u,v) - 1 = 4\int_0^1\int_0^1 uv\,du\,dv - 1 = 0. \qquad \blacksquare$$

Now let C be an arbitrary copula. Because Q is the difference of two probabilities, $Q(C,C) \in [-1,1]$; and as a consequence of part 2 of Corollary 5.1.2 and the values of Q in the above example, it also follows that

$$Q(C,M) \in [0,1], \quad Q(C,W) \in [-1,0],$$
$$\text{and } Q(C,\Pi) \in [-1/3, 1/3]. \tag{5.1.6}$$

In Fig. 5.1, we see a representation of the set **C** of copulas partially ordered by $\prec$ (only seven copulas are shown, C_1, C_2, C_α, and C_β are "typical" copulas), and four "concordance axes," each of which, in a sense, locates the position of each copula C within the partially ordered set $(\mathbf{C},\prec)$.

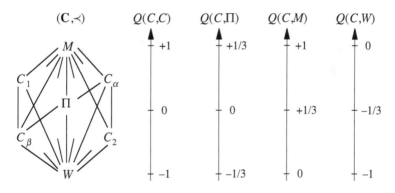

Fig. 5.1. The partially ordered set $(\mathbf{C},\prec)$ and several "concordance axes"

A comparison of (5.1.2), (5.1.3), and (5.1.4) yields

Theorem 5.1.3. *Let X and Y be continuous random variables whose copula is C. Then the population version of Kendall's tau for X and Y (which we will denote by either $\tau_{X,Y}$ or τ_C) is given by*

$$\tau_{X,Y} = \tau_C = Q(C,C) = 4\iint_{\mathbf{I}^2} C(u,v)\,dC(u,v) - 1. \tag{5.1.7}$$

Thus Kendall's tau is the first "concordance axis" in Fig. 5.1. Note that the integral that appears in (5.1.7) can be interpreted as the expected value of the function $C(U,V)$ of uniform $(0,1)$ random variables U and V whose joint distribution function is C, i.e.,

$$\tau_C = 4E\big(C(U,V)\big) - 1. \qquad (5.1.8)$$

When the copula C is a member of a parametric family of copulas (e.g., if C is denoted C_θ or $C_{\alpha,\beta}$), we will write τ_θ and $\tau_{\alpha,\beta}$ rather than τ_{C_θ} and $\tau_{C_{\alpha,\beta}}$, respectively.

Example 5.2. Let C_θ be a member of the Farlie-Gumbel-Morgenstern family (3.2.10) of copulas, where θ is in $[-1,1]$. Because C_θ is absolutely continuous, we have

$$dC_\theta(u,v) = \frac{\partial^2 C_\theta(u,v)}{\partial u \partial v}\, du\, dv = [1 + \theta(1-2u)(1-2v)]\, du\, dv\,,$$

from which it follows that

$$\iint_{I^2} C_\theta(u,v)\, dC_\theta(u,v) = \frac{1}{4} + \frac{\theta}{18}\,,$$

and hence $\tau_\theta = 2\theta/9$. Thus for FGM copulas $\tau_\theta \in [-2/9, 2/9]$ and, as Joe (1997) notes, this limited range of dependence restricts the usefulness of this family for modeling. See Fig. 3.12. ∎

Example 5.3. Let $C_{\alpha,\beta}$ be a member of the Fréchet family of copulas introduced in Exercise 2.4, where $\alpha \ge 0,\ \beta \ge 0,\ \alpha + \beta \le 1$. Then

$$C_{\alpha,\beta} = \alpha M + (1 - \alpha - \beta)\Pi + \beta W\,,$$

and

$$dC_{\alpha,\beta} = \alpha\, dM + (1 - \alpha - \beta)\, d\Pi + \beta\, dW\,,$$

from which it follows from (5.1.7) (using the results of Example 5.1) that

$$\tau_{\alpha,\beta} = \frac{(\alpha - \beta)(\alpha + \beta + 2)}{3}\,. \qquad \blacksquare$$

In general, evaluating the population version of Kendall's tau requires the evaluation of the double integral in (5.1.7). For an Archimedean copula, the situation is simpler, in that Kendall's tau can be evaluated directly from the generator of the copula, as shown in the following corollary (Genest and MacKay 1986a,b). Indeed, one of the reasons that Archimedean copulas are easy to work with is that often expressions with a one-place function (the generator) can be employed rather than expressions with a two-place function (the copula).

Corollary 5.1.4. *Let X and Y be random variables with an Archimed-ean copula C generated by φ in Ω. The population version τ_C of Ken-dall's tau for X and Y is given by*

$$\tau_C = 1 + 4\int_0^1 \frac{\varphi(t)}{\varphi'(t)}\,dt \ . \qquad (5.1.9)$$

Proof. Let U and V be uniform $(0,1)$ random variables with joint dis-tribution function C, and let K_C denote the distribution function of $C(U,V)$. Then from (5.1.8) we have

$$\tau_C = 4E\big(C(U,V)\big) - 1 = 4\int_0^1 t\,dK_C(t) - 1 \qquad (5.1.10)$$

which, upon integration by parts, yields

$$\tau_C = 3 - 4\int_0^1 K_C(t)\,dt \ . \qquad (5.1.11)$$

But as a consequence of Theorem 4.3.4 and Corollary 4.3.6, the distri-bution function K_C of $C(U,V)$ is

$$K_C(t) = t - \frac{\varphi(t)}{\varphi'(t^+)},$$

and hence

$$\tau_C = 3 - 4\int_0^1 \left[t - \frac{\varphi(t)}{\varphi'(t^+)}\right]dt = 1 + 4\int_0^1 \frac{\varphi(t)}{\varphi'(t)}\,dt,$$

where we have replaced $\varphi'(t^+)$ by $\varphi'(t)$ in the denominator of the inte-grand, as concave functions are differentiable almost everywhere.   □

As a consequence of (5.1.10) and (5.1.11), the distribution function K_C of $C(U,V)$ is called the *Kendall distribution function* of the copula C, and is a bivariate analog of the probability integral transform. See (Genest and Rivest, 2001; Nelsen et al. 2001, 2003) for additional de-tails.

Example 5.4. (a) Let C_θ be a member of the Clayton family (4.2.1) of Archimedean copulas. Then for $\theta \geq -1$,

$$\frac{\varphi_\theta(t)}{\varphi'_\theta(t)} = \frac{t^{\theta+1} - t}{\theta} \text{ when } \theta \neq 0, \text{ and } \frac{\varphi_0(t)}{\varphi'_0(t)} = t\ln t;$$

so that

$$\tau_\theta = \frac{\theta}{\theta+2}.$$

(b) Let C_θ be a member of the Gumbel-Hougaard family (4.2.4) of Archimedean copulas. Then for $\theta \geq 1$,

$$\frac{\varphi_\theta(t)}{\varphi'_\theta(t)} = \frac{t \ln t}{\theta},$$

and hence

$$\tau_\theta = \frac{\theta - 1}{\theta}. \qquad \blacksquare$$

The form for τ_C given by (5.1.7) is often not amenable to computation, especially when C is singular or if C has both an absolutely continuous and a singular component. For many such copulas, the expression

$$\tau_C = 1 - 4 \iint_{\mathbf{I}^2} \frac{\partial}{\partial u} C(u,v) \frac{\partial}{\partial v} C(u,v) \, du \, dv \qquad (5.1.12)$$

is more tractable (see Example 5.5 below). The equivalence of (5.1.7) and (5.1.12) is a consequence of the following theorem (Li et al. 2002).

Theorem 5.1.5. *Let C_1 and C_2 be copulas. Then*

$$\iint_{\mathbf{I}^2} C_1(u,v) \, dC_2(u,v) = \frac{1}{2} - \iint_{\mathbf{I}^2} \frac{\partial}{\partial u} C_1(u,v) \frac{\partial}{\partial v} C_2(u,v) \, du \, dv . \qquad (5.1.13)$$

Proof: When the copulas are absolutely continuous, (5.1.13) can be established by integration by parts. In this case the left-hand side of (5.1.13) is given by

$$\iint_{\mathbf{I}^2} C_1(u,v) \, dC_2(u,v) = \int_0^1 \int_0^1 C_1(u,v) \frac{\partial^2 C_2(u,v)}{\partial u \partial v} \, du \, dv .$$

Evaluating the inner integral by parts yields

$$\int_0^1 C_1(u,v) \frac{\partial^2 C_2(u,v)}{\partial u \partial v} \, du$$

$$= C_1(u,v) \frac{\partial C_2(u,v)}{\partial v} \Big|_{u=0}^{u=1} - \int_0^1 \frac{\partial C_1(u,v)}{\partial u} \frac{\partial C_2(u,v)}{\partial v} \, du,$$

$$= v - \int_0^1 \frac{\partial C_1(u,v)}{\partial u} \frac{\partial C_2(u,v)}{\partial v} \, du .$$

Integrating on v from 0 to 1 now yields (5.1.13).

The proof in the general case proceeds by approximating C_1 and C_2 by sequences of absolutely continuous copulas. See (Li et al. 2002) for details. $\qquad \square$

Example 5.5. Let $C_{\alpha,\beta}$ be a member of the Marshall-Olkin family (3.1.3) of copulas for $0 < \alpha,\beta < 1$:

$$C_{\alpha,\beta}(u,v) = \begin{cases} u^{1-\alpha}v, & u^{\alpha} \geq v^{\beta}, \\ uv^{1-\beta}, & u^{\alpha} \leq v^{\beta}. \end{cases}$$

The partials of $C_{\alpha,\beta}$ fail to exist only on the curve $u^{\alpha} = v^{\beta}$, so that

$$\frac{\partial}{\partial u}C_{\alpha,\beta}(u,v)\frac{\partial}{\partial v}C_{\alpha,\beta}(u,v) = \begin{cases} (1-\alpha)u^{1-2\alpha}v, & u^{\alpha} > v^{\beta}, \\ (1-\beta)uv^{1-2\beta}, & u^{\alpha} < v^{\beta}, \end{cases}$$

and hence

$$\iint_{\mathbf{I}^2} \frac{\partial}{\partial u}C_{\alpha,\beta}(u,v)\frac{\partial}{\partial v}C_{\alpha,\beta}(u,v)\,dudv = \frac{1}{4}\left(1 - \frac{\alpha\beta}{\alpha - \alpha\beta + \beta}\right),$$

from which we obtain

$$\tau_{\alpha,\beta} = \frac{\alpha\beta}{\alpha - \alpha\beta + \beta}.$$

It is interesting to note that $\tau_{\alpha,\beta}$ is numerically equal to $S_{\alpha,\beta}(1,1)$, the $C_{\alpha,\beta}$-measure of the singular component of the copula $C_{\alpha,\beta}$ (see Sect. 3.1.1). ∎

Exercises

5.1 Prove Corollary 5.1.2.

5.2 Let X and Y be random variables with the Marshall-Olkin bivariate exponential distribution with parameters λ_1, λ_2, and λ_{12} (see Sect. 3.1.1), i.e., the survival function $\overline{H}$ of X and Y is given by (for $x,y \geq 0$)

$$\overline{H}(x,y) = \exp\left[-\lambda_1 x - \lambda_2 y - \lambda_{12}\max(x,y)\right].$$

(a) Show that the ordinary Pearson product-moment correlation coefficient of X and Y is given by

$$\frac{\lambda_{12}}{\lambda_1 + \lambda_2 + \lambda_{12}}.$$

(b) Show that Kendall's tau and Pearson's product-moment correlation coefficient are numerically equal for members of this family (Edwardes 1993).

5.3   Prove that an alternate expression (Joe 1997) for Kendall's tau
for an Archimedean copula C with generator φ is

$$\tau_C = 1 - 4\int_0^\infty u \left[\frac{d}{du} \varphi^{[-1]}(u) \right]^2 du .$$

5.4   (a) Let C_θ, $\theta \in [0,1]$, be a member of the family of copulas introduced in Exercise 3.9, i.e., the probability mass of C_θ is uniformly distributed on two line segments, one joining $(0,\theta)$ to
$(1-\theta,1)$ and the other joining $(1-\theta,0)$ to $(1,\theta)$, as illustrated in
Fig. 3.7(b). Show that Kendall's tau for a member of this family
is given by

$$\tau_\theta = (1-2\theta)^2.$$

(b) Let C_θ, $\theta \in [0,1]$, be a member of the family of copulas introduced in Example 3.4, i.e., the probability mass of C_θ is uniformly distributed on two line segments, one joining $(0,\theta)$ to $(\theta,0)$
and the other joining $(\theta,1)$ to $(1,\theta)$, as illustrated in Fig. 3.4(a).
Show that Kendall's tau for a member of this family is given by

$$\tau_\theta = -(1-2\theta)^2.$$

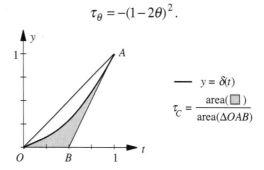

Fig. 5.2. A geometric interpretation of Kendall's tau for diagonal copulas

5.5   Let C be a diagonal copula, that is, let $C(u,v) =$
$\min(u,v,(1/2)[\,\delta(u)+\delta(v)\,])$, where δ satisfies (3.2.21abc).
(a) Show that Kendall's tau is given by

$$\tau_C = 4\int_0^1 \delta(t)\,dt - 1.$$

(b) For diagonal copulas, Kendall's tau has a geometric interpre-
tation. Because $\max(2t-1,0) \le \delta(t) \le t$ for t in $\mathbf{I}$ for any diagonal

δ (Exercise 2.8), then the graph of δ lies in ΔOAB, as illustrated in Fig. 5.2. Show that τ_C is equal to the fraction of the area of ΔOAB that lies below the graph of $y = \delta(t)$.

5.1.2 Spearman's rho

As with Kendall's tau, the population version of the measure of association known as Spearman's rho is based on concordance and discordance. To obtain the population version of this measure (Kruskal 1958; Lehmann 1966), we now let (X_1,Y_1), (X_2,Y_2), and (X_3,Y_3) be three independent random vectors with a common joint distribution function H (whose margins are again F and G) and copula C. The population version $\rho_{X,Y}$ of Spearman's rho is defined to be proportional to the probability of concordance minus the probability of discordance for the two vectors (X_1,Y_1) and (X_2,Y_3)—i.e., a pair of vectors with the same margins, but one vector has distribution function H, while the components of the other are independent:

$$\rho_{X,Y} = 3\left(P[(X_1 - X_2)(Y_1 - Y_3) > 0] - P[(X_1 - X_2)(Y_1 - Y_3) < 0]\right) \quad (5.1.14)$$

(the pair (X_3,Y_2) could be used equally as well). Note that while the joint distribution function of (X_1,Y_1) is $H(x,y)$, the joint distribution function of (X_2,Y_3) is $F(x)G(y)$ (because X_2 and Y_3 are independent). Thus the copula of X_2 and Y_3 is Π, and using Theorem 5.1.1 and part 1 of Corollary 5.1.2, we immediately have

Theorem 5.1.6. *Let X and Y be continuous random variables whose copula is C. Then the population version of Spearman's rho for X and Y (which we will denote by either $\rho_{X,Y}$ or ρ_C) is given by*

$$\rho_{X,Y} = \rho_C = 3Q(C,\Pi), \quad\quad (5.1.15a)$$

$$= 12\iint_{I^2} uv\, dC(u,v) - 3, \quad\quad (5.1.15b)$$

$$= 12\iint_{I^2} C(u,v)\, du\, dv - 3. \quad\quad (5.1.15c)$$

Thus Spearman's rho is essentially the second "concordance axis" in Fig. 5.1. The coefficient "3" that appears in (5.1.14) and (5.1.15a) is a "normalization" constant, because as noted in (5.1.6), $Q(C,\Pi) \in [-1/3,1/3]$. As was the case with Kendall's tau, we will write ρ_θ and $\rho_{\alpha,\beta}$ rather than ρ_{C_θ} and $\rho_{C_{\alpha,\beta}}$, respectively, when the copula C is given by C_θ or $C_{\alpha,\beta}$.

Example 5.6. Let $C_{\alpha,\beta}$ be a member of the Fréchet family of copulas introduced in Exercise 2.4, where $\alpha \geq 0$, $\beta \geq 0$, $\alpha + \beta \leq 1$. Then

$$C_{\alpha,\beta} = \alpha M + (1 - \alpha - \beta)\Pi + \beta W,$$

from which it follows (using (5.1.4) and the results of Example 5.1) that

$$Q(C_{\alpha,\beta}, \Pi) = \alpha Q(M, \Pi) + (1 - \alpha - \beta)Q(\Pi, \Pi) + \beta Q(W, \Pi),$$

$$= \alpha(1/3) + (1 - \alpha - \beta)(0) + \beta(-1/3) = \frac{\alpha - \beta}{3},$$

and hence

$$\rho_{\alpha,\beta} = 3Q(C_{\alpha,\beta}, \Pi) = \alpha - \beta. \qquad \blacksquare$$

Example 5.7. (a) Let C_θ be a member of the Farlie-Gumbel-Morgenstern family (3.2.10) of copulas, where θ is in $[-1,1]$. Then

$$C_\theta(u,v) = uv + \theta uv(1 - u)(1 - v),$$

thus

$$\iint_{\mathbf{I}^2} C_\theta(u,v)\, du\, dv = \frac{1}{4} + \frac{\theta}{36},$$

and hence $\rho_\theta = \theta/3$.

(b) Let $C_{\alpha,\beta}$ be a member of the Marshall-Olkin family (3.1.3) of copulas for $0 < \alpha, \beta < 1$:

$$C_{\alpha,\beta}(u,v) = \begin{cases} u^{1-\alpha} v, & u^\alpha \geq v^\beta, \\ uv^{1-\beta}, & u^\alpha \leq v^\beta. \end{cases}$$

Then

$$\iint_{\mathbf{I}^2} C_{\alpha,\beta}(u,v)\, du\, dv = \frac{1}{2}\left(\frac{\alpha + \beta}{2\alpha - \alpha\beta + 2\beta}\right),$$

so that

$$\rho_{\alpha,\beta} = \frac{3\alpha\beta}{2\alpha - \alpha\beta + 2\beta}.$$

[Cf. Examples 5.2 and 5.5.] $\blacksquare$

Any set of desirable properties for a "measure of concordance" would include those in the following definition (Scarsini 1984).

Definition 5.1.7. A numeric measure κ of association between two continuous random variables X and Y whose copula is C is a *measure of concordance* if it satisfies the following properties (again we write $\kappa_{X,Y}$ or κ_C when convenient):

1. κ is defined for every pair X, Y of continuous random variables;

2. $-1 \leq \kappa_{X,Y} \leq 1$, $\kappa_{X,X} = 1$, and $\kappa_{X,-X} = -1$;

3. $\kappa_{X,Y} = \kappa_{Y,X}$;

4. if X and Y are independent, then $\kappa_{X,Y} = \kappa_{\Pi} = 0$;

5. $\kappa_{-X,Y} = \kappa_{X,-Y} = -\kappa_{X,Y}$;

6. if C_1 and C_2 are copulas such that $C_1 \prec C_2$, then $\kappa_{C_1} \leq \kappa_{C_2}$;

7. if $\{(X_n, Y_n)\}$ is a sequence of continuous random variables with copulas C_n, and if $\{C_n\}$ converges pointwise to C, then $\lim_{n \to \infty} \kappa_{C_n} = \kappa_C$.

As a consequence of Definition 5.1.7, we have the following theorem, whose proof is an exercise.

Theorem 5.1.8. *Let κ be a measure of concordance for continuous random variables X and Y:*

1. *if Y is almost surely an increasing function of X, then $\kappa_{X,Y} = \kappa_M = 1$;*

2. *if Y is almost surely a decreasing function of X, then $\kappa_{X,Y} = \kappa_W = -1$;*

3. *if α and β are almost surely strictly monotone functions on* Ran$X$ *and* Ran$Y$, *respectively, then $\kappa_{\alpha(X),\beta(Y)} = \kappa_{X,Y}$.*

In the next theorem, we see that both Kendall's tau and Spearman's rho are measures of concordance according to the above definition.

Theorem 5.1.9. *If X and Y are continuous random variables whose copula is C, then the population versions of Kendall's tau (5.1.7) and Spearman's rho (5.1.15) satisfy the properties in Definition 5.1.7 and Theorem 5.1.8 for a measure of concordance.*

Proof. For both tau and rho, the first six properties in Definition 5.1.7 follow directly from properties of Q in Theorem 5.1.1, Corollary 5.1.2, and Example 5.1. For the seventh property, we note that the Lipschitz condition (2.2.6) implies that any family of copulas is equicontinuous, thus the convergence of $\{C_n\}$ to C is uniform. $\square$

The fact that measures of concordance, such as ρ and τ, satisfy the sixth criterion in Definition 5.1.7 is one reason that "$\prec$" is called the *concordance ordering*.

Spearman's rho is often called the "grade" correlation coefficient. Grades are the population analogs of ranks—that is, if x and y are observations from two random variables X and Y with distribution functions F and G, respectively, then the *grades* of x and y are given by $u = F(x)$ and $v = G(y)$. Note that the grades (u and v) are observations from the uniform $(0,1)$ random variables $U = F(X)$ and $V = G(Y)$ whose joint distribution function is C. Because U and V each have mean $1/2$ and variance $1/12$, the expression for ρ_C in (5.1.15b) can be re-written in the following form:

$$\rho_{X,Y} = \rho_C = 12\iint_{\mathbf{I}^2} uv \, dC(u,v) - 3 = 12E(UV) - 3,$$
$$= \frac{E(UV) - 1/4}{1/12} = \frac{E(UV) - E(U)E(V)}{\sqrt{\mathrm{Var}(U)}\sqrt{\mathrm{Var}(V)}}.$$

As a consequence, Spearman's rho for a pair of continuous random variables X and Y is identical to Pearson's product-moment correlation coefficient for the grades of X and Y, i.e., the random variables $U = F(X)$ and $V = G(Y)$.

Example 5.8. Let C_θ, $\theta \in \mathbf{I}$, be a member of the family of copulas introduced in Exercise 3.9. If U and V are uniform $(0,1)$ random variables whose joint distribution function is C_θ, then $V = U \oplus \theta$ (where $\oplus$ again denotes addition mod 1) with probability 1, and we have

$$E(UV) = \int_0^1 u(u \oplus \theta) \, du,$$
$$= \int_0^{1-\theta} u(u+\theta) \, du + \int_{1-\theta}^1 u(u+\theta-1) \, du,$$
$$= \frac{1}{3} - \frac{\theta(1-\theta)}{2},$$

and hence

$$\rho_\theta = 12E(UV) - 3 = 1 - 6\theta(1-\theta). \qquad \blacksquare$$

Another interpretation of Spearman's rho can be obtained from its representation in (5.1.15c). The integral in that expression represents the volume under the graph of the copula and over the unit square, and hence ρ_C is a "scaled" volume under the graph of the copula (scaled to lie in the interval $[-1,1]$). Indeed, (5.1.15c) can also be written as

$$\rho_C = 12\iint_{\mathbf{I}^2}[C(u,v) - uv] \, du dv, \qquad (5.1.16)$$

so that ρ_C is proportional to the signed volume between the graphs of the copula C and the product copula Π. Thus ρ_C is a measure of "average distance" between the distribution of X and Y (as represented by C) and independence (as represented by the copula Π). We shall exploit this observation in Sect. 5.3.1 to create and discuss additional measures of association.

Exercises

5.6   Let C_θ, $\theta \in [0,1]$, be a member of the family of copulas introduced in Example 3.3, i.e., the probability mass of C_θ is distributed on two line segments, one joining $(0,0)$ to $(\theta,1)$ and the other
joining $(\theta,1)$ to $(1,0)$, as illustrated in Fig. 3.3(a). Show that Kendall's tau and Spearman's rho for any member of this family are
given by

$$\tau_\theta = \rho_\theta = 2\theta - 1.$$

5.7   Let C_θ, $\theta \in [0,1]$, be a member of the family of copulas introduced in Example 3.4, i.e., the probability mass of C_θ is uniformly distributed on two line segments, one joining $(0,\theta)$ to $(\theta,0)$
and the other joining $(\theta,1)$ to $(1,\theta)$, as illustrated in Fig. 3.5(a).
Show that Spearman's rho for any member of this family is given
by

$$\rho_\theta = 6\theta(1-\theta) - 1.$$

5.8   Let C_θ be a member of the Plackett family of copulas (3.3.3) for
$\theta > 0$. Show that Spearman's ρ for this C_θ is

$$\rho_\theta = \frac{\theta+1}{\theta-1} - \frac{2\theta}{(\theta-1)^2} \ln\theta.$$

There does not appear to be a closed form expression for Kendall's τ for members of this family.

5.9   Let C_θ, $\theta \in \overline{\mathbf{R}}$, be a member of the Frank family (4.2.5) of Archimedean copulas. Show that

$$\tau_\theta = 1 - \frac{4}{\theta}[1 - D_1(\theta)] \quad \text{and} \quad \rho_\theta = 1 - \frac{12}{\theta}[D_1(\theta) - D_2(\theta)],$$

where $D_k(x)$ is the Debye function, which is defined for any
positive integer k by

$$D_k(x) = \frac{k}{x^k} \int_0^x \frac{t^k}{e^t - 1} \, dt.$$

(Genest 1987, Nelsen 1986). For a discussion of estimating the
parameter θ for a Frank copula from a sample using the sample
version of Spearman's rho or Kendall's tau, see (Genest 1987).

5.10 Let C_θ, $\theta \in [-1,1]$, be a member of the Ali-Mikhail-Haq family (4.2.4) of Archimedean copulas.
(a) Show that

$$\tau_\theta = \frac{3\theta - 2}{3\theta} - \frac{2(1-\theta)^2}{3\theta^2} \ln(1-\theta)$$

and

$$\rho_\theta = \frac{12(1+\theta)}{\theta^2} \mathrm{dilog}(1-\theta) - \frac{24(1-\theta)}{\theta^2} \ln(1-\theta) - \frac{3(\theta+12)}{\theta},$$

where $\mathrm{dilog}(x)$ is the dilogarithm function defined by

$$\mathrm{dilog}(x) = \int_1^x \frac{\ln t}{1-t} \, dt.$$

(b) Show that $\rho_\theta \in \left[33 - 48\ln 2, 4\pi^2 - 39\right] \cong [-0.2711,\ 0.4784]$ and $\tau_\theta \in \left[(5 - 8\ln 2)/3, 1/3\right] \cong [-0.1817,\ 0.3333]$.

5.11 Let C_θ, $\theta \in [0,1]$, be a member of the Raftery family of copulas introduced in Exercise 3.6, i.e.,

$$C_\theta(u,v) = M(u,v) + \frac{1-\theta}{1+\theta}(uv)^{1/(1-\theta)}\left\{1 - \left[\max(u,v)\right]^{-(1+\theta)/(1-\theta)}\right\}.$$

Show that

$$\tau_\theta = \frac{2\theta}{3-\theta} \quad \text{and} \quad \rho_\theta = \frac{\theta(4-3\theta)}{(2-\theta)^2}.$$

5.12 (a) Let C_n, n a positive integer, be the ordinal sum of $\{W,W,\cdots,W\}$ with respect to the regular partition $\mathbf{I}_n$ of $\mathbf{I}$ into n subintervals, i.e.,

$$C_n(u,v) = \begin{cases} \max\left(\dfrac{k-1}{n}, u+v-\dfrac{k}{n}\right), & (u,v) \in \left[\dfrac{k-1}{n}, \dfrac{k}{n}\right]^2, \quad k=1,2,\cdots,n, \\ \min(u,v), & \text{otherwise.} \end{cases}$$

The support of C_n consists of n line segments, joining the points $\big((k-1)/n, k/n\big)$ and $\big(k/n, (k-1)/n\big)$, $k = 1,2,\cdots,n$, as illustrated in Fig. 5.3(a) for $n = 4$. Show that

$$\tau_n = 1 - \frac{2}{n} \quad \text{and} \quad \rho_n = 1 - \frac{2}{n^2}.$$

Note that each copula in this family is also a shuffle of M given by $M(n,\mathbf{I}_n,(1,2,\cdots,n),-1)$.

(b) Let C'_n, n a positive integer, be the shuffle of M given by $M(n,\mathbf{I}_n,(n,n-1,\cdots,1),1)$, i.e.,

$$C'_n(u,v) = \begin{cases} \min\left(u-\dfrac{k-1}{n},v-\dfrac{n-k}{n}\right), & (u,v) \in \left[\dfrac{k-1}{n},\dfrac{k}{n}\right] \times \left[\dfrac{n-k+1}{n},\dfrac{n-k}{n}\right], \\ & k=1,2,\cdots,n, \\ \max(u+v-1,0), & \text{otherwise.} \end{cases}$$

The support of C'_n consists of n line segments, joining the points $((k-1)/n,(n-k)/n)$ and $(k/n,(n-k+1)/n)$, $k = 1,2,\cdots,n$, as illustrated in Fig. 5.3(b) for $n = 4$. Show that

$$\tau_n = \frac{2}{n}-1 \quad \text{and} \quad \rho_n = \frac{2}{n^2}-1.$$

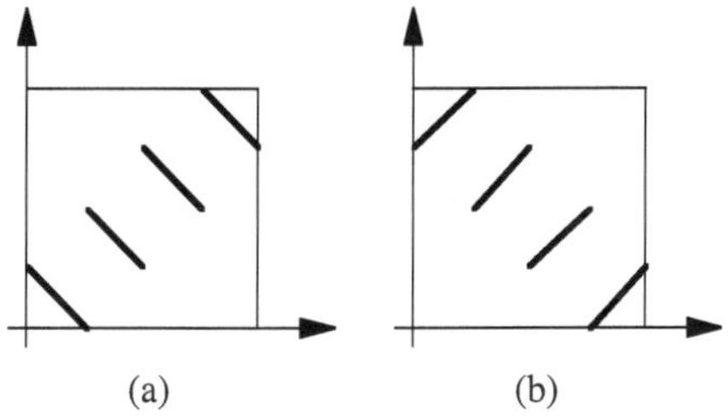

(a)   (b)

Fig. 5.3. Supports of the copulas C_4 and C'_4 in Exercise 5.12

5.13 Let C be a copula with cubic sections in both u and v, i.e., let C be given by
$$C(u,v) = uv + uv(1-u)(1-v)\big[A_1v(1-u) + $$
$$A_2(1-v)(1-u) + B_1uv + B_2u(1-v)\big],$$

where the constants A_1, A_2, B_1, and B_2 satisfy the conditions in Theorem 3.2.9. Show that

$$\rho = \frac{A_1 + A_2 + B_1 + B_2}{12} \quad \text{and} \quad \tau = \frac{A_1 + A_2 + B_1 + B_2}{18} + \frac{A_2B_1 - A_1B_2}{450}.$$

5.14 Let C_0, C_1 be copulas, and let ρ_0, ρ_1, τ_0, τ_1 be the values of Spearman's rho and Kendall's tau for C_0 and C_1, respectively. Let C_θ be the ordinal sum of $\{C_1,C_0\}$ with respect to

$\{[0,\theta],[\theta,1]\}$, for θ in $[0,1]$. Let ρ_θ and τ_θ denote the values of Spearman's rho and Kendall's tau for C_θ. Show that

$$\rho_\theta = \theta^3 \rho_1 + (1-\theta)^3 \rho_0 + 3\theta(1-\theta)$$

and

$$\tau_\theta = \theta^2 \tau_1 + (1-\theta)^2 \tau_0 + 2\theta(1-\theta).$$

5.15 Let C be an extreme value copula given by (3.3.12). Show that

$$\tau_C = \int_0^1 \frac{t(1-t)}{A(t)} dA'(t) \quad \text{and} \quad \rho_C = 12\int_0^1 [A(t)+1]^{-2} dt - 3.$$

(Capéraà et al. 1997).

5.1.3 The Relationship between Kendall's tau and Spearman's rho

Although both Kendall's tau and Spearman's rho measure the probability of concordance between random variables with a given copula, the values of ρ and τ are often quite different. In this section, we will determine just how different ρ and τ can be. In Sect. 5.2, we will investigate the relationship between measures of association and dependence properties in order to partially explain the differences between ρ and τ that we observe here.

We begin with a comparison of ρ and τ for members of some of the families of copulas that we have considered in the examples and exercises in the preceding sections.

Example 5.9. (a) In Exercise 5.6, we have a family of copulas for which $\rho = \tau$ over the entire interval $[-1,1]$ of possible values for these measures.

(b) For the Farlie-Gumbel-Morgenstern family, the results in Examples 5.2 and 5.7(a) yield $3\tau = 2\rho$, but only over a limited range, $|\rho| \le 1/3$ and $|\tau| \le 2/9$ [A similar result holds for copulas with cubic sections which satisfy $A_1 B_2 = A_2 B_1$ (see Exercise 5.13)].

(c) For the Marshall-Olkin family, the results in Examples 5.5 and 5.7(b) yield $\rho = 3\tau/(2+\tau)$ for ρ and τ both in $[0,1]$.

(d) For the Raftery family, the results in Exercise 5.11 yield $\rho = 3\tau(8-5\tau)/(4-\tau)^2$, again for ρ and τ both in $[0,1]$. ∎

Other examples could also be given, but clearly the relationship between ρ and τ varies considerably from family to family. The next theorem, due to Daniels (1950), gives universal inequalities for these measures. Our proof is adapted from Kruskal (1958).

Theorem 5.1.10. *Let X and Y be continuous random variables, and let* τ *and* ρ *denote Kendall's tau and Spearman's rho, defined by (5.1.2) and (5.1.14), respectively. Then*

$$-1 \le 3\tau - 2\rho \le 1. \qquad (5.1.17)$$

Proof. Let (X_1,Y_1), (X_2,Y_2), and (X_3,Y_3) be three independent random vectors with a common distribution. By continuity, (5.1.2) and (5.1.14) are equivalent to

$$\tau = 2P[(X_1 - X_2)(Y_1 - Y_2) > 0] - 1 \qquad (5.1.18)$$

and

$$\rho = 6P[(X_1 - X_2)(Y_1 - Y_3) > 0] - 3.$$

However, the subscripts on X and Y can be permuted cyclically to obtain the following symmetric forms for τ and ρ:

$$\tau = \frac{2}{3}\{P[(X_1 - X_2)(Y_1 - Y_2) > 0] + P[(X_2 - X_3)(Y_2 - Y_3) > 0]$$
$$+ P[(X_3 - X_1)(Y_3 - Y_1) > 0]\} - 1;$$

and

$$\rho = \{P[(X_1 - X_2)(Y_1 - Y_3) > 0] + P[(X_1 - X_3)(Y_1 - Y_2) > 0]$$
$$+ P[(X_2 - X_1)(Y_2 - Y_3) > 0] + P[(X_3 - X_2)(Y_3 - Y_1) > 0]$$
$$+ P[(X_2 - X_3)(Y_2 - Y_1) > 0] + P[(X_3 - X_1)(Y_3 - Y_2) > 0]\} - 3.$$

Because the expressions for τ and ρ above are now invariant under any permutation of the subscripts, we can assume that $X_1 < X_2 < X_3$, in which case

$$\tau = \frac{2}{3}\{P(Y_1 < Y_2) + P(Y_2 < Y_3) + P(Y_1 < Y_3)\} - 1$$

and

$$\rho = \{P(Y_1 < Y_3) + P(Y_1 < Y_2) + P(Y_2 > Y_3)$$
$$+ P(Y_3 > Y_1) + P(Y_2 < Y_1) + P(Y_3 > Y_2)\} - 3,$$
$$= 2[P(Y_1 < Y_3)] - 1.$$

Now let p_{ijk} denote the conditional probability that $Y_i < Y_j < Y_k$ given that $X_1 < X_2 < X_3$. Then the six p_{ijk} sum to one, and we have

$$\tau = \frac{2}{3}\left\{ (p_{123} + p_{132} + p_{312}) + (p_{123} + p_{213} + p_{231}) + (p_{123} + p_{132} + p_{213}) \right\} - 1,$$

$$= p_{123} + \frac{1}{3}(p_{132} + p_{213}) - \frac{1}{3}(p_{231} + p_{312}) - p_{321},$$

and

$$\rho = 2(p_{123} + p_{132} + p_{213}) - 1,$$

$$= p_{123} + p_{132} + p_{213} - p_{231} - p_{312} - p_{321}. \tag{5.1.19}$$

Hence

$$3\tau - 2\rho = p_{123} - p_{132} - p_{213} + p_{231} + p_{312} - p_{321},$$

$$= (p_{123} + p_{231} + p_{312}) - (p_{132} + p_{213} + p_{321}),$$

so that

$$-1 \le 3\tau - 2\rho \le 1. \qquad \square$$

The next theorem gives a second set of universal inequalities relating ρ and τ. It is due to Durbin and Stuart (1951); and again the proof is adapted from Kruskal (1958):

Theorem 5.1.11. *Let X, Y, τ, and ρ be as in Theorem 5.1.9. Then*

$$\frac{1+\rho}{2} \ge \left(\frac{1+\tau}{2}\right)^2 \tag{5.1.20a}$$

and

$$\frac{1-\rho}{2} \ge \left(\frac{1-\tau}{2}\right)^2. \tag{5.1.20b}$$

Proof. Again let (X_1, Y_1), (X_2, Y_2), and (X_3, Y_3) be three independent random vectors with a common distribution function H. If p denotes the probability that some pair of the three vectors is concordant with the third, then, e.g.,

$$p = P[(X_2, Y_2) \text{ and } (X_3, Y_3) \text{ are concordant with } (X_1, Y_1)],$$

$$= \iint_{\mathbf{R}^2} P[(X_2, Y_2) \text{ and } (X_3, Y_3) \text{ are concordant with } (x, y)] \, dH(x, y),$$

$$= \iint_{\mathbf{R}^2} P[(X_2 - x)(Y_2 - y) > 0] P[(X_3 - x)(Y_3 - y) > 0] \, dH(x, y),$$

$$= \iint_{\mathbf{R}^2} \left(P[(X_2 - x)(Y_2 - y) > 0] \right)^2 dH(x, y),$$

$$\ge \left[\iint_{\mathbf{R}^2} P[(X_2 - x)(Y_2 - y) > 0] \, dH(x, y) \right]^2,$$

$$= \left[P[(X_2 - X_1)(Y_2 - Y_1) > 0] \right]^2 = \left(\frac{1+\tau}{2} \right)^2,$$

where the inequality results from $E(Z^2) \geq [E(Z)]^2$ for the (conditional) random variable $Z = P[(X_2 - X_1)(Y_2 - Y_1) > 0 | (X_1, Y_1)]$, and the final equality is from (5.1.18). Permuting subscripts yields

$$p = \frac{1}{3} \{ P[(X_2, Y_2) \text{ and } (X_3, Y_3) \text{ are concordant with } (X_1, Y_1)]$$

$$+ P[(X_3, Y_3) \text{ and } (X_1, Y_1) \text{ are concordant with } (X_2, Y_2)]$$

$$+ P[(X_1, Y_1) \text{ and } (X_2, Y_2) \text{ are concordant with } (X_3, Y_3)] \}.$$

Thus, if $X_1 < X_2 < X_3$ and if we again let p_{ijk} denote the conditional probability that $Y_i < Y_j < Y_k$ given that $X_1 < X_2 < X_3$, then

$$p = \frac{1}{3} \{ (p_{123} + p_{132}) + (p_{123}) + (p_{123} + p_{213}) \},$$

$$= p_{123} + \frac{1}{3} p_{132} + \frac{1}{3} p_{213}.$$

Invoking (5.1.19) yields

$$\frac{1+\rho}{2} = p_{123} + p_{132} + p_{213} \geq p \geq \left(\frac{1+\tau}{2} \right)^2,$$

which completes the proof of (5.1.20a). To prove (5.1.20b), replace "concordant" in the above argument by "discordant." ☐
The inequalities in the preceding two theorems combine to yield

Corollary 5.1.12. *Let X, Y, τ, and ρ be as in Theorem 5.1.9. Then*

$$\frac{3\tau - 1}{2} \leq \rho \leq \frac{1 + 2\tau - \tau^2}{2}, \quad \tau \geq 0,$$

and

$$\frac{\tau^2 + 2\tau - 1}{2} \leq \rho \leq \frac{1 + 3\tau}{2}, \quad \tau \leq 0.$$

(5.1.21)

These bounds for the values of ρ and τ are illustrated in Fig. 5.4. For any pair X and Y of continuous random variables, the values of the population versions of Kendall's tau and Spearman's rho must lie in the shaded region, which we refer to as the *τ-ρ region*.

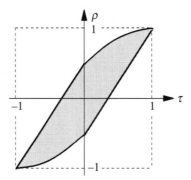

Fig. 5.4. Bounds for ρ and τ for pairs of continuous random variables

Can the bounds in Corollary 5.1.12 be improved? To give a partial answer to this question, we consider two examples.

Example 5.10. (a) Let U and V be uniform $(0,1)$ random variables such that $V = U \oplus \theta$ (where $\oplus$ again denotes addition mod 1)—i.e., the joint distribution function of U and V is the copula C_θ from Exercise 3.9, with $\theta \in [0,1]$. In Example 5.8 we showed that $\rho_\theta = 1 - 6\theta(1-\theta)$, and in Exercise 5.4(a) we saw that $\tau_\theta = (1-2\theta)^2$, hence for this family, $\rho = (3\tau - 1)/2$, $\tau \geq 0$. Thus every point on the linear portion of the lower boundary of the shaded region in Fig. 5.4 is attainable for some pair of random variables.

(b) Similarly, let U and V be uniform $(0,1)$ random variables such that $U \oplus V = \theta$, i.e., the copula of U and V is C_θ from Example 3.4, with $\theta \in [0,1]$. From Exercises 5.4(b) and 5.7, we have $\tau_\theta = -(1-2\theta)^2$ and $\rho_\theta = 6\theta(1-\theta) - 1$, and hence for this family, $\rho = (1+3\tau)/2$, $\tau \leq 0$. Thus every point on the linear portion of the upper boundary of the shaded region in Fig. 5.4 is also attainable. ∎

Example 5.11. (a) Let C_n, n a positive integer, be a member of the family of copulas in Exercise 5.12(a), for which the support consists of n line segments such as illustrated for $n = 4$ in part (a) of Fig. 5.3. When $\tau = (n-2)/n$, we have $\rho = (1+2\tau-\tau^2)/2$. Hence selected points on the parabolic portion of the upper boundary of the shaded region in Fig. 5.4 are attainable.

(b) Similarly let C_n, n a positive integer, be a member of the family of copulas in Exercise 5.12(b), for which the support consists of n line segments such as illustrated for $n = 4$ in part (b) of Fig. 5.3. When $\tau =$

$-(n-2)/n$, we have $\rho = (\tau^2 + 2\tau - 1)/2$. Hence selected points on the parabolic portion of the lower boundary of the shaded region in Fig. 5.4 are also attainable. ∎

In Fig. 5.5, we reproduce Fig. 5.4 with the τ-ρ region augmented by illustrations of the supports of some of the copulas in the preceding two examples, for which ρ and τ lie on the boundary.

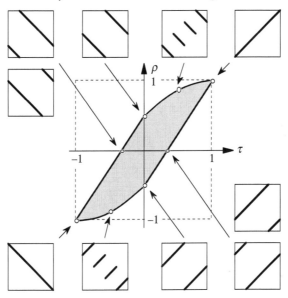

Fig. 5.5. Supports of some copulas for which ρ and τ lie on the boundary of the
τ-ρ region

We conclude this section with several observations.

1. As a consequence of Example 5.10, the linear portion of the boundary of the τ-ρ region cannot be improved. However, the copulas in Example 5.11 do not yield values of ρ and τ at all points on the parabolic portion of the boundary, so it may be possible to improve the inequalities in (5.1.21) at those points.

2. All the copulas illustrated in Fig. 5.5 for which ρ and τ lie on the boundary of the τ-ρ region are shuffles of M.

3. Hutchinson and Lai (1990) describe the pattern in Fig. 5.5 by observing that

> ...[for a given value of τ] very high ρ only occurs with negative correlation locally contrasted with positive overall correlation, and very low ρ only

with negative overall correlation contrasted with positive correlation locally.

4. The relationship between ρ and τ in a one-parameter family of
copulas can be exploited to construct a large sample test of the hypothesis that the copula of a bivariate distribution belongs to a particular family. See (Carriere 1994) for details.

5.1.4 Other Concordance Measures

In the 1910s, Corrado Gini introduced a measure of association g that
he called the *indice di cograduazione semplice*: if p_i and q_i denote the
ranks in a sample of size n of two continuous random variables X and Y,
respectively, then

$$g = \frac{1}{\lfloor n^2/2 \rfloor} \left[\sum_{i=1}^{n} |p_i + q_i - n - 1| - \sum_{i=1}^{n} |p_i - q_i| \right] \qquad (5.1.22)$$

where $\lfloor t \rfloor$ denotes the integer part of t. Let γ denote the population parameter estimated by this statistic, and as usual, let F and G denote the
marginal distribution functions of X and Y, respectively, and set $U =
F(X)$ and $V = G(Y)$. Because p_i/n and q_i/n are observations from dis-
crete uniform distributions on the set $\{1/n, 2/n, \cdots, 1\}$, (5.1.22) can be
re-written as

$$g = \frac{n^2}{\lfloor n^2/2 \rfloor} \left[\sum_{i=1}^{n} \left| \frac{p_i}{n} + \frac{q_i}{n} - \frac{n+1}{n} \right| - \sum_{i=1}^{n} \left| \frac{p_i}{n} - \frac{q_i}{n} \right| \right] \cdot \frac{1}{n}.$$

If we now pass to the limit as n goes to infinity, we obtain $\gamma =
2E(|U + V - 1| - |U - V|)$, i.e.,

$$\gamma = 2 \iint_{\mathbf{I}^2} (|u + v - 1| - |u - v|) dC(u,v) \qquad (5.1.23)$$

(where C is the copula of X and Y). In the following theorem, we show
that γ, like ρ and τ, is a measure of association based upon concordance.

Theorem 5.1.13. *Let X and Y be continuous random variables whose
copula is C. Then the population version of Gini's measure of associa-
tion for X and Y (which we will denote by either $\gamma_{X,Y}$ or γ_C) is given by*

$$\gamma_{X,Y} = \gamma_C = Q(C,M) + Q(C,W). \qquad (5.1.24)$$

Proof. We show that (5.1.24) is equivalent to (5.1.23). Using (5.1.4) and noting that $M(u,v) = (1/2)\big[u+v-|u-v|\big]$, we have

$$Q(C,M) = 4\iint_{\mathbf{I}^2} M(u,v)\, dC(u,v) - 1,$$

$$= 2\iint_{\mathbf{I}^2} \big[u+v-|u-v|\big]\, dC(u,v) - 1.$$

But because any copula is a joint distribution function with uniform $(0,1)$ margins,

$$\iint_{\mathbf{I}^2} u\, dC(u,v) = \frac{1}{2} \quad \text{and} \quad \iint_{\mathbf{I}^2} v\, dC(u,v) = \frac{1}{2},$$

and thus

$$Q(C,M) = 1 - 2\iint_{\mathbf{I}^2} |u-v|\, dC(u,v).$$

Similarly, because $W(u,v) = (1/2)\big[u+v-1+|u+v-1|\big]$, we have

$$Q(C,W) = 4\iint_{\mathbf{I}^2} W(u,v)\, dC(u,v) - 1,$$

$$= 2\iint_{\mathbf{I}^2} \big[u+v-1+|u+v-1|\big]\, dC(u,v) - 1,$$

$$= 2\iint_{\mathbf{I}^2} |u+v-1|\, dC(u,v) - 1,$$

from which the conclusion follows.   □

In a sense, Spearman's $\rho = 3Q(C,\Pi)$ measures a concordance relationship or "distance" between the distribution of X and Y as represented by their copula C and independence as represented by the copula Π. On the other hand, Gini's $\gamma = Q(C,M) + Q(C,W)$ measures a concordance relationship or "distance" between C and monotone dependence, as represented by the copulas M and W. Also note that γ_C is equivalent to the sum of the measures on the third and fourth "concordance axes" in Fig. 5.1.

Using the symmetry of Q from the first part of Corollary 5.1.2 yields the following expression for γ, which shows that γ depends on the copula C only through its diagonal and secondary diagonal sections:

Corollary 5.1.14. *Under the hypotheses of Theorem 5.1.13,*

$$\gamma_C = 4\left[\int_0^1 C(u,1-u)\, du - \int_0^1 [u - C(u,u)]\, du\right]. \qquad (5.1.25)$$

Proof. The result follows directly from

$$Q(C,M) = 4\iint_{\mathbf{I}^2} C(u,v)\, dM(u,v) - 1 = 4\int_0^1 C(u,u)\, du - 1$$

and

$$Q(C,W) = 4\iint_{I^2} C(u,v)\,dW(u,v) - 1 = 4\int_0^1 C(u,1-u)\,du - 1. \qquad \square$$

Note that there is a geometric interpretation of the integrals in (5.1.25)—the second is the area between the graphs of the diagonal sections $\delta_M(u) = M(u,u) = u$ of the Fréchet-Hoeffding upper bound and $\delta_C(u) = C(u,u)$ of the copula C; and the first is the area between the graphs of the secondary diagonal sections $C(u,1-u)$ of C and $W(u,1-u) = 0$ of the Fréchet-Hoeffding lower bound.

We conclude this section with one additional measure of association based on concordance. Suppose that, in the expression (5.1.3) for Q, the probability of concordance minus the probability of discordance, we use a random vector and a fixed point, rather than two random vectors. That is, consider

$$P[(X - x_0)(Y - y_0) > 0] - P[(X - x_0)(Y - y_0) < 0]$$

for some choice of a point (x_0, y_0) in $\mathbf{R}^2$. Blomqvist (1950) proposed and studied such a measure using population medians for x_0 and y_0. This measure, often called the *medial correlation coefficient*, will be denoted β, and is given by

$$\beta = \beta_{X,Y} = P[(X - \tilde{x})(Y - \tilde{y}) > 0] - P[(X - \tilde{x})(Y - \tilde{y}) < 0] \quad (5.1.26)$$

where $\tilde{x}$ and $\tilde{y}$ are medians of X and Y, respectively. But if X and Y are continuous with joint distribution function H and margins F and G, respectively, and copula C, then $F(\tilde{x}) = G(\tilde{y}) = 1/2$ and we have

$$\begin{aligned}
\beta &= 2P[(X - \tilde{x})(Y - \tilde{y}) > 0] - 1, \\
&= 2\{P[X < \tilde{x}, Y < \tilde{y}] + P[X > \tilde{x}, Y > \tilde{y}]\} - 1, \\
&= 2\{H(\tilde{x},\tilde{y}) + [1 - F(\tilde{x}) - G(\tilde{y}) + H(\tilde{x},\tilde{y})]\} - 1, \\
&= 4H(\tilde{x},\tilde{y}) - 1.
\end{aligned}$$

But $H(\tilde{x},\tilde{y}) = C(1/2,1/2)$, and thus

$$\beta = \beta_C = 4C\left(\frac{1}{2},\frac{1}{2}\right) - 1. \qquad (5.1.27)$$

Although Blomqvist's β depends on the copula only through its value at the center of $\mathbf{I}^2$, it can nevertheless often provide an accurate approximation to Spearman's ρ and Kendall's τ, as the following example illustrates.

Example 5.12. Let C_θ, $\theta \in [-1,1]$, be a member of the Ali-Mikhail-Haq family (4.2.3) of Archimedean copulas. In Exercise 5.10 we obtained expressions, involving logarithms and the dilogarithm function, for ρ and τ for members of this family. Blomqvist's β is readily seen to be

$$\beta = \beta_\theta = \frac{\theta}{4-\theta}.$$

Note that $\beta \in [-1/5, 1/3]$. If we reparameterize the expressions in Exercise 5.10 for ρ_θ and τ_θ by replacing θ by $4\beta/(1+\beta)$, and expand each of the results in a Maclaurin series, we obtain

$$\rho = \frac{4}{3}\beta + \frac{44}{75}\beta^3 + \frac{8}{25}\beta^4 + \cdots,$$

$$\tau = \frac{8}{9}\beta + \frac{8}{15}\beta^3 + \frac{16}{45}\beta^4 + \cdots.$$

Hence $4\beta/3$ and $8\beta/9$ are reasonable second-order approximations to ρ and τ, respectively, and higher-order approximations are also possible. ∎

Like Kendall's τ and Spearman's ρ, both Gini's γ and Blomqvist's β are also measures of concordance according to Definition 5.1.7. The proof of the following theorem is analogous to that of Theorem 5.1.9.

Theorem 5.1.15. *If X and Y are continuous random variables whose copula is C, then the population versions of Gini's γ (5.1.24) and Blomqvist's β (5.1.27) satisfy the properties in Definition 5.1.7 and Theorem 5.1.8 for a measure of concordance.*

In Theorem 3.2.3 we saw how the Fréchet-Hoeffding bounds—which are universal—can be narrowed when additional information (such as the value of the copula at a single point in $(0,1)^2$) is known. The same is often true when we know the value of a measure of association.

For any t in $[-1,1]$, let $\mathbf{T}_t$ denote the set of copulas with a common value t of Kendall's τ, i.e.,

$$\mathbf{T}_t = \left\{ C \in \mathbf{C} \mid \tau(C) = t \right\}. \tag{5.1.28}$$

Let $\underline{T}_t$ and $\overline{T}_t$ denote, respectively, the pointwise infimum and supremum of $\mathbf{T}_t$, i.e., for each (u,v) in $\mathbf{I}^2$,

$$\underline{T}_t(u,v) = \inf\left\{ C(u,v) \mid C \in \mathbf{T}_t \right\} \tag{5.1.29a}$$

and

$$\overline{T}_t(u,v) = \sup\{C(u,v)|C \in \mathbf{T}_t\}. \tag{5.1.29b}$$

Similarly, let $\mathbf{P}_t$ and $\mathbf{B}_t$ denote the sets of copulas with a common value t of Spearman's ρ and Blomqvist's β, respectively, i.e.,

$$\mathbf{P}_t = \{C \in \mathbf{C}|\rho(C) = t\} \text{ and } \mathbf{B}_t = \{C \in \mathbf{C}|\beta(C) = t\}, \tag{5.1.30}$$

and define $\underline{P}_t$, $\overline{P}_t$, $\underline{B}_t$ and $\overline{B}_t$ analogously to (5.1.29a) and (5.2.29b). These bounds can be evaluated explicitly, see (Nelsen et al. 2001; Nelsen and Úbeda Flores, 2004) for details.

Theorem 5.1.16. *Let $\underline{T}_t$, $\overline{T}t$, $\underline{P}_t$, $\overline{P}t$, $\underline{B}_t$ and $\overline{B}_t$ denote, respectively, the pointwise infimum and supremum of the sets $\mathbf{T}_t$, $\mathbf{P}_t$ and $\mathbf{B}_t$ in (5.1.28) and (5.1.30). Then for any (u,v) in $\mathbf{I}^2$,*

$$\underline{T}_t(u,v) = \max\left(W(u,v), \frac{1}{2}\left[(u+v) - \sqrt{(u-v)^2 + 1 - t}\right]\right),$$

$$\overline{T}t(u,v) = \min\left(M(u,v), \frac{1}{2}\left[(u+v-1) + \sqrt{(u+v-1)^2 + 1 + t}\right]\right),$$

$$\underline{P}_t(u,v) = \max\left(W(u,v), \frac{u+v}{2} - p(u-v, 1-t)\right),$$

$$\overline{P}t(u,v) = \min\left(M(u,v), \frac{u+v-1}{2} + p(u+v-1, 1+t)\right),$$

$$\underline{B}_t(u,v) = \max\left(W(u,v), \frac{t+1}{4} - \left(\frac{1}{2} - u\right)^+ - \left(\frac{1}{2} - v\right)^+\right), \text{ and}$$

$$\overline{B}t(u,v) = \min\left(M(u,v), \frac{t+1}{4} + \left(u - \frac{1}{2}\right)^+ + \left(v - \frac{1}{2}\right)^+\right),$$

where $p(a,b) = \dfrac{1}{6}\left[\left(9b + 3\sqrt{9b^2 - 3a^6}\right)^{1/3} + \left(9b - 3\sqrt{9b^2 - 3a^6}\right)^{1/3}\right]$. The above bounds are copulas, and hence if X and Y are continuous random variables with joint distribution function H and marginal distribution functions F and G, respectively, and such that $\tau_{X,Y} = t$, then

$$\underline{T}_t(F(x), G(y)) \le H(x,y) \le \overline{T}t(F(x), G(y))$$

for all (x,y) in $\mathbf{R}^2$, and these bounds are joint distributions functions (and similarly when $\rho_{X,Y} = t$ and $\beta_{X,Y} = t$).

For further details, including properties of the six bounds in Theorem 5.1.16 and a comparison of their relative effectiveness in narrowing the Fréchet-Hoeffding bounds, see (Nelsen et al. 2001; Nelsen and Úbeda Flores 2004).

Exercises

5.16 Let X and Y be continuous random variables with copula C. Show that an alternate expression for Spearman's rho for X and Y is

$$\rho = 3\iint_{\mathbf{I}^2}\left([u+v-1]^2 -[u-v]^2\right)dC(u,v).$$

[Cf. (5.1.23).] Gini referred to this expression for rho as the *indice di cograduazione quadratico*.

5.17 Let X and Y be continuous random variables with copula C. Establish the following inequalities between Blomqvist's β and Kendall's τ, Spearman's ρ, and Gini's γ:

$$\frac{1}{4}(1+\beta)^2 -1\leq \tau \leq 1-\frac{1}{4}(1-\beta)^2,$$

$$\frac{3}{16}(1+\beta)^3 -1\leq \rho \leq 1-\frac{3}{16}(1-\beta)^3,$$

$$\frac{3}{8}(1+\beta)^2 -1\leq \gamma \leq 1-\frac{3}{8}(1-\beta)^2.$$

[Hint: Use Theorem 3.2.3.]

5.18 Let C_θ be a member of the Plackett family of copulas (3.3.3) for $\theta > 0$. (a) Show that Blomqvist's β for this family is

$$\beta_\theta = \frac{\sqrt{\theta}-1}{\sqrt{\theta}+1}.$$

Recall (see Sect. 3.3.1) that for the Plackett family, θ represents an "odds ratio." When θ is an odds ratio in a 2×2 table, the expression $\left(\sqrt{\theta}-1\right)/\left(\sqrt{\theta}+1\right)$ is known as "Yule's Y," or "Yule's coefficient of colligation."
(b) Show that $4\beta_\theta/3$ is a second-order approximation to ρ_θ for this family (see Exercise 5.8).

5.19 Let C_θ, $\theta \in \mathbf{R}$, be a member of the Frank family (4.2.5) of Archimedean copulas. In Exercise 5.9, we obtained expressions involving Debye functions for ρ_θ and τ_θ for members of this family.
(a) Show that Blomqvist's β is

$$\beta = \beta_\theta = \frac{4}{\theta}\ln\cosh\frac{\theta}{4}.$$

(b) Show that Maclaurin series expansions for ρ_θ, τ_θ and β_θ are

$$\rho_\theta = \frac{1}{6}\theta - \frac{1}{450}\theta^3 + \frac{1}{23520}\theta^5 - \cdots,$$

$$\tau_\theta = \frac{1}{9}\theta - \frac{1}{900}\theta^3 + \frac{1}{52920}\theta^5 - \cdots,$$

$$\beta_\theta = \frac{1}{8}\theta - \frac{1}{768}\theta^3 + \frac{1}{46080}\theta^5 - \cdots,$$

so that for moderate values of the parameter θ, $4\beta/3$ and $8\beta/9$ are reasonable approximations to ρ and τ, respectively.

5.20 Let X and Y be continuous random variables whose copula C satisfies one (or both) of the functional equations in (2.8.1) for joint symmetry. Show that

$$\tau_{X,Y} = \rho_{X,Y} = \gamma_{X,Y} = \beta_{X,Y} = 0.$$

5.21 Another measure of association between two variates is *Spearman's foot-rule*, for which the sample version is

$$f = 1 - \frac{3}{n^2 - 1}\sum_{i=1}^{n}|p_i - q_i|,$$

where p_i and q_i again denote the ranks of a sample of size n of two continuous random variables X and Y, respectively.
(a) Show that the population version of the footrule, which we will denote ϕ, is given by

$$\phi = 1 - 3\iint_{I^2}|u - v|\,dC(u,v) = \frac{1}{2}[3Q(C,M)-1].$$

where C is again the copula of X and Y.
(b) Show that ϕ fails to satisfy properties 2 and 5 in Definition 5.1.7, and hence is not a "measure of concordance" according to that definition.

5.2 Dependence Properties

Undoubtedly the most commonly encountered dependence property is actually a "lack of dependence" property—independence. If X and Y

are continuous random variables with joint distribution function H, then the independence of X and Y is a property of the joint distribution function H—namely, that it factors into the product of its margins. Thus X and Y are independent precisely when H belongs to a particular subset of the set of all joint distribution functions, the subset characterized by the copula Π (see Theorem 2.4.2). In Sect. 2.5, we observed that one random variable is almost surely a monotone function of the other whenever the joint distribution function is equal to one of its Fréchet-Hoeffding bounds, i.e., the copula is M or W. Hence a "dependence property" for pairs of random variables can be thought of as a subset of the set of all joint distribution functions. Just as the property of independence corresponds to the subset all of whose members have the copula Π (and similarly for monotone functional dependence and the copulas M and W), many dependence properties can be described by identifying the copulas, or simple properties of the copulas, which correspond to the distribution functions in the subset. In this section, we will examine properties of copulas that "describe" other forms of dependence—dependence that lies "between" the extremes of independence and monotone functional dependence.

We begin with some "positive" and "negative" dependence properties—positive dependence properties expressing the notion that "large" (or "small") values of the random variables tend to occur together, and negative dependence properties expressing the notion that "large" values of one variable tend to occur with "small" values of the other. See (Barlow and Proschan 1981; Drouet Mari and Kotz 2001; Hutchinson and Lai 1990; Joe 1997; Tong 1980) and the references therein for further discussion of many of the dependence properties that we present in this section.

5.2.1 Quadrant Dependence

Definition 5.2.1 (Lehmann 1966). Let X and Y be random variables. X and Y are *positively quadrant dependent* (PQD) if for all (x,y) in $\mathbf{R}^2$,

$$P[X \leq x, Y \leq y] \geq P[X \leq x]P[Y \leq y]. \tag{5.2.1}$$

or equivalently (see Exercise 5.22),

$$P[X > x, Y > y] \geq P[X > x]P[Y > y]. \tag{5.2.2}$$

When (5.2.1) or (5.2.2) holds, we will write PQD(X,Y). Negative quadrant dependence is defined analogously by reversing the sense of the inequalities in (5.2.1) and (5.2.2), and we write NQD(X,Y).

Intuitively, X and Y are PQD if the probability that they are simultaneously small (or simultaneously large) is at least as great as it would be were they independent.

Example 5.13 (Barlow and Proschan 1981). Although in many studies of reliability, components are assumed to have independent lifetimes, it may be more realistic to assume some sort of dependence among components. For example, a system may have components that are subject to the same set of stresses or shocks, or in which the failure of one component results in an increased load on the surviving components. In such a two-component system with lifetimes X and Y, we may wish to use a model in which (regardless of the forms of the marginal distributions of X and Y) small values of X tend to occur with small values of Y, i.e., a model for which X and Y are PQD. ∎

If X and Y have joint distribution function H, with continuous margins F and G, respectively, and copula C, then (5.2.1) is equivalent to

$$H(x,y) \geq F(x)G(y) \text{ for all } (x,y) \text{ in } \mathbf{R}^2, \tag{5.2.3}$$

and to

$$C(u,v) \geq uv \text{ for all } (u,v) \text{ in } \mathbf{I}^2. \tag{5.2.4}$$

In the sequel, when continuous random variables X and Y are PQD, we will also say that their joint distribution function H, or their copula C, is PQD.

Note that, like independence, quadrant dependence (positive or negative) is a property of the copula of continuous random variables, and consequently is invariant under strictly increasing transformations of the random variables. Also note that there are other interpretations of (5.2.4). First, if X and Y are PQD, then the graph of the copula of X and Y lies on or above the graph of the independence copula Π. Secondly, (5.2.4) is the same as $C \succ \Pi$—i.e., C is larger than Π (recall Sect. 2.8). Indeed, the concordance ordering $\succ$ is sometimes called the "more PQD" ordering.

Many of the totally ordered one-parameter families of copulas that we encountered in Chapters 2 and 3 include Π and hence have subfamilies of PQD copulas and NQD copulas. For example, If C_θ is a member of the Mardia family (2.2.9), the FGM family (3.2.10), the Ali-Mikhail-Haq family (3.3.7), or the Frank family (4.2.5), then C_θ is PQD for $\theta \geq 0$ and NQD for $\theta \leq 0$ because each family is positively ordered and $C_0 = \Pi$.

Some of the important consequences for measures of association for continuous positively quadrant dependent random variables are summarized in the following theorem.

Theorem 5.2.2. *Let X and Y be continuous random variables with joint distribution function H, margins F and G, respectively, and copula C. If X and Y are PQD, then*

$$3\tau_{X,Y} \geq \rho_{X,Y} \geq 0, \, \gamma_{X,Y} \geq 0, \, and \, \beta_{X,Y} \geq 0.$$

Proof. The first inequality follows from $Q(C,C) \geq Q(C,\Pi) \geq Q(\Pi,\Pi)$; the remaining parts from Theorem 5.1.14. $\square$

Although PQD is a "global" property—(5.2.3) must hold at every point in $\mathbf{R}^2$—we can think of the inequality "locally." That is, at points (x,y) in $\mathbf{R}^2$ where $H(x,y) - F(x)G(y) \geq 0$, X and Y are "locally" PQD; whereas at points (x,y) in $\mathbf{R}^2$ where $H(x,y) - F(x)G(y) \leq 0$, X and Y are "locally" NQD. Equivalently, at points (u,v) in $\mathbf{I}^2$ where $C(u,v) - uv \geq 0$, X and Y are locally PQD; whereas at points (u,v) in $\mathbf{I}^2$ where $C(u,v) - uv \leq 0$, X and Y are locally NQD. But recall from (5.1.16) that one form of Spearman's rho is

$$\rho_C = 12\iint_{\mathbf{I}^2} [C(u,v) - uv] \, du \, dv,$$

and hence ρ_C (or, to be precise, $\rho_C/12$) can be interpreted as a measure of "average" quadrant dependence (both positive and negative) for random variables whose copula is C.

Exercises

5.22 (a) Show that (5.2.1) and (5.2.2) are equivalent.
(b) Show that (5.2.3) is equivalent to

$$\overline{H}(x,y) \geq \overline{F}(x)\overline{G}(y) \text{ for all } (x,y) \text{ in } \mathbf{R}^2.$$

5.23 (a) Let X and Y be random variables with joint distribution function H and margins F and G. Show that PQD(X,Y) if and only if for any (x,y) in $\mathbf{R}^2$,

$$H(x,y)[1 - F(x) - G(y) + H(x,y)] \geq [F(x) - H(x,y)][G(y) - H(x,y)],$$

that is, the product of the two probabilities corresponding to the two shaded quadrants in Fig. 5.6 is at least as great as the product of the two probabilities corresponding to the two unshaded quadrants.
(b) Give an interpretation of quadrant dependence in terms of the cross product ratio (3.3.1) for continuous random variables.
(c) Show that the copula version of this result for continuous random variables is: PQD(X,Y) if and only if for any (u,v) in $\mathbf{I}^2$,

$$C(u,v)[1 - u - v + C(u,v)] \geq [u - C(u,v)][v - C(u,v)];$$

and give an interpretation similar to that in Fig. 5.6.

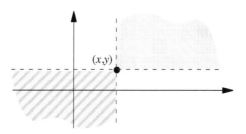

Fig. 5.6. A "product of probabilities" interpretation of PQD(X,Y)

5.24 (a) Show that if X and Y are PQD, then $-X$ and Y are NQD, X and $-Y$ are NQD, and $-X$ and $-Y$ are PQD.
(b) Show that if C is the copula of PQD random variables, then so is $\hat{C}$.

5.25 Let X and Y be continuous random variables with joint distribution function H and margins F and G, and copula C. Consider the random variable $Z = H(X,Y) - F(X)G(Y)$.
(a) Show that $E(Z) = (3\tau_C - \rho_C)/12$.
(b) Show that $\omega_C = 6E(Z) = (3\tau_C - \rho_C)/2$ can be interpreted as a measure of "expected" quadrant dependence for which $\omega_M = 1$, $\omega_\Pi = 0$, and $\omega_W = -1$.
(c) Show that ω_C fails to be a measure of concordance according to Definition 5.1.7.

5.26 *Hoeffding's lemma* (Hoeffding 1940; Lehmann 1966; Shea 1983). Let X and Y be random variables with joint distribution function H and margins F and G, such that $E(|X|)$, $E(|Y|)$, and $E(|XY|)$ are all finite. Prove that

$$\text{Cov}(X,Y) = \iint_{\mathbf{R}^2}[H(x,y) - F(x)G(y)]\,dxdy.$$

5.27 Let X and Y be random variables. Show that if PQD(X,Y), then $\text{Cov}(X,Y) \geq 0$, and hence Pearson's product-moment correlation coefficient is nonnegative for positively quadrant dependent random variables.

5.28 Show that X and Y are PQD if and only if $\text{Cov}[f(X),g(Y)] \geq 0$ for all functions f, g that are nondecreasing in each place and for which the expectations $E[f(X)]$, $E[g(Y)]$, and $E[f(X)g(Y)]$ exist (Lehmann 1966).

5.29 Prove that if the copula of X and Y is max-stable, then PQD(X,Y).

5.2.2 Tail Monotonicity

The expression (5.2.1) for positive quadrant dependence can be written as

$$P[Y \le y | X \le x] \ge P[Y \le y],$$

or as

$$P[Y \le y | X \le x] \ge P[Y \le y | X \le \infty].$$

A stronger condition would be to require that for each y in $\mathbf{R}$, the conditional distribution function $P[Y \le y | X \le x]$ is a nonincreasing function of x. If X and Y represent lifetimes of components in a reliability context, then this says that probability that Y has a short lifetime decreases (to be precise, does not increase) as the lifetime of X increases. This behavior of the left tails of the distributions of X and Y (and a similar behavior for the right tails based on (5.2.2)) is captured in the following definition (Esary and Proschan 1972).

Definition 5.2.3. Let X and Y be random variables.

1. Y is *left tail decreasing* in X [which we denote LTD($Y|X$)] if

$$P[Y \le y | X \le x] \text{ is a nonincreasing function of } x \text{ for all } y. \quad (5.2.5)$$

2. X is *left tail decreasing* in Y [which we denote LTD($X|Y$)] if

$$P[X \le x | Y \le y] \text{ is a nonincreasing function of } y \text{ for all } x. \quad (5.2.6)$$

3. Y is *right tail increasing* in X [which we denote RTI($Y|X$)] if

$$P[Y > y | X > x] \text{ is a nondecreasing function of } x \text{ for all } y. \quad (5.2.7)$$

4. Y is *right tail increasing* in X [which we denote RTI($X|Y$)] if

$$P[X > x | Y > y] \text{ is a nondecreasing function of } y \text{ for all } x. \quad (5.2.8)$$

Of course, if the joint distribution of X and Y is H, with margins F and G, respectively, then we can write $H(x,y)/F(x)$ (when $F(x) > 0$) rather than $P[Y \le y | X \le x]$ in (5.2.5) and similarly in (5.2.6); $\overline{H}(x,y)/\overline{F}(x)$ rather than $P[Y > y | X > x]$ in (5.2.7) and similarly in (5.2.8). The terms "left tail decreasing" and "right tail increasing" are from (Esary and Proschan 1972) where, as is often the case, "decreasing" means nonincreasing and "increasing" means nondecreasing.

There are similar negative dependence properties, known as *left tail increasing* and *right tail decreasing*, defined analogously by exchanging the words "nonincreasing" and "nondecreasing" in Definition 5.2.3.

Each of the four tail monotonicity conditions implies positive quadrant dependence. For example, if LTD($Y|X$), then

$$P[Y \le y|X \le x] \ge P[Y \le y|X \le \infty] = P[Y \le y],$$

and hence

$$P[X \le x, Y \le y] = P[X \le x]P[Y \le y|X \le x] \ge P[X \le x]P[Y \le y],$$

so that PQD(X,Y). Similarly, if RTI($Y|X$),

$$P[Y > y|X > x] \ge P[Y > y|X > -\infty] = P[Y > y],$$

and hence

$$P[X > x, Y > y] = P[X > x]P[Y > y|X > x] \ge P[X > x]P[Y > y],$$

and thus PQD(X,Y) by Exercise 5.22(a). Thus we have

Theorem 5.2.4. *Let X and Y be random variables. If X and Y satisfy any one of the four properties in Definition 5.2.3, then X and Y are positively quadrant dependent.*

However, positive quadrant dependence does not imply any of the four tail monotonicity properties—see Exercise 5.30.

The next theorem shows that, when the random variables are continuous, tail monotonicity is a property of the copula. The proof follows immediately from the observation that univariate distribution functions are nondecreasing.

Theorem 5.2.5. *Let X and Y be continuous random variables with copula C. Then*

1. LTD($Y|X$) *if and only if for any v in* **I**, $C(u,v)/u$ *is nonincreasing in u,*

2. LTD($X|Y$) *if and only if for any u in* **I**, $C(u,v)/v$ *is nonincreasing in v,*

3. RTI($Y|X$) *if and only if for any v in* **I**, $[1-u-v+C(u,v)]/(1-u)$ *is nondecreasing in u, or equivalently, if* $[v-C(u,v)]/(1-u)$ *is nonincreasing in u;*

4. RTI($X|Y$) *if and only if for any u in* **I**, $[1-u-v+C(u,v)]/(1-v)$ *is nondecreasing in v, or equivalently, if* $[u-C(u,v)]/(1-v)$ *is nonincreasing in v.*

Verifying that a given copula satisfies one or more of the conditions in Theorem 5.2.5 can often be tedious. As a consequence of Theorem 2.2.7, we have the following criteria for tail monotonicity in terms of the partial derivatives of C.

Corollary 5.2.6. *Let X and Y be continuous random variables with copula C. Then*

1. LTD($Y|X$) *if and only if for any v in* **I**, *$\partial C(u,v)/\partial u \leq C(u,v)/u$ for almost all u*;

2. LTD($X|Y$) *if and only if for any u in* **I**, *$\partial C(u,v)/\partial v \leq C(u,v)/v$ for almost all v*;

3. RTI($Y|X$) *if and only if for any v in* **I**, *$\partial C(u,v)/\partial u \geq [v - C(u,v)]/(1-u)$ for almost all u*;

4. RTI($X|Y$) *if and only if for any u in* **I**, *$\partial C(u,v)/\partial v \geq [u - C(u,v)]/(1-v)$ for almost all v*.

In the preceding section, we saw that there was a geometric interpretation for the copula of positive quadrant dependent random variables—the graph of the copula must lie on or above the graph of Π. There are similar geometric interpretations of the graph of the copula when the random variables satisfy one or more of the tail monotonicity properties—interpretations that involve the shape of regions determined by the horizontal and vertical sections of the copula.

To illustrate this, we first introduce some new notation. For each u_0 in **I**, let $S_1(u_0) = \left\{(u_0,v,x) \in \mathbf{I}^3 \middle| 0 \leq v \leq 1, 0 \leq z \leq C(u_0,v)\right\}$, i.e., $S_1(u_0)$ consists of the points in the unit cube $\mathbf{I}^3$ below the graph of the vertical section at $u = u_0$, i.e., lying in a plane perpendicular to the u-axis on or below the graph $z = C(u_0,v)$. Similarly, for each v_0 in **I**, we let $S_2(v_0) = \left\{(u,v_0,x) \in \mathbf{I}^3 \middle| 0 \leq u \leq 1, 0 \leq z \leq C(u,v_0)\right\}$, i.e., $S_2(v_0)$ consists of the points in the unit cube $\mathbf{I}^3$ below the graph of the horizontal section at $v = v_0$, i.e., lying in a plane perpendicular to the v-axis on or below the graph $z = C(u,v_0)$. The shaded region in Fig. 5.7 represents $S_2(v_0)$ for $C = M$ and $v_0 = 0.4$.

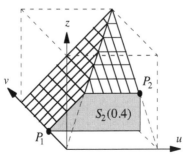

Fig. 5.7. An example of $S_2(v_0)$ for $C = M$ and $v_0 = 0.4$

Furthermore, we say that a plane region S is *starlike with respect to the point P in S* if for every point Q in S, all points on the line segment

PQ are in *S*. In Fig. 5.7, the shaded region $S_2(0.4)$ is starlike with respect to both $P_1 = (0,0.4,0)$ and $P_2 = (1,0.4,0.4)$.

The next theorem expresses the criteria for tail monotonicity in terms of the shapes of the regions $S_1(u)$ and $S_2(v)$ determined by the vertical and horizontal sections of the copula.

Theorem 5.2.7. *Let X and Y be continuous random variables with copula C. Then*

1. LTD($Y|X$) *if and only if for any v in* **I***, the region* $S_2(v)$ *is starlike with respect to the point* $(0,v,0)$ *in* $S_2(v)$.

2. LTD($X|Y$) *if and only if for any u in* **I***, the region* $S_1(u)$ *is starlike with respect to the point* $(u,0,0)$ *in* $S_1(u)$.

3. RTI($Y|X$) *if and only if for any v in* **I***, the region* $S_2(v)$ *is starlike with respect to the point* $(1,v,v)$ *in* $S_2(v)$.

4. RTI($X|Y$) *if and only if for any u in* **I***, the region* $S_1(u)$ *is starlike with respect to the point* $(u,1,u)$ *in* $S_1(u)$.

Proof. We prove part 1, leaving the proof of the remaining parts as an exercise. Assume LTD($Y|X$) and fix *v* in **I**. To show that $S_2(v)$ is starlike with respect to the point $(0,v,0)$, we will show that for $0 < t < 1$, the line segment joining $(0,v,0)$ to $(t,v,C(t,v))$ lies inside $S_2(v)$. Consider the points λt and t for $0 < \lambda < 1$. Because $\lambda t < t$, $C(\lambda t,v)/\lambda t \geq C(t,v)/t$ (because $C(u,v)/u$ is nonincreasing in *u*), or equivalently, $C(\lambda t,v) \geq \lambda C(t,v)$. Hence $C(\lambda t+(1-\lambda)0,v) \geq \lambda C(t,v) + (1-\lambda)C(0,v)$, so that every point on the line segment joining $(0,v,0)$ to $(t,v,C(t,v))$ lies inside $S_2(v)$. Conversely, assume that $S_2(v)$ is starlike with respect to $(0,v,0)$. Let $0 \leq u_1 < u_2 \leq 1$. Because the line segment joining $(0,v,0)$ to $(u_2,v,C(u_2,v))$ lies inside $S_2(v)$, we have

$$C(u_1,v) = C\left(\frac{u_1}{u_2}u_2 + \left(1-\frac{u_1}{u_2}\right)0,v\right)$$

$$\geq \frac{u_1}{u_2}C(u_2,v) + \left(1-\frac{u_1}{u_2}\right)C(0,v) = \frac{u_1}{u_2}C(u_2,v),$$

and thus $C(u_1,v)/u_1 \geq C(u_2,v)/u_2$. Hence LTD($Y|X$).   □

An important consequence of tail monotonicity is the following theorem (Capéraà and Genest 1993), in which the bounds for Spearman's rho and Kendall's tau in Corollary 5.1.11 can be narrowed when one random variable is simultaneously left tail decreasing and right tail increasing in the other. The proof proceeds along lines similar to those in the proofs of Theorems 5.1.9 and 5.1.10, and can be found in (Capéraà and Genest 1993).

Theorem 5.2.8. *Let X and Y be continuous random variables. If* LTD$(Y|X)$ *and* RTI$(Y|X)$, *then* $\rho_{X,Y} \geq \tau_{X,Y} \geq 0$ (*and similarly if* LTD$(X|Y)$ *and* RTI$(X|Y)$).

Because any one of the four tail monotonicity properties implies positive quadrant dependence, Theorem 5.2.2 can be invoked to strengthen the inequality in the preceding theorem to $3\tau_{X,Y} \geq \rho_{X,Y} \geq \tau_{X,Y} \geq 0$. However, positive quadrant dependence alone is insufficient to guarantee $\rho_{X,Y} \geq \tau_{X,Y}$, as the following example shows.

Example 5.14. Let U and V be uniform $(0,1)$ random variables whose joint distribution function is the diagonal copula constructed from the diagonal $\delta(t) = t^2$, i.e., $C(u,v) = \min(u,v,(u^2+v^2)/2)$ [see Examples 3.5 and 3.17(c)]. Because $u \geq uv$, $v \geq uv$, and $(u^2+v^2)/2 \geq uv$, U and V are positively quadrant dependent. However, it is easy to check that

$$P[U \leq 1/2 | V \leq 1/2] = 1/2 \quad \text{and} \quad P\left[U \leq 1/2 | V \leq \sqrt{3}/2\right] = \sqrt{3}/3 \cong .577$$

so that U is not left tail decreasing in V, and

$$P\left[U > 1/2 | V > 1 - \sqrt{3}/2\right] = \sqrt{3}/3 \cong .577 \quad \text{and} \quad P[U > 1/2 | V > 1/2] = 1/2,$$

so that U is not right tail increasing in V. By symmetry, V is not left tail decreasing in U, nor is V is not right tail increasing in U. Furthermore, from Exercise 5.4,

$$\tau = 4\int_0^1 t^2 \, dt - 1 = \frac{1}{3},$$

and from (5.1.15c),

$$\rho = 12\iint_{I^2} C(u,v) \, du dv - 3 = 5 - \frac{3\pi}{2} \cong .288,$$

and thus $\rho < \tau$.   ∎

5.2.3 Stochastic Monotonicity, Corner Set Monotonicity, and Likelihood Ratio Dependence

In the preceding section, we studied the monotonicity of $P[Y > y | X > x]$ and similar expressions. Replacing $P[Y > y | X > x]$ by $P[Y > y | X = x]$ yields further forms of dependence collectively known as "stochastic monotonicity":

Definition 5.2.9. Let X and Y be random variables.

1. Y is *stochastically increasing* in X [which we denote SI($Y|X$)] if

$P[Y > y|X = x]$ is a nondecreasing function of x for all y.   (5.2.9)

2. X is *stochastically increasing* in Y [which we denote SI($X|Y$)] if

$P[X > x|Y = y]$ is a nondecreasing function of y for all x.   (5.2.10)

Two negative dependence properties, SD($Y|X$) (Y is *stochastically decreasing* in X) and SD($X|Y$) (X is *stochastically decreasing* in Y) are defined analogously by replacing "nondecreasing" by "nonincreasing" in (5.2.9) and (5.2.10).

Example 5.15 (Barlow and Proschan 1981). Suppose X and Y are random variables with the Marshall-Olkin bivariate exponential distribution with parameters λ_1, λ_2, and λ_{12}, as presented in Sect. 3.1.1. The conditional survival probability $P[Y > y|X = x]$ is

$$P[Y > y|X = x] = \begin{cases} \dfrac{\lambda_1}{\lambda_1 + \lambda_{12}} \exp(-\lambda_{12}(y - x) - \lambda_2 y), & x \le y, \\ \exp(-\lambda_2 y), & x > y. \end{cases}$$

Because this conditional survival probability is nondecreasing in x, SI($Y|X$).   ∎

The term "stochastically increasing" is from (Shaked 1977; Barlow and Proschan 1981). However, in (Lehmann 1966) this property is called *positive regression dependence*, a term used by other authors as well with the notation PRD($Y|X$) and PRD($X|Y$) rather than SI($Y|X$) and SI($X|Y$). Although we obtained the two SI properties from the RTI properties, they can also be obtained from the LTD properties, because $P[Y \le y|X = x] = 1 - P[Y > y|X = x]$. Hence SI($Y|X$) if $P[Y \le y|X = x]$ is a nonincreasing function of x for all y and similarly for SI($X|Y$).

In the next theorem, we show that when the random variables are continuous, stochastic monotonicity, like tail monotonicity and quadrant dependence, is a property of the copula.

Theorem 5.2.10. *Let X and Y be continuous random variables with copula C. Then*

1. *SI($Y|X$) if and only if for any v in $\mathbf{I}$ and for almost all u, $\partial C(u,v)/\partial u$ is nonincreasing in u;*

2. *SI($X|Y$) if and only if for any u in $\mathbf{I}$ and for almost all v, $\partial C(u,v)/\partial v$ is nonincreasing in v.*

Proof: Because the marginal distribution functions F and G, respectively, of X and Y are nondecreasing, $P[Y \le y | X = x]$ is a nonincreasing function of x for all y if and only if $P[V \le v | U = u]$ is a nonincreasing function of u for all v, where $U = F(X)$ and $V = G(Y)$ are uniform $(0,1)$ random variables whose joint distribution function is the copula C. But, as shown in (2.9.1), $P[V \le v | U = u] = \partial C(u,v)/\partial u$. $\qquad\square$

The following geometric interpretation of stochastic monotonicity now follows [see (Roberts and Varberg 1973)]:

Corollary 5.2.11. *Let X and Y be continuous random variables with copula C. Then*

1. $SI(Y | X)$ *if and only if for any v in $\mathbf{I}$, $C(u,v)$ is a concave function of u,*

2. $SI(X | Y)$ *if and only if for any u in $\mathbf{I}$, $C(u,v)$ is a concave function of v.*

Example 5.16. Let C_θ be a member of the Plackett family (3.3.3). Then

$$\frac{\partial^2}{\partial u^2} C_\theta(u,v) = \frac{-2\theta(\theta-1)v(1-v)}{\left([1+(\theta-1)(u+v)]^2 - 4uv\theta(\theta-1)\right)^{3/2}},$$

so that $\partial^2 C_\theta(u,v)/\partial u^2 \le 0$ for $\theta \ge 1$, thus $C_\theta(u,v)$ is a concave function of u for $\theta \ge 1$. It follows that if X and Y are continuous random variables with copula C_θ, then $SI(Y | X)$ (and by symmetry, $SI(X | Y)$ as well) for $\theta \ge 1$ [Recall that for this family, $C_1 = \Pi$, $C_\infty = M$]. $\qquad\blacksquare$

The stochastic monotonicity properties imply the tail monotonicity properties:

Theorem 5.2.12. *Let X and Y be continuous random variables with copula C. Then*

1. *if $SI(Y | X)$, then $LTD(Y | X)$ and $RTI(Y | X)$,*

2. *if $SI(X | Y)$, then $LTD(X | Y)$ and $RTI(X | Y)$.*

Proof. Assume $SI(Y | X)$, fix v, and again let $S_2(v)$ be the set of points in the unit cube $\mathbf{I}^3$ lying in a plane perpendicular to the v-axis on or below the graph of the copula. Because $C(u,v)$ is a concave function of u, it follows that $S_2(v)$ is starlike with respect to both $(0,v,0)$ and $(1,v,v)$ in $S_2(v)$, hence $LTD(Y | X)$ and $RTI(Y | X)$. The second part of the proof is analogous. $\qquad\square$

The converse to Theorem 5.2.12 is false—the tail monotonicity properties do not imply stochastic monotonicity—see Exercise 5.32.

Another pair of dependence properties, also derived from tail monotonicity, are the "corner set monotonicity" properties, introduced in (Harris 1970). Because LTD($Y|X$) is defined in terms of $P[Y \leq y|X \leq x]$, and LTD($X|Y$) in terms of $P[X \leq x|Y \leq y]$, we are led to considering the behavior of the joint probability $P[X \leq x, Y \leq y|X \leq x', Y \leq y']$ (and similarly for the right tail properties):

Definition 5.2.13. Let X and Y be continuous random variables.

1. X and Y are *left corner set decreasing* [which we denote LCSD(X,Y)] if

$$P[X \leq x, Y \leq y|X \leq x', Y \leq y'] \text{ is nonincreasing in } x' \text{ and in } y' \qquad (5.2.11)$$

for all x and y;

2. X and Y are *right corner set increasing* [which we denote RCSI(X,Y)] if

$$P[X > x, Y > y|X > x', Y > y'] \text{ is nondecreasing in } x' \text{ and in } y' \qquad (5.2.12)$$

for all x and y.

Two negative dependence properties, LCSI(X,Y) (X and Y are *left corner set increasing*) and RCSD($X|Y$) (X and Y are *right corner set decreasing*) are defined analogously by exchanging the words "nondecreasing" and "nonincreasing" in (5.2.11) and (5.2.12).

As an immediate consequence, we have that the corner set monotonicity properties imply the corresponding tail monotonicity properties:

Theorem 5.2.14. *Let X and Y be continuous random variables.*

1. *If* LCSD(X,Y)]*, then* LTD($Y|X$) *and* LTD($X|Y$);

2. *If* RCSI(X,Y)]*, then* RTI($Y|X$) *and* RTI($X|Y$).

Proof. For part 1, set $x = \infty$ and $y' = \infty$ to obtain LTD($Y|X$), and set $y = \infty$ and $x' = \infty$ to obtain LTD($X|Y$). Part 2 is similar. $\square$

The converse to Theorem 5.2.14 is false—the tail monotonicity properties do not imply corner set monotonicity—see Exercise 5.33.

The following theorem gives simple criteria for LCSD(X,Y) and RCSI(X,Y) in terms of inequalities for the joint distribution and survival function of X and Y:

Theorem 5.2.15. *Let X and Y be continuous random variables with joint distribution function H:*

1. LCSD(X,Y) *if and only if*

$$H(x,y)H(x',y') \geq H(x,y')H(x',y) \qquad (5.2.13)$$

for all x, y, x', y' in $\overline{\mathbf{R}}$ such that $x \leq x'$ and $y \leq y'$.

2. RCSI(X,Y) *if and only if*

$$\overline{H}(x,y)\overline{H}(x',y') \geq \overline{H}(x,y')\overline{H}(x',y) \qquad (5.2.14)$$

for all x, y, x′, y′ in **R** *such that x ≤ x′ and y ≤ y′.*

Proof: We prove part 1, the proof of part 2 is similar. First assume LCSD(X,Y). Thus $P[X \leq x, Y \leq y | X \leq x', Y \leq y']$ is nonincreasing in x' and in y' for all x and y, so that for $y = \infty$, $P[X \leq x | X \leq x', Y \leq y']$ is nonincreasing in x' and in y' for all x. Hence if $x \leq x'$, then

$$\frac{P[X \leq x, Y \leq y']}{P[X \leq x', Y \leq y']}$$

is nonincreasing in y'. Thus, for $y \leq y'$,

$$\frac{P[X \leq x, Y \leq y]}{P[X \leq x', Y \leq y]} \geq \frac{P[X \leq x, Y \leq y']}{P[X \leq x', Y \leq y']}, \qquad (5.2.15)$$

which is equivalent to (5.2.13).

Conversely, assume that (5.2.15) holds. It follows that $P[X \leq x | X \leq x', Y \leq y']$ is nonincreasing in x' and in y' for all x, and that $P[Y \leq y | X \leq x', Y \leq y']$ is nonincreasing in x' and in y' for all y. If $x \leq x'$ and $y \leq y'$, then $P[X \leq x, Y \leq y | X \leq x', Y \leq y']$ is trivially nonincreasing in x' and in y'. If $x > x'$ and $y \leq y'$, then $P[X \leq x, Y \leq y | X \leq x', Y \leq y'] = P[Y \leq y | X \leq x', Y \leq y']$, which we have just shown is nonincreasing in x' and in y'. The case $x \leq x'$ and $y > y'$ is similar, and when $x > x'$ and $y > y'$, $P[X \leq x, Y \leq y | X \leq x', Y \leq y'] = 1$. Hence LCSD(X,Y), as claimed. □

The criteria in Theorem 5.2.14 for LCSD(X,Y) and RCSI(X,Y) can be succinctly expressed using the notion of "totally positive" functions. A function f from $\overline{\mathbf{R}}^2$ to **R** is *totally positive of order two* (Karlin 1968), abbreviated TP$_2$, if $f(x,y) \geq 0$ on $\overline{\mathbf{R}}^2$ and whenever $x \leq x'$ and $y \leq y'$,

$$\begin{vmatrix} f(x,y) & f(x,y') \\ f(x',y) & f(x',y') \end{vmatrix} \geq 0 \qquad (5.2.16)$$

("order two" refers to the size of the matrix). When the inequality in (5.2.16) is reversed, we say that f is *reverse regular of order two* (or *reverse rule of order two*), abbreviated RR$_2$. In terms of total positivity, we have

Corollary 5.2.16. *Let X and Y be continuous random variables with joint distribution function H. Then LCSD(X,Y) if and only if H is* TP$_2$, *and* RCSI(X,Y) *if and only if* $\overline{H}$ *is* TP$_2$.

Example 5.17. Let X and Y be random variables with the Marshall-Olkin bivariate exponential distribution presented in Sect. 3.1.1, with parameters λ_1, λ_2, and λ_{12}. If $\overline{H}$ denotes the joint survival function of X and Y, then

$$\overline{H}(x,y) = \exp\left[-\lambda_1 x - \lambda_2 y - \lambda_{12}\max(x,y)\right].$$

Thus

$$\overline{H}(x,y)\overline{H}(x',y') =$$
$$\exp\left[-\lambda_1(x+x') - \lambda_2(y+y') - \lambda_{12}[\max(x,y)+\max(x',y')]\right]$$

and

$$\overline{H}(x',y)\overline{H}(x,y') =$$
$$\exp\left[-\lambda_1(x+x') - \lambda_2(y+y') - \lambda_{12}[\max(x',y)+\max(x,y')]\right].$$

So if $0 \le x \le x'$ and $0 \le y \le y'$, then $\max(x,y) + \max(x',y') \le \max(x',y) + \max(x,y')$. It follows that $\overline{H}(x,y)\overline{H}(x',y') \ge \overline{H}(x',y)\overline{H}(x,y')$, and hence RCSI$(X,Y)$. ∎

In terms of the copula and survival copula of X and Y, we have:

Corollary 5.2.17. *Let X and Y be continuous random variables with copula C. Then LCSD(X,Y) if and only if C is* TP$_2$, *and RCSI(X,Y) if and only if $\hat{C}$ is* TP$_2$.

Theorems 5.2.5, 5.2.12, and 5.2.14 yield the implications illustrated in Fig. 5.8 among the various dependence properties presented so far. Exercises 5.30, 5.32, and 5.33 show that none of the implications are equivalences.

$$\text{SI}(Y\,|\,X) \;\Rightarrow\; \text{RTI}(Y\,|\,X) \;\Leftarrow\; \text{RCSI}(X,Y)$$
$$\Downarrow \qquad\qquad \Downarrow \qquad\qquad \Downarrow$$
$$\text{LTD}(Y\,|\,X) \;\Rightarrow\; \text{PQD}(X,Y) \;\Leftarrow\; \text{RTI}(X\,|\,Y)$$
$$\Uparrow \qquad\qquad \Uparrow \qquad\qquad \Uparrow$$
$$\text{LCSD}(X,Y) \;\Rightarrow\; \text{LTD}(X\,|\,Y) \;\Leftarrow\; \text{SI}(X\,|\,Y)$$

Fig. 5.8. Implications among the various dependence properties

The final dependence property that we discuss in this section is likelihood ratio dependence (Lehmann 1966). It differs from those considered above in that it is defined in terms of the joint density function rather than conditional distribution functions.

Definition 5.2.18. Let X and Y be continuous random variables with joint density function $h(x,y)$. Then X and Y are *positively likelihood ratio dependent* [which we denote PLR(X,Y)] if h satisfies

$$h(x,y)h(x',y') \ge h(x,y')h(x',y) \qquad (5.2.17)$$

for all x, y, x', y' in $\mathbf{R}$ such that $x \leq x'$ and $y \leq y'$, i.e., h is TP_2.

Some authors use the notation $\text{TP}_2(X,Y)$ rather than $\text{PLR}(X,Y)$. This property derives its name from the fact that the inequality in (5.2.17) is equivalent to the requirement that the conditional density of Y given x has a monotone likelihood ratio. Negative likelihood ratio dependence is defined analogously, by reversing the sense of the inequality in (5.2.17) (i.e., h is RR_2).

Of the dependence properties discussed so far, positive likelihood ratio dependence is the strongest, implying all of the properties in Fig. 5.8. To prove this statement, we need only prove Theorem 5.2.19.

Theorem 5.2.19. *Let X and Y be random variables with an absolutely continuous distribution function. If $\text{PLR}(X,Y)$, then $\text{SI}(Y|X)$, $\text{SI}(X|Y)$, $\text{LCSD}(X,Y)$, and $\text{RCSI}(X,Y)$.*

Proof (Barlow and Proschan 1981). Let h, f, and g denote the joint and respective marginal densities of X and Y, and assume $\text{PLR}(X,Y)$. Then if $x \leq x'$ and $t \leq t'$, $h(x,t)h(x',t') \geq h(x,t')h(x',t)$, so that if we divide both sides of the inequality by $f(x)f(x')$ and integrate with respect to t from $-\infty$ to y and with respect to t' from y to ∞ (where y is arbitrary), we have $P[Y \leq y|X = x]P[Y > y|X = x'] \geq P[Y \leq y|X = x'] \cdot P[Y > y|X = x']$. Adding $P[Y > y|X = x']P[Y > y|X = x]$ to both sides, the inequality simplifies to $P[Y > y|X = x'] \geq P[Y > y|X = x]$, i.e., $P[Y > y|X = x]$ is nondecreasing in x for all y, so that $\text{SI}(Y|X)$. The proof that $\text{PLR}(X,Y)$ implies $\text{SI}(X|Y)$ is similar.

To show that $\text{PLR}(X,Y)$ implies $\text{LCSD}(X,Y)$, we first note that if $x \leq x'$ and $y \leq y'$, then $P[X \leq x, Y \leq y|X \leq x', Y \leq y']$ is trivially nonincreasing in x' and in y', and if $x > x'$ and $y > y'$, then $P[X \leq x, Y \leq y|X \leq x', Y \leq y'] = 1$. So assume $x > x'$ and $y \leq y'$, in which case $P[X \leq x, Y \leq y|X \leq x', Y \leq y'] = P[Y \leq y|X \leq x', Y \leq y']$. As this is clearly nonincreasing in y', we need only show that $P[Y \leq y|X \leq x', Y \leq y'] = P[X \leq x', Y \leq y]/P[X \leq x', Y \leq y']$ is nonincreasing in x', i.e., that for $x' \leq x''$,

$$\frac{P[X \leq x', Y \leq y]}{P[X \leq x', Y \leq y']} \geq \frac{P[X \leq x'', Y \leq y]}{P[X \leq x'', Y \leq y']}. \qquad (5.2.18)$$

Assume $\text{PLR}(X,Y)$, so that if $s \leq s'$ and $t \leq t'$, $h(s,t)h(s',t') \geq h(s,t')h(s',t)$. Integrating both sides of this inequality with respect to s from $-\infty$ to x', with respect to s' from x' to x'', with respect to t from

$-\infty$ to y, and with respect to t' from y to y' yields $P[X \le x', Y \le y]P[x' < X \le x'', y < Y \le y'] \ge P[X \le x', y < Y \le y'] \cdot P[x' < X \le x'', Y \le y]$. If we add $P[X \le x', Y \le y] P[x' < X \le x'', Y \le y]$ to both sides, the inequality simplifies to $P[X \le x', Y \le y] \cdot P[x' < X \le x'', Y \le y'] \ge P[X \le x', Y \le y']P[x' < X \le x'', Y \le y]$. Now adding $P[X \le x', Y \le y]P[X \le x', Y \le y']$ to both sides yields $P[X \le x', Y \le y]P[X \le x'', Y \le y'] \ge P[X \le x', Y \le y']P[X \le x'', Y \le y]$, which is equivalent to (5.2.18). The case $x \le x'$ and $y > y'$ is similar, which completes the proof that PLR(X,Y) implies LCSD(X,Y). The proof that PLR(X,Y) implies RCSI(X,Y) is analogous. $\square$

Although positive likelihood ratio dependence is a "global" property, one can view it "locally," as we did with positive quadrant dependence (see the paragraph following Theorem 5.2.2). That is, for any x, y, x', y' in **R** such that $x \le x'$ and $y \le y'$, we evaluate the density h at the four points (x,y), (x,y'), (x',y), and (x',y'), and when $h(x,y)h(x',y') - h(x,y')h(x',y) \ge 0$, (X,Y) is "locally positively likelihood ratio dependent" (or h is "locally" TP$_2$). When the inequality is reversed, (X,Y) is "locally negatively likelihood ratio dependent" (or h is "locally" RR$_2$). In the next theorem, we relate local likelihood ratio dependence to Kendall's tau:

Theorem 5.2.20. *Let X and Y be random variables with joint density function h, and let*

$$T = \int_{-\infty}^{\infty}\int_{-\infty}^{\infty}\int_{-\infty}^{y'}\int_{-\infty}^{x'} [h(x,y)h(x',y') - h(x,y')h(x',y)] \, dxdydx'dy'.$$

Then Kendall's tau for X and Y is given by $\tau_{X,Y} = 2T$.

Proof: Let C, F, G, f, and g denote the copula, the marginal distribution functions, and the marginal densities of X and Y, respectively, and set $u = F(x)$, $u' = F(x')$, $v = G(y)$, and $v' = G(y')$. Also let $c(u,v) = \partial^2 C(u,v)/\partial u \partial v$, so that $h(x,y) = c(F(x),G(y))f(x)g(y)$. Then

$$T = \int_0^1\int_0^1\int_0^{v'}\int_0^{u'} [c(u,v)c(u',v') - c(u,v')c(u',v)] \, dudvdu'dv'. \quad (5.2.19)$$

The inner double integral is readily evaluated to yield

$$T = \int_0^1\int_0^1 \left[C(u',v') \frac{\partial^2}{\partial u' \partial v'} C(u',v') - \frac{\partial}{\partial u'} C(u',v') \frac{\partial}{\partial v'} C(u',v') \right] du'dv',$$

$$= \iint_{\mathbf{I}^2} C(u,v) \, dC(u,v) - \iint_{\mathbf{I}^2} \frac{\partial}{\partial u} C(u,v) \frac{\partial}{\partial v} C(u,v) \, dudv. \quad (5.2.20)$$

But from (5.1.7) the first integral is $(\tau_{X,Y}+1)/4$, and from (5.1.9) the second integral is $(1-\tau_{X,Y})/4$, and the conclusion now follows.   □

Recall that in Sect. 5.2.1 we observed that Spearman's rho can be interpreted as a measure of "average" quadrant dependence, and from the above discussion we see that Kendall's tau can be interpreted as a measure of "average" likelihood ratio dependence. Of the dependence properties discussed in this chapter, quadrant dependence is the weakest, whereas likelihood ratio dependence is the strongest. Thus the two most commonly used measures of association are related to two rather different dependence properties, a fact that may partially explain the differences between the values of rho and tau that we observed in several of the examples and exercises in earlier sections of this chapter.

The notion of positive likelihood ratio dependence can be extended to random variables whose joint distribution function fails to be absolutely continuous. To do so, we need some new notation: Let J and K denote intervals in $\overline{\mathbf{R}}$. Then we will write $H(J,K)$ for $P[X\in J, Y\in K]$. We also write $J < K$ whenever $s\in J$ and $t\in K$ implies $s < t$.

Definition 5.2.21 (Block et al. 1982; Kimeldorf and Sampson 1987). Let X and Y be continuous random variables with joint distribution function $H(x,y)$. Then X and Y are *positively likelihood ratio dependent* if H satisfies

$$H(J_2,K_2)H(J_1,K_1) \geq H(J_1,K_2)H(J_2,K_1) \qquad (5.2.21)$$

for all intervals J_1,J_2,K_1,K_2 in $\overline{\mathbf{R}}$ such that $J_1 < J_2$ and $K_1 < K_2$.

It is easy to verify that when H has a density h, then Definitions 5.2.18 and 5.2.21 are equivalent. Furthermore, (5.2.21) can be expressed in terms of the copula C of X and Y, because $H(J,K) = C(F(J),G(K))$, where for any two intervals $[u_1,u_2]$ and $[v_1,v_2]$, $C([u_1,u_2],[v_1,v_2]) = C(u_2,v_2) - C(u_1,v_2) - C(u_2,v_1) + C(u_1,v_1)$.

Using Definition 5.2.21, it is possible to prove an extension of Theorem 5.2.20 without the assumption that X and Y have a joint density—they need only be continuous. To do so, we use the copula C of X and Y to construct a measure of "local" positive likelihood ratio dependence analogous to that which appears in the integrand of (5.2.19). We begin with partitioning the interval [0,1] on the u-axis in the usual manner: choose points $\{u_p\}_{p=0}^{n}$ such that $0 = u_0 < u_1 < \cdots < u_n = 1$ and let $J_p = [u_{p-1},u_p]$. Similarly partition [0,1] on the v-axis into intervals $K_q = [v_{q-1},v_q]$ for $1 \leq q \leq m$; thus generating a partition P of $\mathbf{I}^2$ into mn rectangles $J_p \times K_q$. Let $\|P\|$ denote the norm of P. For each of the $\binom{n}{2}\binom{m}{2}$ choices of intervals J_r, J_p, K_s, and K_q, with $1 \leq r < p \leq n$ and $1 \leq s < q \leq m$, the quantity

$$C(J_p,K_q)C(J_r,K_s) - C(J_p,K_s)C(J_r,K_q)$$

measures "local" positive (or negative) likelihood ratio dependence. Analogous to (5.2.19), we now define

$$T = \lim_{\|P\|\to 0} \sum_{p=2}^{n} \sum_{q=2}^{m} \sum_{r=1}^{p-1} \sum_{s=1}^{q-1} \left[C(J_p,K_q)C(J_r,K_s) - C(J_p,K_s)C(J_r,K_q) \right].$$

The inner double summation in T readily telescopes to yield

$$T =$$

$$\lim_{\|P\|\to 0} \sum_{p=2}^{n} \sum_{q=2}^{m} \left[C(u_p,v_q)C(u_{p-1},v_{q-1}) - C(u_{p-1},v_q)C(u_p,v_{q-1}) \right].(5.2.22)$$

It can be shown (Nelsen 1992) that T exists and $\tau_{XY} = 2T$, as in Theorem 5.2.20.

Exercises

5.30 This exercise shows that positive quadrant dependence does not imply any of the tail monotonicity properties. Let C be the ordinal sum of $\{M,W,M\}$ with respect to the partition $\{[0,\theta],[\theta,1-\theta],[1-\theta,1]\}$, for any θ in $(1/4,1/2)$ [this copula is also a shuffle of M]. Show that if X and Y are random variables with copula C, then PQD(X,Y) but none of the four tail monotonicity properties hold.

5.31 Let X and Y be continuous random variables whose copula is C.
(a) Show that if $C = \hat{C}$, then LTD($Y|X$) if and only if RTI($Y|X$), and LTD($X|Y$) if and only if RTI($X|Y$).
(b) Show that is C is symmetric [i.e., $C(u,v) = C(v,u)$], then LTD($Y|X$) if and only if LTD($X|Y$), and RTI($Y|X$) if and only if RTI($X|Y$).

5.32 This exercise shows that the tail monotonicity properties do not imply the stochastic monotonicity properties. Let C be given by

$$C(u,v) = \begin{cases} \dfrac{3uv}{2} - \dfrac{u+v-1}{2}, & \dfrac{1}{3} \le v \le 1 - u \le \dfrac{2}{3}, \\[2mm] \dfrac{3uv}{2}, & \dfrac{1}{3} \le 1 - u \le v \le \dfrac{2}{3}, \\[2mm] M(u,v), & \text{otherwise.} \end{cases}$$

(a) Let X and Y be continuous random variables whose copula is C. Show that LTD$(Y|X)$, LTD$(X|Y)$, RTI$(Y|X)$, and RTI$(X|Y)$.

(b) Show that SI$(Y|X)$, and SI$(X|Y)$ both fail.

5.33 This exercise shows that the tail monotonicity properties do not imply the corner set monotonicity properties. Let C be the copula whose probability mass is uniformly distributed on two line segments, one joining $(0,0)$ to $(1,1/2)$, and the other joining $(0,1/2)$ to $(1,1)$.

(a) Show that C is given by

$$C(u,v) = \min\left(u,v,\frac{u}{2}+(v-1/2)^+ \right).$$

(b) Let X and Y be continuous random variables whose copula is C. Show that LTD$(Y|X)$ and RTI$(Y|X)$.

(c) Show that LCSD(X,Y) and RCSI(X,Y) both fail. [Hint: Consider the points $u_i = v_i = i/3$, $i = 1,2$; and note that $C = \hat{C}$.]

5.34 Let X and Y be random variables whose copula C is Archimedean with a strict generator φ in Ω. Prove that

(a) PQD(X,Y) if and only if $-\ln\varphi^{-1}$ is subadditive on $(0,\infty)$.

(b) If φ^{-1} is differentiable, then SI$(Y|X)$ or SI$(X|Y)$ if and only if $\ln(-d\varphi^{-1}(t)/dt)$ is convex on $(0,\infty)$ (Capéraà and Genest 1993).

(c) The LTD and LCSD properties are equivalent.

(d) If φ^{-1} is completely monotone, then LCSD(X,Y) (Alsina et al. 2005).

5.35 Let X and Y be continuous random variables whose copula is C. Show that

(a) LTD$(Y|X)$ and LTD$(X|Y)$ if and only if for all u, u', v, v' in $\mathbf{I}$ such that $0 < u \le u' \le 1$ and $0 < v \le v' \le 1$,

$$\frac{C(u,v)}{uv} \ge \frac{C(u',v')}{u'v'}.$$

(b) Conclude that LTD$(Y|X)$ and LTD$(X|Y)$ if and only if for every point (u',v') in $(0,1]^2$, the graph of $z = C(u,v)$ lies above the graph of the hyperbolic paraboloid through the four points $(0,0,0)$, $(u',0,0)$, $(0,v',0)$, and $(u',v',C(u',v'))$, i.e., $z = [C(u',v')/u'v'] \cdot uv$.

5.36 Let X and Y be continuous random variables whose copula is C. Show that

(a) if the function $u - C(u,v)$ is TP_2, then $LTD(Y|X)$ and $RTI(X|Y)$;

(b) if the function $v - C(u,v)$ is TP_2, then $LTD(X|Y)$ and $RTI(Y|X)$;

(c) the function $1 - u - v + C(u,v)$ is TP_2 if and only if $\hat{C}$ is TP_2.

5.37 Let X and Y be continuous random variables whose copula C has cubic sections in both u and v, i.e., C satisfies the conclusion of Theorem 3.2.8. Show that (Nelsen et al. 1997):

(a) $PQD(X,Y)$ if and only if A_1, A_2, B_1, B_2 are all nonnegative;

(b) $LTD(Y|X)$ if and only if $2A_1 \geq B_1 \geq 0, 2A_2 \geq B_2 \geq 0$;

(c) $RTI(Y|X)$ if and only if $2B_1 \geq A_1 \geq 0, 2B_2 \geq A_2 \geq 0$;

(d) $SI(Y|X)$ if and only if $2A_1 \geq B_1, 2B_1 \geq A_1, 2A_2 \geq B_2$, $2B_2 \geq A_2$;

(e) $LTD(X|Y)$ if and only if $2A_2 \geq A_1 \geq 0, 2B_2 \geq B_1 \geq 0$;

(f) $RTI(X|Y)$ if and only if $2A_1 \geq A_2 \geq 0, 2B_1 \geq B_2 \geq 0$;

(g) $SI(X|Y)$ if and only if $2A_1 \geq A_2, 2A_2 \geq A_1, 2B_1 \geq B_2$, $2B_2 \geq B_1$.

5.38 Let X and Y be random variables such that $SI(Y|X)$ and $SI(X|Y)$. Hutchinson and Lai (1990) conjectured that for such random variables, $\rho \leq 3\tau/2$. Let θ be in $[0,1/4]$, and let C_θ be the copula

$$C_\theta(u,v) = uv + 2\theta uv(1-u)(1-v)(1+u+v-2uv).$$

Note that C_θ is cubic in u and in v, so that C_θ is given by (3.2.19) with $A_1 = B_2 = 4\theta$, $A_2 = B_1 = 2\theta$. Suppose the copula of X and Y is C_θ.

(a) Show that $SI(Y|X)$ and $SI(X|Y)$.

(b) Show (see Exercise 5.13) that

$$\rho_\theta = \theta \quad \text{and} \quad \tau_\theta = \frac{2}{3}\theta - \frac{2}{75}\theta^2,$$

so that $\rho > 3\tau/2$ for θ in $(0,1/4]$, and hence that the conjecture is false. Hutchinson and Lai also conjectured that when $SI(Y|X)$ and $SI(X|Y)$,

$$-1+\sqrt{1+3\tau} \le \rho \le 2\tau - \tau^2,$$

but this conjecture remains to be proven or disproven. However, when C is an extreme value copula (see Sect. 3.3.4), Hürlimann (2003) has shown that

$$-1+\sqrt{1+3\tau} \le \rho \le \min\{3\tau/2, 2\tau - \tau^2\}.$$

5.39 This exercise provides an alternate proof (Joe 1997) that PLR(X,Y) implies LCSD(X,Y). Let X, Y, and h be as in Theorem 5.2.19 and let H be the joint distribution function of X and Y. Suppose that $x \le x'$ and $y \le y'$. Show that PLR(X,Y) implies that

$$\int_{-\infty}^{x} \int_{-\infty}^{y} \int_{x}^{x'} \int_{y}^{y'} [h(s,t)h(s',t') - h(s,t')h(s',t)]\, dt'ds'dtds \ge 0,$$

which in turn implies

$$H(x,y)[H(x',y') - H(x',y) - H(x,y') + H(x,y)]$$
$$\ge [H(x,y') - H(x,y)][H(x',y) - H(x,y)],$$

a condition equivalent to LCSD(X,Y). There is an analogous proof that PLR(X,Y) implies RCSI(X,Y).

5.3 Other Measures of Association

In Sect. 5.1, we discussed four measures of association based on the notion of concordance. There are many other nonparametric measures of association that depend on the copula of the random variables. In Sect. 5.3.1, we discuss a class of measures, known as measures of dependence, which are based on a "distance" between the copula of a pair of random variables X and Y and the "independence" copula Π. In Sect. 5.3.2, we consider measures derived from Gini's γ, measures that are based on "distances" between the copula of X and Y and the copulas M and W of the Fréchet bounds.

5.3.1 Measures of Dependence

In Definition 5.1.7, we presented a list of seven properties for one class of measures of association—those known as "measures of concordance." In Theorems 5.1.9 and 5.1.14, we saw that Kendall's τ, Spearman's ρ, Gini's γ, and Blomqvist's β are measures of concordance. But one defect of such measures is that for the fourth property in Definition

5.1.7, which states that if the random variables are independent then the measure equals zero, the converse fails to hold. Examples abound in which a measure of concordance is zero but the random variables are dependent.

In this section, we will consider measures of association that are commonly called "measures of dependence" (Rényi 1959; Schweizer and Wolff 1981; Jogdeo 1982; Lancaster 1982). Recall that measures of concordance measure how "large" values of one variable tend be associated with "large" values of the other (and "small" with "small"), and consequently they attain their extreme values when the copula of the random variables is either M (where the measure is +1) or W (where the measure is –1). On the other hand, in the words of Lancaster (1982), "a measure of dependence indicates in some defined way, how closely X and Y are related, with extremes at mutual independence and (monotone) dependence." Here then is a minimal set of desirable properties for a nonparametric measure of dependence. It is adapted from sets of properties discussed in (Rényi 1959; Schweizer and Wolff 1981; Jogdeo 1982; Lancaster 1982):

Definition 5.3.1. A numeric measure δ of association between two continuous random variables X and Y whose copula is C is a *measure of dependence* if it satisfies the following properties (where we write $\delta_{X,Y}$ or δ_C if convenient):

1. δ is defined for every pair of continuous random variables X and Y;

2. $\delta_{X,Y} = \delta_{Y,X}$;

3. $0 \le \delta_{X,Y} \le 1$;

4. $\delta_{X,Y} = 0$ if and only if X and Y are independent;

5. $\delta_{X,Y} = 1$ if and only if each of X and Y is almost surely a strictly monotone function of the other;

6. if α and β are almost surely strictly monotone functions on Ran$X$ and Ran$Y$, respectively, then $\delta_{\alpha(X),\beta(Y)} = \delta_{X,Y}$,

7. if $\{(X_n, Y_n)\}$ is a sequence of continuous random variables with copulas C_n, and if $\{C_n\}$ converges pointwise to C, then $\lim_{n \to \infty} \delta_{C_n} = \delta_C$.

Our first example of such a measure is closely related to Spearman's rho. Recall from (5.1.16) that for continuous random variables X and Y with copula C, Spearman's rho can be written as

$$\rho_{X,Y} = \rho_C = 12 \iint_{I^2} [C(u,v) - uv] \, du \, dv .$$

As noted before, ρ_C is proportional to the signed volume between the graphs of the copula C and the product copula Π. If in the integral above, we replace the difference $[C(u,v) - uv]$ between C and Π by the absolute difference $|C(u,v) - uv|$, then we have a measure based upon the L_1 distance between the graphs of C and Π. This measure (Schweizer and Wolff 1981; Wolff 1977), which is known as *Schweizer and Wolff's* σ, is given by

$$\sigma_{X,Y} = \sigma_C = 12 \iint_{\mathbf{I}^2} |C(u,v) - uv| \, du \, dv. \qquad (5.3.1)$$

Theorem 5.3.2. *Let X and Y be continuous random variables with copula C. Then the quantity σ_C defined in (5.3.1) is a measure of dependence, i.e., it satisfies the seven properties in Definition 5.3.1.*

Proof (Schweizer and Wolff 1981). It is easy to see from its definition that σ satisfies the first two properties. The third property is also easily established for σ by first showing that for any copula C,

$$\iint_{\mathbf{I}^2} |C(u,v) - uv| \, du \, dv \le \frac{1}{12}. \qquad (5.3.2)$$

The fourth property follows from Theorem 2.4.2, and the fifth from Theorems 2.5.4 and 2.5.5 and the observation the equality holds in (5.3.2) if and only if C is M or W. If both α and β are almost surely strictly increasing, σ satisfies the sixth property as a consequence of Theorem 2.4.3. If α is almost surely strictly increasing, and β almost surely strictly decreasing, σ satisfies the sixth property as a consequence of Theorem 2.4.4 and the observation that $C_{\alpha(X),\beta(Y)}(u,v) - \Pi(u,v) = \Pi(1 - u,v) - C_{X,Y}(1 - u,v)$. The remaining cases (for the sixth property) are similar. For the seventh property, we note that the Lipschitz condition (2.2.5) implies that any family of copulas is equicontinuous, thus the convergence of $\{C_n\}$ to C is uniform.   □

Of course, it is immediate that if X and Y are PQD, then $\sigma_{X,Y} = \rho_{X,Y}$; and that if X and Y are NQD, then $\sigma_{X,Y} = -\rho_{X,Y}$. Hence for many of the totally ordered families of copulas presented in earlier chapters (e.g., Plackett, Farlie-Gumbel-Morgenstern, and many families of Archimedean copulas), $\sigma_{X,Y} = |\rho_{X,Y}|$. But for random variables X and Y that are neither PQD nor NQD, i.e., random variables whose copulas are neither larger nor smaller than Π, σ is often a better measure than ρ, as the following examples (Wolff 1977; Schweizer and Wolff 1981) illustrate.

Example 5.18. Let X and Y be random variables with the circular uniform distribution presented in Sect. 3.1.2. Because X and Y are jointly symmetric, the measures of concordance τ, ρ, β, and γ are all 0 (see Exercise 5.20). But clearly X and Y are not independent, and hence a measure of dependence such as σ will be positive. The copula C of X and Y is given by (3.1.5), from which it is easy to see that $C(u,v) \geq uv$ for (u,v) in $[0,1/2]^2 \cup [1/2,1]^2$, and $C(u,v) \leq uv$ elsewhere in $\mathbf{I}^2$. Evaluating the integral in (5.3.1) yields $\sigma_{XY} = 1/4$.   ∎

Example 5.19. Let X and Y be continuous random variables whose copula $C_\theta, \theta \in [0,1]$, is a member of the family of copulas introduced in Example 3.3. Recall that the probability mass of C_θ is distributed on two line segments, one joining $(0,0)$ to $(\theta,1)$ and the other joining $(\theta,1)$ to $(1,0)$, as illustrated in Fig. 3.2(a). In Exercise 5.6, we saw that $\tau_\theta = \rho_\theta = 2\theta - 1$ so that when $\theta = 1/2$, $\tau_{1/2} = \rho_{1/2} = 0$. However, X and Y are not independent—indeed, Y is a function of X. For each θ in $[0,1]$, $C_\theta(u,v) \geq uv$ for u in $[0,\theta]$ and $C_\theta(u,v) \leq uv$ for u in $[\theta,1]$, and it follows from (5.3.1) that

$$\sigma_\theta = 1 - 2\theta(1-\theta) = \frac{1}{2}\left[1 + (2\theta - 1)^2\right] = \frac{1}{2}\left(1 + \rho_\theta^2\right).$$

Note that $\sigma_0 = \sigma_1 = 1$ and $\sigma_{1/2} = 1/2$.   ∎

As Schweizer and Wolff (1981) note, "...any suitably normalized measure of distance between the surfaces $z = C(u,v)$ and $z = uv$, i.e., any L_p distance, should yield a symmetric nonparametric measure of dependence." For any p, $1 \leq p < \infty$, the L_p distance between C and Π is given by

$$\left(k_p \iint_{\mathbf{I}^2} |C(u,v) - uv|^p \, du dv\right)^{1/p}, \tag{5.3.3}$$

where k_p is a constant chosen so that the quantity in (5.3.3) is 1 when $C = M$ or W (so that properties 3 and 5 in Definition 5.3.1 are satisfied). The same techniques used in the proof of Theorem 5.3.2 can be used to show that for each p, $1 < p < \infty$, the quantity in (5.3.3) is a measure of dependence. For example, when $p = 2$, we have

$$\Phi_{XY} = \Phi_C = \left(90 \iint_{\mathbf{I}^2} |C(u,v) - uv|^2 \, du dv\right)^{1/2}. \tag{5.3.4}$$

The square of this measure of dependence, i.e., Φ_{XY}^2, is called the "dependence index" (Hoeffding 1940), while Φ_{XY} (but without the normalizing constant 90) was discussed in (Blum et al. 1961). When $p = \infty$, the L_∞ distance between C and Π is

$$\Lambda_{X,Y} = \Lambda_C = 4 \sup_{u,v \in \mathbf{I}} |C(u,v) - uv|. \tag{5.3.5}$$

It can be shown that this quantity satisfies all the properties in Definition 5.3.1 except the fifth (see Exercise 5.40).

Example 5.20. Let X, Y and C_θ, $\theta \in [0,1]$, be as in Example 5.19. Computations analogous to those in that example for σ_θ yield

$$\Phi_\theta = \sqrt{1 - 3\theta(1-\theta)} = \frac{1}{2}\left[1 + 3(2\theta - 1)^2\right]^{1/2} = \frac{1}{2}\left(1 + 3\rho_\theta^2\right)^{1/2}.$$

Note that $\Phi_0 = \Phi_1 = 1$ and $\Phi_{1/2} = 1/2$. To evaluate Λ_θ for this family, we first note that for u in $[0,\theta]$, $C_\theta(u,v) \geq uv$, and that $C_\theta(u,v) - uv$ attains its maximum on the line $v = u/\theta$. Elementary calculus then yields a maximum value θ at the point $(\theta/2, 1/2)$. For u in $[\theta,1]$, $C_\theta(u,v) \leq uv$, and $uv - C_\theta(u,v)$ has a maximum value $1 - \theta$ at the point $((1+\theta)/2, 1/2)$. Hence

$$\Lambda_\theta = \max(\theta, 1-\theta) = \frac{1}{2}\left[1 + |2\theta - 1|\right] = \frac{1}{2}\left(1 + |\rho_\theta|\right).$$

As with σ_θ and Φ_θ, $\Lambda_0 = \Lambda_1 = 1$ and $\Lambda_{1/2} = 1/2$. ∎

5.3.2 Measures Based on Gini's Coefficient

As a consequence of the expression $|C(u,v) - uv|$ in (5.3.1), (5.3.4), and (5.3.5), measures of dependence such as σ_C, Φ_C, and Λ_C measure "distances" from independence, i.e., distances between C and Π. Alternatively, we could look at a "distance" from complete monotone dependence, i.e., distances between C and either M or W, or both.

Let X and Y be continuous random variables with joint distribution function H and margins F and G, and copula C. In Sect. 5.1.4, we saw in (5.1.23) that the measure of concordance between X and Y known as Gini's γ can be expressed as

$$\gamma_C = 2\iint_{\mathbf{I}^2} \left(|u + v - 1| - |u - v|\right) dC(u,v).$$

In the derivation of this result, we noted that if we set $U = F(X)$ and $V = G(Y)$, then U and V are uniform $(0,1)$ random variables whose joint distribution function is C, and

$$\gamma_C = 2E\big(|U+V-1|-|U-V|\big). \tag{5.3.6}$$

There is a natural interpretation of (5.3.6) in terms of expected distances between (U,V) and the two diagonals of $\mathbf{I}^2$. Recall that for $p \geq 1$, the ℓ_p distance between points (a,b) and (c,d) in $\mathbf{R}^2$ is $\big(|a-c|^p+|b-d|^p\big)^{1/p}$. The principal diagonal of $\mathbf{I}^2$ is $\{(t,t)|t\in\mathbf{I}\}$, i.e., the support of M; while the secondary diagonal is $\{(t,1-t)|t\in\mathbf{I}\}$, the support of W. So if (u,v) represents a point in $\mathbf{I}^2$, then $|u-v|$ is the ℓ_1 distance between (u,v) and the foot $\big((u+v)/2,(u+v)/2\big)$ of the perpendicular from (u,v) to the principal diagonal of $\mathbf{I}^2$, and $|u+v-1|$ is the ℓ_1 distance between (u,v) and the foot $\big((u-v+1)/2,(v-u+1)/2\big)$ of the perpendicular from (u,v) to the secondary diagonal of $\mathbf{I}^2$. Thus γ_C in (5.3.6) is twice the difference of the expected ℓ_1 distances of (U,V) from "perfect" positive and "perfect" negative dependence.

In Exercise 5.16, we saw that Spearman's rho can be written as

$$\rho_C = 3\iint_{\mathbf{I}^2}\big([u+v-1]^2-[u-v]^2\big)dC(u,v),$$

from which it follows that, for $U = F(X)$ and $V = G(Y)$, then

$$\rho_C = 3E\big([U+V-1]^2-[U-V]^2\big). \tag{5.3.7}$$

Thus ρ_C is proportional to the difference of the expected squared ℓ_2 distances of (U,V) from "perfect" positive and "perfect" negative dependence (Long and Krzysztofowicz 1996). Other ℓ_p distances yield other measures of association (Conti 1993).

Another form for Gini's γ is given by (5.1.25):

$$\gamma_C = 4\left[\int_0^1 C(u,1-u)\,du - \int_0^1[u-C(u,u)]\,du\right]. \tag{5.3.8}$$

The second integral above is the L_1 distance between the diagonal section $\delta_M(u) = M(u,u) = u$ of the Fréchet upper bound copula and the diagonal section $\delta_C(u) = C(u,u)$ of C, while the first integral is the L_1 distance between the secondary diagonal section $W(u,1-u) = 0$ of the Fréchet lower bound copula and the secondary diagonal section of C. Employing other L_p distances between the diagonal sections of C and M and the secondary diagonal sections of C and W yields other measures. For example, using L_∞ distances yields the population version of the rank correlation coefficient R_g in (Gideon and Hollister 1987):

$$R_g = 2 \sup_{0<u<1} C(u,1-u) - 2 \sup_{0<u<1} \left[u - C(u,u) \right].$$

Exercises

5.40 Let X and Y be continuous random variables whose copula C is a member of a totally ordered (with respect to the concordance ordering $\prec$) family that includes Π. Show that $\sigma_{X,Y} = |\rho_{X,Y}|$.

5.41 Let X and Y be random variables with the circular uniform distribution presented in Sect. 3.1.2. In Example 5.18, we saw that $\sigma_{X,Y} = 1/4$. Show that $\Phi_{X,Y} = \Lambda_{X,Y} = 1/4$.

5.42 Show that $\Lambda_{X,Y}$ defined in (5.3.5) satisfies all the properties in Definition 5.3.1 for a measure of dependence except the fifth property. [Hint: to construct a counterexample to the fifth property, consider the copula C_θ from Example 3.4 with $\theta = 1/2$.]

5.43 Show that Gideon and Hollister's R_g satisfies the properties in Definition 5.1.7 for a measure of concordance.

5.44 Show that k_p in (5.3.3) is given by

$$k_p = \frac{\Gamma(2p+3)}{2\Gamma^2(p+1)}.$$

5.45 Show that, for p in $[1,\infty)$, the " ℓ_p " generalization of γ_C and ρ_C in (5.3.6) and (5.3.7), respectively, leads to measures of association given by

$$(p+1)\iint_{\mathbf{I}^2} \left(|u+v-1|^p - |u-v|^p \right) dC(u,v).$$

5.46 Show that, for p in $[1,\infty)$, the " L_p " generalization of γ_C in (5.3.8) leads to measures of association given by

$$2^p(p+1)\left[\int_0^1 [C(u,1-u)]^p \, du - \int_0^1 [u - C(u,u)]^p \, du \right].$$

5.4 Tail Dependence

Many of the dependence concepts introduced in Sect. 5.2 are designed to describe how large (or small) values of one random variable appear with large (or small) values of the other. Another such concept is tail dependence, which measures the dependence between the variables in the upper-right quadrant and in the lower-left quadrant of $\mathbf{I}^2$.

Definition 5.4.1. Let X and Y be continuous random variables with distribution functions F and G, respectively. The *upper tail dependence parameter* λ_U is the limit (if it exists) of the conditional probability that Y is greater than the $100t$-th percentile of G given that X is greater than the $100t$-th percentile of F as t approaches 1, i.e.

$$\lambda_U = \lim_{t \to 1^-} P\left[Y > G^{(-1)}(t) \,\middle|\, X > F^{(-1)}(t)\right]. \qquad (5.4.1)$$

Similarly, the *lower tail dependence parameter* λ_L is the limit (if it exists) of the conditional probability that Y is less than or equal to the $100t$-th percentile of G given that X is less than or equal to the $100t$-th percentile of F as t approaches 0, i.e.

$$\lambda_L = \lim_{t \to 0^+} P\left[Y \le G^{(-1)}(t) \,\middle|\, X \le F^{(-1)}(t)\right]. \qquad (5.4.2)$$

These parameters are nonparametric and depend only on the copula of X and Y, as the following theorem demonstrates.

Theorem 5.4.2. *Let X, Y, F, G, λ_U, and λ_L be as in Definition 5.4.1, and let C be the copula of X and Y, with diagonal section δ_C. If the limits in (5.4.1) and (5.4.2) exist, then*

$$\lambda_U = 2 - \lim_{t \to 1^-} \frac{1 - C(t,t)}{1-t} = 2 - \delta_C'(1^-) \qquad (5.4.3)$$

and

$$\lambda_L = \lim_{t \to 0^+} \frac{C(t,t)}{t} = \delta_C'(0^+). \qquad (5.4.4)$$

Proof: We establish (5.4.3), the proof of (5.4.4) is similar.

$$\lambda_U = \lim_{t \to 1^-} P\left[Y > G^{(-1)}(t) \,\middle|\, X > F^{(-1)}(t)\right]$$
$$= \lim_{t \to 1^-} P\left[G(Y) > t \,\middle|\, F(X) > t\right]$$
$$= \lim_{t \to 1^-} \frac{\overline{C}(t,t)}{1-t} = \lim_{t \to 1^-} \frac{1 - 2t + C(t,t)}{1-t}$$

$$= 2 - \lim_{t \to 1^-} \frac{1-C(t,t)}{1-t} = 2 - \delta'_C(1^-). \qquad \square$$

If λ_U is in $(0,1]$, we say C has upper tail tependence; if $\lambda_U = 0$, we say C has no upper tail dependence; and similarly for λ_L.

Example 5.21. The tail dependence parameters λ_U and λ_L are easily evaluated for some of the families of copulas encountered earlier:

Family	λ_L	λ_U
Fréchet (Exercise 2.4)	α	α
Cuadras-Augé (2.2.10)	0	θ
Marshall-Olkin (3.1.3)	0	$\min(\alpha,\beta)$
Raftery (Exercise 3.6)	$2\theta/(\theta+1)$	0
Plackett (3.3.3)	0	0

For an Archimedean copula, the tail dependence parameters can be expressed in terms of limits involving the generator and its inverse (Nelsen 1997):

Corollary 5.4.3. *Let C be an Archimedean copula with generator $\varphi \in \Omega$. Then*

$$\lambda_U = 2 - \lim_{t \to 1^-} \frac{1-\varphi^{[-1]}(2\varphi(t))}{1-t} = 2 - \lim_{x \to 0^+} \frac{1-\varphi^{[-1]}(2x)}{1-\varphi^{[-1]}(x)}$$

and

$$\lambda_L = \lim_{t \to 0^+} \frac{\varphi^{[-1]}(2\varphi(t))}{t} = \lim_{x \to \infty} \frac{\varphi^{[-1]}(2x)}{\varphi^{[-1]}(x)}.$$

Example 5.22. Using Theorem 5.4.2 and Corollary 5.4.3, the tail dependence parameters λ_U and λ_L can be evaluated for all of the families of Archimedean copulas in Table 4.1 (for the values of θ in the "$\theta \in$" column):

Family (4.2.#)	λ_L	λ_U
3, 5, 7-11, 13, 17, 22	0	0
2, 4, 6, 15, 21	0	$2-2^{1/\theta}$
18	0	1
1 ($\theta \geq 0$)	$2^{-1/\theta}$	0
12	$2^{-1/\theta}$	$2-2^{1/\theta}$
16	$1/2$	0
14	$1/2$	$2-2^{1/\theta}$
19, 20	1	0

The values of the parameters can be different for a limiting case. For example, the copula denoted $\Pi/(\Sigma-\Pi)$ has $\lambda_L = 1/2$ although it is a limiting case in families (4.2.3), (4.2.8), and (4.2.19); and M has $\lambda_U = 1$ although it is a limiting case in families (4.2.1), (4.2.5), (4.2.13), etc.∎

Tail dependence can be observed in several of the scatterplots of the Archimedean copula simulations in Figs. 4.2 through 4.9. It appears as a pronounced "spike" in the data points in the upper-right or lower-left corner of the plot.

When a two-parameter family of Archimedean copulas is an interior or exterior power family associated with a generator φ in Ω, the tail dependence parameters are determined by the parameters of the copula generated by φ. The proof of the following theorem can be found in (Nelsen 1997).

Theorem 5.4.4. *Let φ in Ω generate the copula C with upper and lower tail dependence parameters λ_U and λ_L, and let $C_{\alpha,1}$ and $C_{1,\beta}$ denote the copulas generated by $\varphi_{\alpha,1}(t) = \varphi(t^{\alpha})$ and $\varphi_{1,\beta}(t) = [\varphi(t)]^{\beta}$, respectively. Then the upper and lower tail dependence parameters of $C_{\alpha,1}$ are λ_U and $\lambda_L^{1/\alpha}$, respectively, and the upper and lower tail dependence parameters of $C_{1,\beta}$ are $2 - (2 - \lambda_U)^{1/\beta}$ and $\lambda_L^{1/\beta}$, respectively.*

Example 5.23. In Example 4.22, we constructed a two-parameter family $C_{\alpha,\beta}$ in (4.5.3) from the generator $\varphi(t) = (1/t) - 1$, which generates the copula denoted by $\Pi/(\Sigma - \Pi)$ with $\lambda_U = 0$ and $\lambda_L = 1/2$. Hence the upper tail dependence parameter $\lambda_{U\alpha,\beta}$ for $C_{\alpha,\beta}$ is $\lambda_{U\alpha,\beta} = 2 \div 2^{1/\beta}$ and the lower tail dependence parameter $\lambda_{L\alpha,\beta}$ for $C_{\alpha,\beta}$ is $\lambda_{L\alpha,\beta} = 2^{-1/(\alpha\beta)}$. This pair of equations is invertible, so to find a member of the family (4.5.3) with a predetermined upper tail dependence parameter λ_U^* and a predetermined lower tail dependence parameter λ_L^*, set $\alpha = -\ln(2 - \lambda_U^*)/\ln\lambda_L^*$ and $\beta = \ln 2/\ln(2 - \lambda_U^*)$.   ∎

Exercises

5.47 Verify the entries for λ_U and λ_L in Examples 5.21 and 5.22.

5.48 Write $\lambda_U(C)$ and $\lambda_L(C)$ to specify the copula under consideration. Prove that $\lambda_U(\hat{C}) = \lambda_L(C)$ and $\lambda_L(\hat{C}) = \lambda_U(C)$.

5.49 Let C be an extreme value copula given by (3.3.12). Prove that

$$\lambda_L = \begin{cases} 0, & A(1/2) > 1/2, \\ 1, & A(1/2) = 1/2, \end{cases} \text{ and } \lambda_U = 2[1 - A(1/2)].$$

[Note that $A(1/2) = 1/2$ if and only if $C = M$.]

5.50 Let $C(u,v) = \min\big(uf(v), vf(u)\big)$ where f is an increasing function on **I** with $f(1) = 1$ and $t \mapsto f(t)/t$ decreasing on $(0,1]$ [this is the symmetric case of the copulas in Exercise 3.3, see (Durante 2005)]. Show that $\lambda_U = 1 - f'(1^-)$ and $\lambda_L = f(0)$.

5.5 Median Regression

In addition to measures of association and dependence properties, regression is a method for describing the dependence of one random variable on another. For random variables X and Y, the regression curve $y = E(Y|x)$ specifies a "typical" (the mean) value of Y for each value of X, and the regression curve $x = E(X|y)$ specifies a "typical" value of X for each value of Y. In general, however, $E(Y|x)$ and $E(X|y)$ are parametric and thus do not have simple expressions in terms of distribution functions and copulas.

An alternative to the mean for specifying "typical" values of Y for each value of X is the median, which leads to the notion of median regression (Mardia 1970; Conway 1986):

Definition 5.5.1. Let X and Y be random variables. For x in Ran$X$, let $y = \tilde{y}(x)$ denote a solution to the equation $P[Y \le y|X = x] = 1/2$. Then the graph of $y = \tilde{y}(x)$ is the *median regression curve* of Y on X.

Of course, the median regression curve $x = \tilde{x}(y)$ of X on Y is defined analogously in terms of $P[X \le x|Y = y]$, so in the rest of this section we present results only for median regression of Y on X.

Now suppose that X and Y are continuous, with joint distribution function H, marginal distribution functions F and G, respectively, and copula C. Then $U = F(X)$ and $V = G(Y)$ are uniform $(0,1)$ random variables with joint distribution function C. As a consequence of (2.9.1), we have

$$P[Y \le y|X = x] = P[V \le G(y)|U = F(x)] = \frac{\partial C(u,v)}{\partial u}\bigg|_{\substack{u=F(x) \\ v=G(y)}}, \quad (5.5.1)$$

which yields the following algorithm for finding median regression curves for continuous random variables. To find the median regression curve $y = \tilde{y}(x)$ of Y on X:

1. Set $\partial C(u,v)/\partial u = 1/2$;
2. Solve for the regression curve $v = \tilde{v}(u)$ (of V on U);
3. Replace u by $F(x)$ and v by $G(y)$.

Example 5.24. Let U and V be uniform random variables whose joint distribution function C_θ is a member of the Plackett family (3.3.3) for θ in $[0,\infty]$. Thus

$$\frac{\partial C(u,v)}{\partial u} = \frac{\theta v + (1-\theta)C(u,v)}{1+(\theta-1)[u+v-2C(u,v)]},$$

so that setting $\partial C(u,v)/\partial u = 1/2$ and simplifying yields

$$(\theta+1)v = 1+(\theta-1)u.$$

Thus the median regression curve of V on U is the line in $\mathbf{I}^2$ connecting the points $\left(0,1/(\theta+1)\right)$ and $\left(1,\theta/(\theta+1)\right)$. Note the special cases: when $\theta = 0$, $C_0 = W$ and the median regression line is $v = 1-u$, the support of W; when $\theta = \infty$, $C_\infty = M$ and the median regression line is $v = u$, the support of M; and when $\theta = 1$, $C_1 = \Pi$ and the median regression line is $v = 1/2$. The slope of the median regression line is $(\theta-1)/(\theta+1)$. Recall (Sect. 3.3.1, also see Exercise 5.17) that for the Plackett family, θ represents an "odds ratio." When θ is an odds ratio in a 2×2 table, the expression $(\theta-1)/(\theta+1)$ is known as "Yule's Q," or "Yule's coefficient of association."

If X and Y are continuous random variables with distribution functions F and G, respectively, and copula C_θ, then the median regression curve is linear in $F(x)$ and $G(y)$:

$$(\theta+1)G(y) = 1+(\theta-1)F(x). \qquad \blacksquare$$

Example 5.25. Let C be an Archimedean copula with generator φ in Ω. From $\varphi(C) = \varphi(u) + \varphi(v)$ we obtain $\varphi'(C)\partial C(u,v)/\partial u = \varphi'(u)$. Setting $\partial C(u,v)/\partial u = 1/2$ and solving for v yields the median regression curve of V on U for Archimedean copulas:

$$v = \varphi^{[-1]}\left\{\varphi\left[(\varphi')^{(-1)}(2\varphi'(u))\right] - \varphi(u)\right\}.$$

For example, for the Clayton family (4.2.1) with $\varphi(t) = (t^{-\theta}-1)/\theta$, $\theta > -1$, $\theta \neq 0$, we have

$$v = u\left[(2^{\theta/(\theta+1)}-1) + u^\theta\right]^{1/\theta}. \qquad \blacksquare$$

5.6 Empirical Copulas

In this section, we will show that there are expressions for the sample versions of several measures of association analogous to those whose population versions were discussed in Sects. 5.1 and 5.2. The population versions can be expressed in terms of copulas—the sample versions will now be expressed in terms of empirical copulas and the corresponding empirical copula frequency function:

Definition 5.6.1. Let $\{(x_k,y_k)\}_{k=1}^n$ denote a sample of size n from a continuous bivariate distribution. The *empirical copula* is the function C_n given by

$$C_n\left(\frac{i}{n},\frac{j}{n}\right) = \frac{\text{number of pairs } (x,y) \text{ in the sample with } x \le x_{(i)}, y \le y_{(j)}}{n}.$$

where $x_{(i)}$ and $y_{(j)}$, $1 \le i,j \le n$, denote order statistics from the sample. The *empirical copula frequency* c_n is given by

$$c_n\left(\frac{i}{n},\frac{j}{n}\right) = \begin{cases} 1/n, & \text{if } (x_{(i)},y_{(j)}) \text{ is an element of the sample,} \\ 0, & \text{otherwise.} \end{cases}$$

Note that C_n and c_n are related via

$$C_n\left(\frac{i}{n},\frac{j}{n}\right) = \sum_{p=1}^{i}\sum_{q=1}^{j} c_n\left(\frac{p}{n},\frac{q}{n}\right)$$

and

$$c_n\left(\frac{i}{n},\frac{j}{n}\right) = C_n\left(\frac{i}{n},\frac{j}{n}\right) - C_n\left(\frac{i-1}{n},\frac{j}{n}\right) - C_n\left(\frac{i}{n},\frac{j-1}{n}\right) + C_n\left(\frac{i-1}{n},\frac{j-1}{n}\right).$$

Empirical copulas were introduced and first studied by Deheuvels (1979), who called them *empirical dependence functions*.

Recall the population versions of Spearman's ρ, Kendall's τ, and Gini's γ from (5.1.16), (5.2.19), and (5.1.25), respectively, for continuous random variables X and Y with copula C:

$$\rho = 12\iint_{\mathbf{I}^2}[C(u,v) - uv]\,dudv,$$

$$\tau = 2\int_0^1\int_0^1\int_0^{v'}\int_0^{u'}[c(u,v)c(u',v') - c(u,v')c(u',v)]\,dudvdu'dv',$$

and

$$\gamma = 4\left[\int_0^1 C(u,1-u)\,du - \int_0^1[u - C(u,u)]\,du\right].$$

In the next theorem, we present the corresponding version for a sample (we use Latin letters for the sample statistics):

Theorem 5.6.2. *Let C_n and c_n denote, respectively, the empirical copula and the empirical copula frequency function for the sample* $\{(x_k, y_k)\}_{k=1}^{n}$. *If r, t and g denote, respectively, the sample versions of Spearman's rho, Kendall's tau, and Gini's gamma, then*

$$r = \frac{12}{n^2 - 1} \sum_{i=1}^{n} \sum_{j=1}^{n} \left[C_n\left(\frac{i}{n}, \frac{j}{n}\right) - \frac{i}{n} \cdot \frac{j}{n} \right], \tag{5.6.1}$$

$$t = \frac{2n}{n-1} \sum_{i=2}^{n} \sum_{j=2}^{n} \sum_{p=1}^{i-1} \sum_{q=1}^{j-1} \left[c_n\left(\frac{i}{n}, \frac{j}{n}\right) c_n\left(\frac{p}{n}, \frac{q}{n}\right) - c_n\left(\frac{i}{n}, \frac{q}{n}\right) c_n\left(\frac{p}{n}, \frac{j}{n}\right) \right], \tag{5.6.2}$$

and

$$g = \frac{2n}{\lfloor n^2/2 \rfloor} \left\{ \sum_{i=1}^{n-1} C_n\left(\frac{i}{n}, 1 - \frac{i}{n}\right) - \sum_{i=1}^{n} \left[\frac{i}{n} - C_n\left(\frac{i}{n}, \frac{i}{n}\right) \right] \right\}. \tag{5.6.3}$$

Proof. We will show that the above expressions are equivalent to the expressions for r, t, and g that are usually encountered in the literature. The usual expression for r is (Kruskal 1958; Lehmann 1975)

$$r = \frac{12}{n(n^2 - 1)} \left[\sum_{k=1}^{n} kR_k - \frac{n(n+1)^2}{4} \right], \tag{5.6.4}$$

where $R_k = m$ whenever $(x_{(k)}, y_{(m)})$ is an element of the sample. To show that (5.6.1) is equivalent to (5.6.4), we need only show that

$$\sum_{i=1}^{n} \sum_{j=1}^{n} C_n\left(\frac{i}{n}, \frac{j}{n}\right) = \frac{1}{n} \sum_{k=1}^{n} kR_k . \tag{5.6.5}$$

Observe that a particular pair $(x_{(k)}, y_{(m)})$ in the sample contributes $1/n$ to the double sum in (5.6.5) for each pair of subscripts (i,j) with $i \geq k$ and $j \geq m$. That is, the total contribution to the double sum in (5.6.5) by a particular pair $(x_{(k)}, y_{(m)})$ is $1/n$ times $(n - k + 1)(n - m + 1)$, the total number of pairs (i,j) such that $i \geq k$ and $j \geq m$. Hence, writing R_k for m and summing on k, we have

$$\sum_{i=1}^{n} \sum_{j=1}^{n} C_n\left(\frac{i}{n}, \frac{j}{n}\right) = \frac{1}{n} \sum_{k=1}^{n} (n - k + 1)(n - R_k + 1) = \frac{1}{n} \sum_{k=1}^{n} kR_k ,$$

as claimed.

Next we show that (5.6.2) is equivalent to Kendall's tau in (5.1.1), i.e., the difference between number of concordant and discordant pairs in the sample divided by the total number $\binom{n}{2}$ of pairs of elements from the sample. Note that the summand in (5.6.2) reduces to $(1/n)^2$ whenever the sample contains both $(x_{(p)},y_{(q)})$ and $(x_{(i)},y_{(j)})$, a concordant pair because $x_{(p)} < x_{(i)}$ and $y_{(q)} < y_{(j)}$; reduces to $-(1/n)^2$ whenever the sample contains both $(x_{(p)},y_{(j)})$ and $(x_{(i)},y_{(q)})$, a discordant pair; and is 0 otherwise. Thus the quadruple sum in (5.6.2) is $(1/n)^2$ times the difference between the number of concordant and discordant pairs, which is equivalent to (5.1.1). Evaluating the inner double summation in (5.6.2) yields

$$t = \frac{2n}{n-1} \sum_{i=2}^{n} \sum_{j=2}^{n} \left[C_n\left(\frac{i}{n},\frac{j}{n}\right) C_n\left(\frac{i-1}{n},\frac{j-1}{n}\right) - C_n\left(\frac{i}{n},\frac{j-1}{n}\right) C_n\left(\frac{i-1}{n},\frac{j}{n}\right) \right],$$

a sample version of (5.2.20) and (5.2.22).

To show that (5.6.3) is equivalent to the sample version of Gini's gamma in (5.1.19), we need only show that

$$\sum_{i=1}^{n} |p_i - q_i| = 2n \sum_{i=1}^{n} \left[\frac{i}{n} - C_n\left(\frac{i}{n},\frac{i}{n}\right) \right] \tag{5.6.6}$$

and

$$\sum_{i=1}^{n} |n+1 - p_i - q_i| = 2n \sum_{i=1}^{n-1} C_n\left(\frac{i}{n},\frac{n-i}{n}\right). \tag{5.6.7}$$

where, recall, p_i and q_i denote the ranks of x_i and y_i, respectively. The sample $\{(x_k,y_k)\}_{k=1}^{n}$ can be written $\{(x_{(p_i)},y_{(q_i)})\}_{i=1}^{n}$. Because $nC_n(i/n,i/n)$ is the number of points $(x_{(p_i)},y_{(q_i)})$ in the sample for which $p_i \le i$ and $q_i \le i$, the sample point $(x_{(p_i)},y_{(q_i)})$ is counted $n - \max(p_i,q_i) + 1$ times in the sum $n\sum_{1}^{n} C_n(i/n,i/n)$. Thus

$$2n \sum_{i=1}^{n} \left[\frac{i}{n} - C_n\left(\frac{i}{n},\frac{i}{n}\right) \right] = n(n+1) - 2\sum_{i=1}^{n} [(n+1) - \max(p_i,q_i)]$$

$$= \left[2\sum_{i=1}^{n} \max(p_i,q_i) \right] - n(n+1) = \sum_{i=1}^{n} [2\max(p_i,q_i) - (p_i + q_i)].$$

But $2\max(u,v) - (u + v) = |u - v|$, and hence

$$2n\sum_{i=1}^{n}\left[\frac{i}{n}-C_n\left(\frac{i}{n},\frac{i}{n}\right)\right]=\sum_{i=1}^{n}\left|p_i-q_i\right|.$$

The verification of (5.6.7) is similar.　　　　　　　　□

Empirical copulas can also be used to construct nonparametric tests for independence. See (Deheuvels 1979, 1981a,b) for details.

5.7 Multivariate Dependence

As Barlow and Proschan (1981) note, "the notions of positive dependence in the multivariate case are more numerous, more complex, and their interrelationships are less well understood." This is true as well of the role played by n-copulas in the study of multivariate dependence. However, many of the dependence properties encountered in earlier sections of this chapter have natural extensions to the multivariate case. We shall examine only a few and provide references for others.

In three or more dimensions, rather than quadrants we have "orthants," and the generalization of quadrant dependence is known as *orthant dependence*:

Definition 5.7.1. Let $\mathbf{X} = \left(X_1,X_2,\cdots,X_n\right)$ be an n-dimensional random vector.

1. $\mathbf{X}$ is *positively lower orthant dependent* (PLOD) if for all $\mathbf{x} = \left(x_1,x_2,\cdots,x_n\right)$ in $\mathbf{R}^n$,

$$P[\mathbf{X} \leq \mathbf{x}] \geq \prod_{i=1}^{n} P[X_i \leq x_i]. \tag{5.7.1}$$

2. $\mathbf{X}$ is *positively upper orthant dependent* (PUOD) if for all $\mathbf{x} = \left(x_1,x_2,\cdots,x_n\right)$ in $\mathbf{R}^n$,

$$P[\mathbf{X} > \mathbf{x}] \geq \prod_{i=1}^{n} P[X_i > x_i], \tag{5.7.2}$$

3. $\mathbf{X}$ is *positively orthant dependent* (POD) if for all $\mathbf{x}$ in $\mathbf{R}^n$, both (5.7.1) and (5.7.2) hold.

Negative lower orthant dependence (NLOD), negative upper orthant dependence (NUOD), and negative orthant dependence (NOD) are defined analogously, by reversing the sense of the inequalities in (5.7.1) and (5.7.2).

For $n = 2$, (5.7.1) and (5.7.2) are equivalent to (5.2.1) and (5.2.2), respectively. As a consequence of Exercise 5.21, PLOD and PUOD are the same for $n = 2$. However, this is not the case for $n \geq 3$.

Example 5.26. Let $\mathbf{X}$ be a three-dimensional random vector that assumes the four values (1,1,1), (1,0,0), (0,1,0), and (0,0,1) each with probability $1/4$. It is now easy to verify that $\mathbf{X}$ is PUOD but not PLOD. Note that $P[\mathbf{X} \leq \mathbf{0}] = 0$ while $P[X_1 \leq 0]P[X_2 \leq 0]P[X_3 \leq 0] = 1/8$. $\blacksquare$

If X has a joint n-dimensional distribution function H, continuous margins $F_1, F_2, \cdots, F_n$, and n-copula C, then (5.7.1) is equivalent to

$$H(x_1,x_2,\cdots,x_n) \geq F_1(x_1)F_2(x_2)\cdots F_n(x_n) \text{ for all } x_1,x_2,\cdots,x_n \text{ in } \mathbf{R},$$

and to

$$C(u_1,u_2,\cdots,u_n) \geq u_1 u_2 \cdots u_n \text{ for all } u_1,u_2,\cdots,u_n \text{ in } \mathbf{I},$$

i.e., $C(\mathbf{u}) \geq \Pi^n(\mathbf{u})$ for all $\mathbf{u} = (u_1,u_2,\cdots,u_n)$ in $\mathbf{I}^n$. Analogously, (5.7.2) is equivalent to

$$\overline{H}(x_1,x_2,\cdots,x_n) \geq \overline{F}_1(x_1)\overline{F}_2(x_2)\cdots \overline{F}_n(x_n) \text{ for all } x_1,x_2,\cdots,x_n \text{ in } \mathbf{R},$$

and to (where $\overline{C}$ denotes the n-dimensional joint survival function corresponding to C)

$$\overline{C}(u_1,u_2,\cdots,u_n) \geq (1-u_1)(1-u_2)\cdots(1-u_n) \text{ for all } u_1,u_2,\cdots,u_n \text{ in } \mathbf{I},$$

i.e., $\overline{C}(\mathbf{u}) \geq \overline{\Pi}^n(\mathbf{u})$ for all $\mathbf{u}$ in $\mathbf{I}^n$.

Closely related to the notion of orthant dependence is multivariate concordance. Recall (see Sect. 2.9) that in the bivariate case, a copula C_1 is more concordant than (or more PQD than) C_2 if $C_1(u,v) \geq C_2(u,v)$ for all (u,v) in $\mathbf{I}^2$. The multivariate version is similar:

Definition 5.7.2. Let C_1 and C_2 be an n-copulas, and let $\overline{C}_1$ and $\overline{C}_2$ denote the corresponding n-dimensional joint survival functions.

1. C_1 is *more PLOD than* C_2 if for all $\mathbf{u}$ in $\mathbf{I}^n$, $C_1(\mathbf{u}) \geq C_2(\mathbf{u})$;

2. C_1 is *more PUOD than* C_2 if for all $\mathbf{u}$ in $\mathbf{I}^n$, $\overline{C}_1(\mathbf{u}) \geq \overline{C}_2(\mathbf{u})$;

3. C_1 is *more POD than* C_2, or C_1 is *more concordant than* C_2, if

for all $\mathbf{u}$ in $\mathbf{I}^n$, both $C_1(\mathbf{u}) \geq C_2(\mathbf{u})$ and $\overline{C}_1(\mathbf{u}) \geq \overline{C}_2(\mathbf{u})$ hold.

In the bivariate case, parts 1 and 2 of the above definition are equivalent (see Exercise 2.30), however, that is not the case in higher dimensions.

Many of the measures of concordance in Sect. 5.1 have multivariate versions. In general, however, each measure of bivariate concordance has several multidimensional versions. See (Joe 1990; Nelsen 1996) for details. There are also multivariate versions of some of the measures of dependence in Sect. 5.3.1. For example, the n-dimensional version of Schweizer and Wolff's σ for an n-copula C, which we denote by σ_C^n, is given by

$$\sigma_C^n = \frac{2^n(n+1)}{2^n-(n+1)} \iint \cdots \int_{\mathbf{I}^n} \left| C(\mathbf{u}) - u_1 u_2 \cdots u_n \right| du_1 du_2 \cdots du_n .$$

See (Wolff 1977, 1981) for details.

Extensions of some of the other dependence properties to the multivariate case are similar. A note on notation: for $\mathbf{x}$ in $\mathbf{R}^n$, a phrase such as "nondecreasing in $\mathbf{x}$" means nondecreasing in each component x_i, $i = 1, 2, \cdots, n$; and if A and B are nonempty disjoint subsets of $\{1, 2, \cdots, n\}$, then $\mathbf{X}_A$ and $\mathbf{X}_B$ denote the vectors $(X_i | i \in A)$ and $(X_i | i \in B)$, respectively, where $X_i \in \mathbf{X}$. The following definitions are from (Brindley and Thompson 1972; Harris 1970; Joe 1997).

Definition 5.7.3. Let $\mathbf{X} = (X_1, X_2, \cdots, X_n)$ be an n-dimensional random vector, and let the sets A and B partition $\{1, 2, \cdots, n\}$.

1. LTD($\mathbf{X}_B \mid \mathbf{X}_A$) if $P\left[\mathbf{X}_B \leq \mathbf{x}_B \middle| \mathbf{X}_A \leq \mathbf{x}_A\right]$ is nonincreasing in $\mathbf{x}_A$ for all $\mathbf{x}_B$;

2. RTI($\mathbf{X}_B \mid \mathbf{X}_A$) if $P\left[\mathbf{X}_B > \mathbf{x}_B \middle| \mathbf{X}_A > \mathbf{x}_A\right]$ is nondecreasing in $\mathbf{x}_A$ for all $\mathbf{x}_B$;

3. SI($\mathbf{X}_B \mid \mathbf{X}_A$) if $P\left[\mathbf{X}_B > \mathbf{x}_B \middle| \mathbf{X}_A = \mathbf{x}_A\right]$ is nondecreasing in $\mathbf{x}_A$ for all $\mathbf{x}_B$;

4. LCSD($\mathbf{X}$) if $P\left[\mathbf{X} \leq \mathbf{x} \middle| \mathbf{X} \leq \mathbf{x}'\right]$ is nonincreasing in $\mathbf{x}'$ for all $\mathbf{x}$;

5. RCSI($\mathbf{X}$) if $P\left[\mathbf{X} > \mathbf{x} \middle| \mathbf{X} > \mathbf{x}'\right]$ is nondecreasing in $\mathbf{x}'$ for all $\mathbf{x}$.

Two additional multivariate dependence properties are expressible in terms of the stochastic increasing property (Joe 1997). When SI($\mathbf{X}_B \mid \mathbf{X}_A$) holds for all singleton sets A, i.e., $A = \{i\}$, $i = 1,2,\cdots,n$; then $\mathbf{X}$ is *positive dependent through the stochastic ordering* (PDS); and when SI($\mathbf{X}_B \mid \mathbf{X}_A$) holds for all singleton sets $B = \{i\}$ and $A = \{1,2,\cdots,i - 1\}$, $i = 2,3,\cdots,n$, then $\mathbf{X}$ is *conditional increasing in sequence* (CIS). Note that for $n = 2$, both PDS and CIS are equivalent to SI($Y \mid X$) and SI($X \mid Y$).

In the bivariate case, the corner set monotonicity properties were expressible in terms of total positivity (see Corollary 5.2.16). The same is true in the multivariate case with the following generalization of total positivity: A function f from $\overline{\mathbf{R}}^n$ to $\mathbf{R}$ is *multivariate totally positive of order two* (MTP$_2$) if

$$f(\mathbf{x} \vee \mathbf{y}) f(\mathbf{x} \wedge \mathbf{y}) \geq f(\mathbf{x}) f(\mathbf{y})$$

for all $\mathbf{x}, \mathbf{y}$ in $\overline{\mathbf{R}}^n$ where

$$\mathbf{x} \vee \mathbf{y} = \left(\max(x_1, y_1), \max(x_2, y_2), \cdots, \max(x_n, y_n)\right),$$

$$\mathbf{x} \wedge \mathbf{y} = \left(\min(x_1, y_1), \min(x_2, y_2), \cdots, \min(x_n, y_n)\right).$$

Lastly, $\mathbf{X}$ is positively likelihood ratio dependent if its joint n-dimensional density h is MTP$_2$.

For implications among these (and other) dependence concepts, see (Block and Ting 1981; Joe 1997; Block et al. 1997).

We close this section with the important observation that the symmetric relationship between positive and negative dependence properties that holds in two dimensions does not carry over to the n-dimensional case. In two dimensions, if the vector (X,Y) satisfies some positive dependence property, then the vector $(X,-Y)$ satisfies the corresponding negative dependence property, and similarly for the vector $(-X,Y)$. Furthermore, as a consequence of Theorem 2.4.4, if C is the copula of (X,Y), then $C'(u,v) = u - C(u,1-v)$ is the copula of $(X,-Y)$, and it is easy to show that for all u,v in $\mathbf{I}$,

$$C(u,v) - \Pi(u,v) = \Pi(u,1-v) - C'(u,1-v)$$

and

$$M(u,v) - C(u,v) = C'(u,1-v) - W(u,1-v);$$

that is, the graph of C' is a "twisted reflection" in Π of the graph of C, and C' has a relationship with W analogous to the relationship between C and M.

There is no analog to these relationships in n dimensions, $n \geq 3$. Indeed, as n increases, the graphs of $z = W^n(\mathbf{u})$ and $z = \Pi^n(\mathbf{u})$ are much closer to one another than are the graphs of $z = M^n(\mathbf{u})$ and $z = \Pi^n(\mathbf{u})$. It is an exercise in multivariable calculus to show that the n-volume between the graphs of M^n and Π^n is given by

$$a_n = \iint \cdots \int_{\mathbf{I}^n} \left[M^n(\mathbf{u}) - \Pi^n(\mathbf{u})\right] du_1 du_2 \cdots du_n = \frac{1}{n+1} - \frac{1}{2^n},$$

and the n-volume between the graphs of Π^n and W^n is given by

$$b_n = \iint \cdots \int_{\mathbf{I}^n} \left[\Pi^n(\mathbf{u}) - W^n(\mathbf{u})\right] du_1 du_2 \cdots du_n = \frac{1}{2^n} - \frac{1}{(n+1)!},$$

and hence

$$\lim_{n \to \infty} \frac{b_n}{a_n} = 0.$$

For a further discussion of negative multivariate dependence concepts, see (Block et al. 1982).

6 Additional Topics

In this chapter, we consider four topics related to copulas. The first, distributions with fixed margins, dates back to the early history of the subject. The original question whose answer leads to the Fréchet-Hoeffding bounds (2.5.1) is: Of all joint distribution functions H constrained to have *fixed margins* F and G, which is the "largest," and which the "smallest"? Another example, which also involves optimization when the margins are fixed, is the following. In studying "distances" between distributions, Dall'Aglio (1956, 1991) considered the following problem (see Exercise 6.5): What is the minimum value of

$$E|X-Y|^\alpha = \iint_{\mathbf{R}^2} |x-y|^\alpha \, dH(x,y),$$

given that the margins of H are fixed to be F and G, respectively?

Secondly, we introduce quasi-copulas—functions closely related to copulas—which arise when finding bounds on sets of copulas. They also occur in the third section of this chapter, where we employ copulas and quasi-copulas to study some aspects of the relationship between operations on distribution functions and corresponding functions of random variables.

The final topic in this chapter is an application of copulas to Markov processes and leads to new interpretations of and approaches to these stochastic processes.

6.1 Distributions with Fixed Margins

It is common in statistics and probability to know (or assume to know) the distributions of each of two random variables X and Y but not to know their joint distribution function H, or equivalently, their copula. For example, one of the central problems in statistics concerns testing the hypothesis that two random variables are independent, i.e., that their copula is Π. In such situations, it is often either assumed that the margins are normal, or no assumption at all is made concerning the margins. In this section, we will be concerned with problems in which it is assumed that the marginal distributions of X and Y are *known*, that is, the margins of H are given distribution functions F_X and F_Y, respectively. In this situation, we say that the joint distribution has "fixed margins."

The following problem, attributed to A. N. Kolmogorov in (Makarov 1981), is typical: Suppose X and Y are random variables with distribution functions F_X and F_Y, respectively. Let G denote the distribution function of the sum $X+Y$, i.e., $G(z) = P[X+Y \leq z]$. Find $G^\wedge(z) = \sup G(z)$ and $G^\vee(z) = \inf G(z)$, where the supremum and infimum are taken over the Fréchet-Hoeffding class $\mathbf{H}(F_X, F_Y)$ of all joint distribution functions H with marginals F_X and F_Y. This problem leads to copulas naturally, for if H is the joint distribution function of X and Y, then $H(x,y) = C(F_X(x), F_Y(y))$ for at least one copula C (exactly one if X and Y are continuous). The problem can be solved without copulas (Makarov 1981), but the arguments are cumbersome and nonintuitive. The following theorem and proof are from (Frank et al. 1987). Another proof can be found in (Rüschendorf 1982).

Theorem 6.1.1. *Let X and Y be random variables with distribution functions F_X and F_Y. Let G denote the distribution function of $X+Y$. Then*

$$G^\vee(z) \leq G(z) \leq G^\wedge(z) \qquad (6.1.1)$$

where

$$G^\vee(z) = \sup_{x+y=z} \left\{ W\left(F_X(x), F_Y(y)\right) \right\} \qquad (6.1.2a)$$

and

$$G^\wedge(z) = \inf_{x+y=z} \left\{ \tilde{W}\left(F_X(x), F_Y(y)\right) \right\}, \qquad (6.1.2b)$$

and $\tilde{W}(u,v) = u+v-W(u,v) = \min(u+v,1)$ is the dual of W.

Proof. Fix z in $\mathbf{R}$, and let H denote the (unknown) joint distribution function of X and Y. Then $G(z)$ is the H-volume over the half-plane $\{(x,y) \in \mathbf{R}^2 | x+y \leq z\}$, i.e., the H-volume of the region below and to the left of the line $x+y=z$ in Fig. 6.1. If (x_1, y_1) is any point on the line $x+y=z$, then $H(x_1, y_1) \leq G(z)$ because the H-volume of the rectangle $(-\infty, x_1] \times (-\infty, y_1]$ cannot exceed $G(z)$. But $H(x_1, y_1)$ is bounded below by its Fréchet-Hoeffding lower bound $W(F_X(x_1), F_Y(y_1))$, and thus

$$W\left(F_X(x_1), F_Y(y_1)\right) \leq H\left(x_1, y_1\right) \leq G(z).$$

Because this inequality holds for every (x_1, y_1) on the line $x+y=z$, the left-hand inequality in (6.1.1) follows. Similarly, if (x_2, y_2) is any point on the line $x+y=z$, then

$$G(z) \leq 1 - \overline{H}(x_2, y_2),$$
$$\leq 1 - \overline{W}\big(F_X(x_2), F_Y(y_2)\big),$$
$$\leq F_X(x_2) + F_Y(y_2) - W\big(F_X(x_2), F_Y(y_2)\big),$$
$$= \widetilde{W}\big(F_X(x_2), F_Y(y_2)\big),$$

from which the right-hand inequality in (6.1.1) follows. $\square$

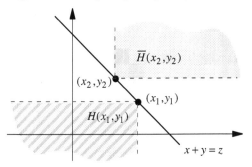

Fig. 6.1. Illustrating the inequalities in the proof of Theorem 6.1.1

Viewed as an inequality among all possible distribution functions, (6.1.1) cannot be improved, for it is easy to show (see Exercise 6.1) that if either F_X or F_Y is the unit step function ε_a for some finite a (see Example 2.4), then for all z in $\mathbf{R}$, $G^{\wedge}(z) = G(z) = G^{\vee}(z)$, i.e., we have equality throughout (6.1.1).

But more is true. Given *any* pair of distribution functions F_X and F_Y, (6.1.1) cannot be improved, that is, the bounds $G^{\wedge}(z)$ and $G^{\vee}(z)$ for $G(z)$ are pointwise best-possible.

Theorem 6.1.2. *Let F_X and F_Y be any two distribution functions, and let $G^{\wedge}(z)$ and $G^{\vee}(z)$ be given by (6.1.2ab). Let X and Y be random variables whose distribution functions are F_X and F_Y, respectively, and let G denote the distribution function of $X + Y$. Let z_0 be any number in $\mathbf{R}$, and set $s = G^{\wedge}(z_0)$ and $t = G^{\vee}(z_0^-) = \lim_{t \to z_0^-} G^{\vee}(t)$. Then:*

1. There exists a copula C_s, dependent only on s, such that if the joint distribution function of X and Y is $C_s(F_X(x), F_Y(y))$, then

$$G(z_0) = G^{\wedge}(z_0) = s.$$

2. There exists a copula C_t, dependent only on t, such that if the joint distribution function of X and Y is $C_t(F_X(x), F_Y(y))$, then

$$G(z_0^-) = G^\vee(z_0^-) = t. \tag{6.1.3}$$

The proof of this theorem (Frank et al. 1987) is long and technical, hence we will only present an outline here. In part 1, we need to show $G(z_0) \geq s$, as from (6.1.1) we have $G(z_0) \leq s$. For the copula C_s in part 1, we use

$$C_s(u,v) = \begin{cases} \max(u+v-s,0), & (u,v) \in [0,s]^2, \\ \min(u,v), & \text{elsewhere.} \end{cases}$$

This copula is the ordinal sum of $\{W,M\}$ with respect to the partition $\{[0,s],[s,1]\}$, whose support is illustrated in Fig. 6.2(a). If we set $H_s(x,y) = C_s(F_X(x),F_Y(y))$, then, as a consequence of Theorems 2.5.4 and 2.5.5, the support of H_s consists of two components: a nonincreasing set in the quadrant $(-\infty,x_0] \times (-\infty,y_0]$ and a nondecreasing set in the quadrant $[x_0,\infty) \times [y_0,\infty)$, where $x_0 = F_X^{(-1)}(s)$ and $y_0 = F_Y^{(-1)}(s)$, as illustrated in Fig. 6.2(b). Note that the H_s-measure of the nonincreasing component is s. In the proof, one proceeds to show that every point in the nonincreasing component of the support in $(-\infty,x_0] \times (-\infty,y_0]$ lies on or below the line $x+y=z_0$. Because $G(z_0)$ equals the H_s-volume of the half-plane $\{(x,y) \in \mathbf{R}^2 | x+y \leq z_0\}$, it now follows that $G(z_0) \geq s$.

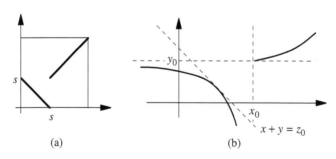

(a)   (b)

Fig. 6.2. The supports of (a) C_s and (b) H_s

The procedure for proving part 2 is similar. Because $G(z_0^-) \geq t$, we need only show that $G(z_0^-) \leq t$. For the copula C_t, we will use a member of the family of copulas from Exercise 2.10:

$$C_t(u,v) = \begin{cases} \max(u+v-1,t), & (u,v) \in [t,1]^2, \\ \min(u,v), & \text{elsewhere.} \end{cases}$$

This copula is the ordinal sum of $\{M,W\}$ with respect to the partition $\{[0,t],[t,1]\}$, whose support is illustrated in Fig. 6.3(a). If we now set $H_t(x,y) = C_t(F_X(x), F_Y(y))$, then, as with H_s, the support of H_t consists of two components: a nondecreasing set in the quadrant $(-\infty,x_1] \times (-\infty,y_1]$ and a nonincreasing set in the quadrant $[x_1,\infty) \times [y_1,\infty)$, where $x_1 = F_X^{(-1)}(t)$ and $y_1 = F_Y^{(-1)}(t)$, as illustrated in Fig. 6.3(b). Note that the H_t-measure of the nonincreasing component is $1-t$. One then proceeds to show that every point in the nonincreasing component of the support in $[x_1,\infty) \times [y_1,\infty)$ lies on or above the line $x+y=z_0$, from which it follows that the H_t-measure of the half-plane $\{(x,y) \in \mathbf{R}^2 \,|\, x+y \ge z_0\}$ is at least $1-t$. Because $G(z_0^-)$ is the H_t-measure of the (open) half-plane $\{(x,y) \in \mathbf{R}^2 \,|\, x+y < z_0\}$, we have $G(z_0^-) \le t$.

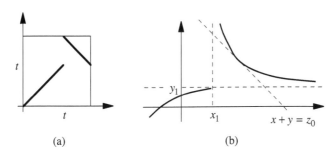

Fig. 6.3. The supports of (a) C_t and (b) H_t

The second part of Theorem 6.1.2 cannot be strengthened to read $G(z_0) = G^\vee(z_0) = t$ in (6.1.3), not even when the distribution functions F_X and F_Y are continuous. To see this, suppose that both F_X and F_Y are $U_{0,1}$, the uniform distribution function on $(0,1)$. Then $G^\vee = U_{1,2}$ (see Example 6.1 below), and for any t in $(0,1)$, $G(1+t) = P[X+Y \le 1+t] = 1$, and thus $t = G^\vee(1+t) < G(1+t) = 1$.

Also note that the crucial property of the copula C_s in the proof of part 1 is the fact that for $u + v = s$, the C_s-volume of any rectangle of the form $[u,1] \times [v,1]$ is $1-s$. Hence C_s is not unique—indeed, we could choose for C_s the ordinal sum of $\{W,C\}$ with respect to the partition $\{[0,s],[s,1]\}$ for any copula C.

Example 6.1 (Alsina 1981). Let X and Y be random variables with uniform distributions, i.e., for $a \le b$ and $c \le d$, let $F_X = U_{ab}$ and $F_Y = U_{cd}$. Let G, $G^{\vee}$, and $G^{\wedge}$ be as in Theorem 6.1.1. Then for the lower bound $G^{\vee}(z)$, from (6.1.2a) we have

$$G^{\vee}(z) = \sup_{x+y=z} W\big(U_{a,b}(x), U_{c,d}(y)\big)$$

$$= \begin{cases} 0, & z \le a+c, \\ \displaystyle\sup_{x+y=z} \max\left(\frac{x-a}{b-a} + \frac{y-c}{d-c} - 1, 0\right), & a+c \le z \le b+d, \\ 1, & z \ge b+d. \end{cases}$$

If $a+d \le b+c$, then for $a+c \le z \le a+d$, the middle line in the last display simplifies to 0; and when $a+d \le z \le b+d$, it simplifies to $[z-(a+d)]/(b-a)$. Hence in this case, $G^{\vee} = U_{a+d,b+d}$. When $b+c \le a+d$, a similar analysis yields $G^{\vee} = U_{b+c,b+d}$, and thus

$$G^{\vee} = U_{\min(a+d,b+c),b+d}.$$

The evaluation of the upper bound $G^{\wedge}$ is analogous, and yields

$$G^{\wedge} = U_{a+c,\max(a+d,b+c)}. \qquad \blacksquare$$

Example 6.2. Let X and Y be normally distributed random variables, with means μ_X, μ_Y and variances σ_X^2, σ_Y^2, respectively; so that if Φ denotes the standard normal distribution function, then

$$F_X(x) = \Phi\left(\frac{x-\mu_X}{\sigma_X}\right) \quad \text{and} \quad F_Y(y) = \Phi\left(\frac{y-\mu_Y}{\sigma_Y}\right).$$

Let G, $G^{\vee}$, and $G^{\wedge}$ be as in Theorem 6.1.1. There are two cases to consider:

(1) If $\sigma_X^2 = \sigma_Y^2 = \sigma^2$, then (6.1.2ab) yields

$$G^{\vee}(z) = \sup_{x+y=z} \max\left\{\Phi\left(\frac{x-\mu_X}{\sigma}\right) + \Phi\left(\frac{y-\mu_Y}{\sigma}\right) - 1, 0\right\}$$

and

$$G^{\wedge}(z) = \inf_{x+y=z} \min\left\{\Phi\left(\frac{x-\mu_X}{\sigma}\right) + \Phi\left(\frac{y-\mu_Y}{\sigma}\right), 1\right\}.$$

Using the method of Lagrange multipliers to find the extrema of $\Phi((x-\mu_X)/\sigma) + \Phi((y-\mu_Y)/\sigma)$ subject to the constraint $x+y=z$, we obtain

$$G^\vee(z) = \begin{cases} 0, & z \leq \mu_X + \mu_Y, \\ 2\Phi\left(\dfrac{z-\mu_X-\mu_Y}{2\sigma}\right)-1, & z \geq \mu_X + \mu_Y, \end{cases}$$

and

$$G^\wedge(z) = \begin{cases} 2\Phi\left(\dfrac{z-\mu_X-\mu_Y}{2\sigma}\right), & z \leq \mu_X + \mu_Y, \\ 1, & z \geq \mu_X + \mu_Y. \end{cases}$$

(2) If $\sigma_X^2 \neq \sigma_Y^2$, then the same procedure yields

$$G^\vee(z) = \Phi\left(\frac{-\sigma_X\theta-\sigma_Y\varphi}{\sigma_Y^2-\sigma_X^2}\right)+\Phi\left(\frac{\sigma_Y\theta-\sigma_X\varphi}{\sigma_Y^2-\sigma_X^2}\right)-1$$

and

$$G^\wedge(z) = \Phi\left(\frac{-\sigma_X\theta+\sigma_Y\varphi}{\sigma_Y^2-\sigma_X^2}\right)+\Phi\left(\frac{\sigma_Y\theta+\sigma_X\varphi}{\sigma_Y^2-\sigma_X^2}\right),$$

where $\theta = z - \mu_X - \mu_Y$ and $\varphi = \left[\theta^2 + 2(\sigma_Y^2-\sigma_X^2)\ln(\sigma_Y/\sigma_X)\right]^{1/2}$. ∎

Analogous results may be obtained for operations other than addition—i.e., in Theorem 6.1.1, the sum $X+Y$ can be replaced by $L(X,Y)$ where L is a function from $\mathbf{R}\times\mathbf{R}$ to $\mathbf{R}$, which is nondecreasing and continuous in each place. Similar results hold in higher dimensions as well [see (Frank et al. 1987; Li et al. 1996a) for details]. It is also possible to bound the distribution functions of the sum of squares X^2+Y^2 and the so-called radial error $[X^2+Y^2]^{1/2}$ in a similar fashion—see Exercise 6.4 and (Nelsen and Schweizer 1991).

Exercises

6.1 Let X and Y be random variables with distribution functions F_X and F_Y, respectively. Assume F_X equals the unit step function ε_a for some finite a. Show that equality holds throughout (6.1.1) by showing that

$$G^\wedge(z) = G(z) = G^\vee(z) = F_Y(z - a).$$

6.2   Let X and Y be exponentially distributed random variables with means α and β, respectively. Let G, $G^\vee$, and $G^\wedge$ be as in Theorem 6.1.1. Let $\theta = (\alpha + \beta)\ln(\alpha + \beta) - \alpha\ln\alpha - \beta\ln\beta$. Show that

$$G^\vee(z) = \begin{cases} 0, & z \le \theta, \\ 1 - e^{-(z-\theta)/(\alpha+\beta)}, & z \ge \theta, \end{cases}$$

and

$$G^\wedge(z) = \begin{cases} 0, & z \le 0, \\ 1 - e^{-z/\max(\alpha,\beta)}, & z \ge 0. \end{cases}$$

6.3   Let X and Y be random variables with Cauchy distributions with location parameters α_X, α_Y and scale parameters β_X, β_Y, respectively, i.e.,

$$F_X(x) = \frac{1}{2} + \frac{1}{\pi}\arctan\left(\frac{z-\alpha_X}{\beta_X}\right) \text{ and } F_Y(y) = \frac{1}{2} + \frac{1}{\pi}\arctan\left(\frac{z-\alpha_Y}{\beta_Y}\right).$$

Let G, $G^\vee$, and $G^\wedge$ be as in Theorem 6.1.1. Show that
(a) If $\beta_X = \beta_Y = \beta$, then

$$G^\vee(z) = \begin{cases} 0, & z \le \alpha_X + \alpha_Y, \\ \dfrac{2}{\pi}\arctan\left(\dfrac{z-\alpha_X-\alpha_Y}{2\beta}\right), & z \ge \alpha_X + \alpha_Y, \end{cases}$$

and

$$G^\wedge(z) = \begin{cases} 1 + \dfrac{2}{\pi}\arctan\left(\dfrac{z-\alpha_X-\alpha_Y}{2\beta}\right), & z \le \alpha_X + \alpha_Y, \\ 1, & z \ge \alpha_X + \alpha_Y. \end{cases}$$

(b) If $\beta_X \ne \beta_Y$, then

$$G^\vee(z) = \frac{1}{\pi}\left[\arctan\left(\frac{-\theta + \beta_Y\varphi}{\beta_Y - \beta_X}\right) + \arctan\left(\frac{\theta - \beta_X\varphi}{\beta_Y - \beta_X}\right)\right]$$

and

$$G^\wedge(z) = 1 + \frac{1}{\pi}\left[\arctan\left(\frac{-\theta - \beta_Y\varphi}{\beta_Y - \beta_X}\right) + \arctan\left(\frac{\theta + \beta_X\varphi}{\beta_Y - \beta_X}\right)\right],$$

where $\theta = z - \alpha_X - \alpha_Y$ and $\varphi = \left\{\left[\theta^2 - (\beta_Y - \beta_X)^2\right]/\beta_X\beta_Y\right\}^{1/2}$.

6.4   Let X and Y be symmetric (about 0) random variables with a common distribution function F that is concave on $(0,\infty)$. Let G_1 denote the distribution function of $X^2 + Y^2$, and let G_2 denote the distribution function of $[X^2 + Y^2]^{1/2}$. Show that

$$\max\left(4F\left(\sqrt{z/2}\right) - 3,0\right) \le G_1(z) \le 2F\left(\sqrt{z}\right) - 1$$

and

$$\max\left(4F\left(z/\sqrt{2}\right) - 3,0\right) \le G_2(z) \le 2F(z) - 1.$$

These bounds are best-possible. See (Nelsen and Schweizer 1991).

6.5   (Dall'Aglio 1956, 1991). Let X and Y be random variables with distribution functions F and G, respectively, and let

$$E|X - Y|^\alpha = \iint_{\mathbf{R}^2} |x - y|^\alpha \, dH(x,y) \qquad (6.1.4)$$

where H is any joint distribution function with margins F and G. Show that
(a) If $\alpha = 1$, then

$$E|X - Y| = \int_{-\infty}^{\infty} [F(t) + G(t) - 2H(t,t)] \, dt;$$

and if $\alpha > 1$, then

$$E|X - Y|^\alpha = \alpha(\alpha - 1) \int_{-\infty}^{\infty} \int_{-\infty}^{x} [G(y) - H(x,y)](x - y)^{\alpha-2} \, dy \, dx$$

$$+ \alpha(\alpha - 1) \int_{-\infty}^{\infty} \int_{-\infty}^{y} [F(x) - H(x,y)](y - x)^{\alpha-2} \, dx \, dy.$$

(b) When $\alpha > 1$, (6.1.4) attains its minimum value precisely when $H(x,y) = M(F(x),G(y))$; and if $\alpha = 1$, then there is a set of minimizing joint distribution functions, the largest of which is $M(F(x),G(y))$, and the smallest of which is given by

$$H(x,y) = \begin{cases} F(x) - \max\left(\inf_{x \le t \le y} [F(t) - G(t)],0 \right), & x \le y, \\ G(y) - \max\left(\inf_{y \le t \le x} [G(t) - F(t)],0 \right), & x \ge y. \end{cases}$$

[Also see (Bertino 1968).]

6.2 Quasi-copulas

Quasi-copulas are functions from $\mathbf{I}^2$ to $\mathbf{I}$ that mimic many but not all of the properties of copulas.

Definition 6.2.1. A *quasi-copula* is a function $Q: \mathbf{I}^2 \to \mathbf{I}$ that satisfies the same boundary conditions (2.2.2a) and (2.2.2b) as do copulas, but in place of the 2-increasing condition (2.2.3), the weaker conditions of nondecreasing in each variable and the Lipschitz condition (2.2.6).

Clearly every copula is a quasi-copula, and quasi-copulas that are not copulas are called *proper* quasi-copulas. For example, the function $Q(u,v)$ in Exercise 2.11 is a proper quasi-copula.

The conditions of being nondecreasing and Lipschitz in each variable together are equivalent to only requiring that the 2-increasing condition (2.2.3) holds when at least one of u_1, u_2, v_1, v_2 is 0 or 1. Geometrically, this means that at least the rectangles in $\mathbf{I}^2$ that share a portion of their boundary with the boundary of $\mathbf{I}^2$ must have nonnegative Q-volume. See Fig. 6.4.

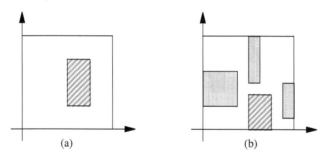

(a)   (b)

Fig. 6.4. Typical rectangles with nonnegative volume for (a) copulas and (b) quasi-copulas

Quasi-copulas were introduced in Alsina et al. (1993) [see also (Nelsen et al. 1996)] in order to characterize operations on univariate distribution functions that can or cannot be derived from corresponding operations on random variables (defined on the same probability space), which we discuss in the next section. The original definition was as follows (Alsina et al. 1993):

Definition 6.2.2. A *quasi-copula* is a function $Q: \mathbf{I}^2 \to \mathbf{I}$ such that for every *track* B in $\mathbf{I}^2$ (i.e., B can be described as $B = \{(\alpha(t), \beta(t)); 0 \le t \le 1\}$ for some continuous and nondecreasing functions α, β with $\alpha(0) =$

$\beta(0) = 0$, $\alpha(1) = \beta(1) = 1$), there exists a copula C_B such that $Q(u,v) = C_B(u,v)$ whenever $(u,v) \in B$.

Genest et al. (1999) established the equivalence of Definitions 6.2.1 and 6.2.2, presented the Q-volume interpretation following Definition 6.2.1, and proved that quasi-copulas also satisfy the Fréchet-Hoeffding bounds inequality (2.2.5). Indeed, any property of copulas that can be established without appealing to the 2-increasing property will also be a property of quasi-copulas. For a copula C, the C-volume of a rectangle $R = [a,b] \times [c,d]$ must be between 0 and 1 as a consequence of the 2-increasing condition (2.2.3). The next theorem (Nelsen et al. 2002b) presents the corresponding result for quasi-copulas.

Theorem 6.2.3. *Let Q be a quasi-copula, and $R = [a,b] \times [c,d]$ any rectangle in $\mathbf{I}^2$. Then $-1/3 \leq V_Q(R) \leq 1$. Furthermore, $V_Q(R) = 1$ if and only if $R = \mathbf{I}^2$, and $V_Q(R) = -1/3$ implies $R = [1/3,2/3]^2$.*

For example, the proper quasi-copula Q in Exercise 2.11 has $V_Q([1/3,2/3]^2) = -1/3$. While Theorem 6.2.3 limits the Q-volume of a rectangle, the lower bound of $-1/3$ does not hold for more general subsets of $\mathbf{I}^2$. Let μ_Q denotes the finitely additive set function on finite unions of rectangles given by $\mu_Q(S) = \sum_i V_Q(R_i)$ where $S = \bigcup_i R_i$ with $\{R_i\}$ nonoverlapping. Analogous to Theorem 3.2.2, the copula Π can be approximated arbitrarily closely by quasi-copulas with as much negative "mass" (i.e., value of μ_Q) as desired:

Theorem 6.2.4. *Let ε, $M > 0$. Then there exists a quasi-copula Q and a set $S \subseteq \mathbf{I}^2$ such that $\mu_Q(S) < -M$ and*

$$\sup_{u,v \in \mathbf{I}} |Q(u,v) - \Pi(u,v)| < \varepsilon.$$

The proof in (Nelsen et al. 2002b) is constructive and can be generalized by replacing Π by any quasi-copula whatsoever.

Quasi-copulas arise in a natural fashion when working with copulas.

Example 6.3. Suppose we try to construct a copula C from two copulas C_1 and C_2 via the construction $C(u,v) = \max\{C_1(u,v), C_2(u,v)\}$ for each (u,v) in $\mathbf{I}^2$. Is such a C always a copula? The answer is no—let C_1 and C_2 be the copulas given by (3.2.2) in Example 3.4 with $\theta = 1/3$ and $2/3$, respectively. Then $C(u,v) = \max\{C_1(u,v), C_2(u,v)\}$ is not a copula but rather the proper quasi-copula in Exercise 2.11. ∎

However, the pointwise infimum and supremum of any nonempty set of quasi-copulas (or copulas) are quasi-copulas:

Theorem 6.2.5 (Nelsen et al. 2004). *Let* **S** *be any nonempty set of quasi-copulas, and define* $\underline{S}$ *and* $\overline{S}$ *by*

$$\underline{S}(u,v) = \inf\left\{S(u,v)\big|S \in \mathbf{S}\right\} \text{ and } \overline{S}(u,v) = \sup\left\{S(u,v)\big|S \in \mathbf{S}\right\}$$

for all (u,v) *in* $\mathbf{I}^2$. *Then* $\underline{S}$ *and* $\overline{S}$ *are quasi-copulas.*

Proof. We show that $\overline{S}$ is a quasi-copula, the proof for $\underline{S}$ is similar. For the boundary conditions (2.2.2a) and (2.2.2b), we have $\overline{S}(u,0) = \sup\left\{0\big|S \in \mathbf{S}\right\} = 0$, and $\overline{S}(u,1) = \sup\left\{u\big|S \in \mathbf{S}\right\} = u$, and similarly $\overline{S}(0,v) = 0$ and $\overline{S}(1,v) = v$. Because each quasi-copula is nondecreasing in its arguments, we have $\overline{S}(u_1,v) = \sup\left\{S(u_1,v)\big|S \in \mathbf{S}\right\} \le \sup\left\{S(u_2,v)\big|S \in \mathbf{S}\right\} \le \overline{S}(u_2,v)$ whenever $u_1 \le u_2$, so that $\overline{S}$ is nondecreasing in u (and similarly in v). To show that $\overline{S}$ is Lipschitz, it will suffice to show that whenever $u_1 \le u_2$, $\overline{S}(u_2,v) - \overline{S}(u_1,v) \le u_2 - u_1$. Let u_2,u_1 be fixed in $\mathbf{I}$ with $u_1 \le u_2$. For any $\varepsilon > 0$, there exists a quasi-copula Q_ε in $\mathbf{S}$ such that $Q_\varepsilon(u_2,v) > \overline{S}(u_2,v) - \varepsilon$. Because $Q_\varepsilon(u_1,v) \le \overline{S}(u_1,v)$, it follows that $\overline{S}(u_2,v) - \overline{S}(u_1,v) \le Q_\varepsilon(u_2,v) + \varepsilon - Q_\varepsilon(u_1,v) \le u_2 - u_1 + \varepsilon$. Because this is true for every $\varepsilon > 0$, we have $\overline{S}(u_2,v) - \overline{S}(u_1,v) \le u_2 - u_1$ as required.□

As a consequence of the preceding example and theorem, the partially ordered set $(\mathbf{C},\prec)$ is not a lattice, as not every pair of copulas has a supremum and infimum in the set $\mathbf{C}$. However, when $\mathbf{Q}$, the set of quasi-copulas, is ordered with the same order $\prec$ in Definition 2.8.1, then $(\mathbf{Q},\prec)$ is a complete lattice (i.e., every subset of $\mathbf{Q}$ has a supremum and infimum in $\mathbf{Q}$). Furthermore, $(\mathbf{Q},\prec)$ is order-isomorphic to the Dedekind-MacNeille completion of $(\mathbf{C},\prec)$ (Nelsen and Úbeda Flores 2005). Thus the set of quasi-copulas is a lattice-theoretic completion of the set of copulas, analogous to Dedekind's construction of the reals as a completion by cuts of the set of rationals.

Thus any nonempty set of copulas or quasi-copulas with a specific property is guaranteed to have bounds in the set of quasi-copulas. These bounds are often copulas, as occurred in Theorems 3.2.3 and 5.1.16. However, the bounds may be proper quasi-copulas, as in the following example.

Example 6.4. Let δ be a diagonal, and consider the set $\mathbf{Q}_\delta$ of quasi-copulas with diagonal section δ, i.e.,

$$\mathbf{Q}_\delta = \left\{Q \in \mathbf{Q}\big|Q(t,t) = \delta(t), \, t \in \mathbf{I}\right\}.$$

Let $\underline{Q}_\delta$ and $\overline{Q}_\delta$ be the pointwise infimum and supremum of $\mathbf{Q}_\delta$, i.e., let $\underline{Q}_\delta(u,v) = \inf\{Q(u,v)|Q \in \mathbf{Q}_\delta\}$ and $\overline{Q}_\delta(u,v) = \sup\{Q(u,v)|Q \in \mathbf{Q}_\delta\}$ for (u,v) in $\mathbf{I}^2$. Then $\underline{Q}_\delta$ and $\overline{Q}_\delta$ are quasi-copulas. Indeed, it can be shown (Nelsen et al. 2004) that $\underline{Q}_\delta$ corresponds to the Bertino copula B_δ in (3.2.23), and that $\overline{Q}_\delta$ is given by

$$\overline{Q}_\delta(u,v) = \begin{cases} \min\{u, v - \max(t - \delta(t)|t \in [u,v])\}, & u \le v, \\ \min\{v, u - \max(t - \delta(t)|t \in [v,u])\}, & v \le u. \end{cases}$$

This may or may not be a copula, depending on δ. For example, if $\delta(t) = (2t-1)^+$, then $\overline{Q}_\delta$ is the shuffle of M given by $M(2, \mathbf{I}_2, (2,1), 1)$, i.e., the copula $C_{1/2}$ from Exercise 3.9; and if $\delta(t) = (t-1/3)^+ + (t-2/3)^+$, then $\overline{Q}_\delta$ is the proper quasi-copula in Exercise 2.11. ∎

The quasi-copula concept can be extended to n dimensions, analogous to the extension of copulas to n dimensions in Sect. 2.10 in equations (2.10.4a)-(2.10.4c). The following definition is from (Cuculescu and Theodorescu 2001), where it is shown to be equivalent to the one in (Nelsen et al. 1996).

Definition 6.2.6. An n-dimensional *quasi-copula* (or n-quasi-copula) is a function $Q: \mathbf{I}^n \to \mathbf{I}$ such that:

1. for every $\mathbf{u}$ in $\mathbf{I}^n$, $Q(\mathbf{u}) = 0$ if at least one coordinate of $\mathbf{u}$ is 0, and $Q(\mathbf{u}) = u_k$ if all coordinates of $\mathbf{u}$ are 1 except u_k;
2. Q is nondecreasing in each variable;
3. Q satisfies the Lipschitz condition

$$|Q(\mathbf{u}) - Q(\mathbf{u}')| \le \sum_{i=1}^{n} |u_i - u_i'|$$

for all $\mathbf{u}$ and $\mathbf{u}'$ in $\mathbf{I}^n$.

It is easy to show that n-quasi-copulas satisfy the n-dimension version of the Fréchet-Hoeffding bounds inequality (2.10.9), and that W^n is a proper n-quasi-copula for $n \ge 3$ (see Exercise 6.8). We conclude this section with a discussion of the distribution of positive and negative "mass" (i.e., Q-volume) in $\mathbf{I}^3$ induced by the proper quasi-copula W^3.

Let $n \ge 2$ and partition $\mathbf{I}^3$ into n^3 3-boxes $B_{ijk} = [(i-1)/n, i/n] \times [(j-1)/n, j/n] \times [(k-1)/n, k/n]$, $1 \le i,j,k \le n$. The W^3-volume of B_{ijk} is given by

$$V_{W^3}\left(B_{ijk}\right) = \begin{cases} +\dfrac{1}{n}, & i+j+k = 2n+1, \\[2mm] -\dfrac{1}{n}, & i+j+k = 2n+2, \\[2mm] 0, & \text{otherwise.} \end{cases}$$

The number of 3-boxes B_{ijk} with positive W^3-volume is equal to the number of integer solutions to $i+j+k = 2n+1$ with $1 \le i,j,k \le n$, i.e., $\binom{n+1}{2}$; while the number of 3-boxes B_{ijk} with negative W^3-volume is equal to the number of integer solutions to $i+j+k = 2n+2$ with $1 \le i,j,k \le n$, i.e., $\binom{n}{2}$. Note that the net W^3-volume is thus $\left[\binom{n+1}{2}-\binom{n}{2}\right]\big/n = 1$. However, the total W^3-volume of the 3-boxes with positive W^3-volume is $\binom{n+1}{2}\big/n = (n+1)/2$, whereas the total W^3-volume of the 3-boxes with negative W^3-volume is $-\binom{n}{2}\big/n = (-n+1)/2$. Consequently there are subsets of $\mathbf{I}^3$ with arbitrarily large W^3-volume and subsets of $\mathbf{I}^3$ with arbitrarily small (i.e., very negative) W^3-volume. Similar results hold for the W^n-volume of subsets of $\mathbf{I}^n$ for $n \ge 4$.

Exercises

6.6 Prove the following extension of Theorem 6.2.3: Let $R = [u_1, u_2] \times [v_1, v_2]$ be a rectangle in $\mathbf{I}^2$. If $V_Q(R) = \theta$ for some quasi-copula Q, then $A(R)$, the area of R, satisfies

$$\theta^2 \le A(R) \le \left((1+\theta)/2\right)^2.$$

Furthermore, when $A(R)$ attains either bound, R must be a square (Nelsen et al. 2002).

6.7 Prove that there are no proper Archimedean 2-quasi-copulas [Hint: Lemma 4.2.3]. This is not the case in higher dimensions; see (Nelsen et al. 2002a).

6.8 (a) Let Q be an n-quasi-copula, $n \ge 2$. Prove that $W^n(\mathbf{u}) \le Q(\mathbf{u}) \le M^n(\mathbf{u})$ for all $\mathbf{u}$ in $\mathbf{I}^n$.

(b) Prove that W^n is a proper n-quasi-copula for $n \geq 3$. [See Exercise 2.36.]

6.3 Operations on Distribution Functions

It is common practice to use operations on distribution functions such as convolution or discrete mixtures to construct new distribution functions. We illustrate with two examples.

Example 6.5. For any two distribution functions F and G, the *convolution $F \otimes G$* of F and G is the function on $\overline{\mathbf{R}}$ defined by

$$(F \otimes G)(x) = \int_{-\infty}^{\infty} F(x - t) dG(t) \quad \text{for } x \text{ in } \mathbf{R},$$

with $(F \otimes G)(-\infty) = 0$ and $(F \otimes G)(\infty) = 1$. If X and Y are independent random variables with distribution functions F and G, respectively, then as is well-known $F \otimes G$ is the distribution function of the sum $X + Y$; and the study of sums of independent random variables plays a central role in probability and statistics. ∎

Example 6.6 (Lawless 1982). Discrete mixture models arise in the theory of reliability when individuals belong to one of n distinct types, with a proportion p_k of the population being of the kth type, where the p_k's satisfy $0 < p_k < 1$, $p_1 + p_2 + \cdots + p_n = 1$. Individuals of type k are assumed to have a lifetime distribution function F_k. An individual randomly selected from the population then has the lifetime distribution function

$$p_1 F_1(t) + p_2 F_2(t) + \cdots + p_n F_n(t).$$

In the simplest case, $n = 2$, the *mixture* of any two distribution functions F and G is the distribution function $pF + (1 - p)G$, where p is a fixed real number in $(0,1)$. ∎

Because the operation of convolution derives from the sum of (independent) random variables, it is natural to ask whether a similar result holds for mixtures. That is, does there exist a two-place function Z such that, for any pair X,Y of random variables with respective distribution functions F and G, the mixture $pF + (1 - p)G$ is the distribution function of the random variable $Z(X,Y)$? If so, we could say that mixtures are "derivable" from a function on random variables. To be precise, we have (Alsina and Schweizer 1988):

Definition 6.3.1. A binary operation ψ on the set of distribution functions is *derivable* from a function on random variables if there exists a Borel-measurable two-place function Z satisfying the following condition: For every pair of distribution functions F and G, there exist random variables X and Y defined on a common probability space, such that F and G are, respectively, the distribution functions of X and Y, and $\psi(F,G)$ is the distribution function of the random variable $Z(X,Y)$.

Mixtures are the binary operations μ_p on distribution functions given by

$$\mu_p(F,G) = pF + (1-p)G, \tag{6.3.1}$$

where p is a fixed number in $(0,1)$. Somewhat surprisingly, mixtures are not derivable (Alsina and Schweizer 1988):

Theorem 6.3.2. *The mixture μ_p is not derivable from any binary operation on random variables.*

Proof. Assume that μ_p is derivable, i.e., that a suitable function Z exists. For any real numbers a and b $(a \neq b)$, let F and G be the unit step functions ε_a and ε_b, respectively (see Example 2.4). Then F and G are, respectively, the distribution functions of random variables X and Y, which are defined on a common probability space and equal, respectively, to a and b almost surely. Hence $Z(X,Y)$ is a random variable defined on the same probability space as X and Y and equal to $Z(a,b)$ almost surely. Thus the distribution function of $Z(X,Y)$ is the unit step $\varepsilon_{Z(a,b)}$. But because μ_p is derivable from Z, the distribution function of $Z(X,Y)$ must be $\mu_p(\varepsilon_a,\varepsilon_b) = p\varepsilon_a + (1-p)\varepsilon_b$; and because $p\varepsilon_a + (1-p)\varepsilon_b \neq \varepsilon_{Z(a,b)}$, we have a contradiction. ◻

This argument can be easily extended to mixtures of any finite collection of distribution functions, such as in Example 6.4. It also shows that if ψ is to be derivable, then for any unit step functions ε_a and ε_b, $\psi(\varepsilon_a,\varepsilon_b)$ must also be a unit step function. But this condition is not sufficient, as shown in the next theorem (Alsina and Schweizer 1988).

Theorem 6.3.3. *The operation of forming the geometric mean $g(F,G)$ $= \sqrt{FG}$ of two distribution functions is not derivable from any binary operation on random variables.*

Proof. Assume to the contrary that a suitable function Z exists. Then proceeding as in Theorem 6.3.2, it follows that $Z(a,b) = \max(a,b)$ (because $\sqrt{\varepsilon_a\varepsilon_b} = \varepsilon_{\max(a,b)}$). Next, for any pair of continuous random variables X and Y with distribution functions F and G, respectively, and copula C, it follows from Exercise 2.16 that the distribution function of $Z(X,Y) = \max(X,Y)$ is $C(F(t),G(t))$. Hence

$$\sqrt{F(t)G(t)} = P[Z(X,Y) \le t] = C(F(t),G(t)) \le M(F(t),G(t)).$$

But this is a contradiction, because $\sqrt{uv} \leq \min(u,v)$ is not true for all u,v in $\mathbf{I}$. □

To proceed we need the following:

Definition 6.3.4. A binary operation ψ on distribution functions is *induced pointwise* by a two-place function Ψ from $\mathbf{I}^2$ into $\mathbf{I}$ if for every pair F, G of distribution functions and all t in $\mathbf{R}$,

$$\psi(F,G)(t) = \Psi\big(F(t),G(t)\big).$$

Thus the mixture $\mu_p(F,G)(t) = pF(t)+(1-p)G(t)$ is induced pointwise by the two-place function $M_p(u,v) = pu+(1-p)v$; and the geometric mean $g(F,G)(t) = \sqrt{F(t)G(t)}$ is induced pointwise by the two-place function $G(u,v) = \sqrt{uv}$. But because the value of the convolution of two distribution functions F and G at a point t generally depends on more than the values of F and G at t, convolution is not induced pointwise. Thus we are led to the question: Do there exist binary operations on distribution functions that are both derivable and induced pointwise?

The answer to this question is given by the following theorem, whose proof can be found in (Alsina et al. 1993) [the *dual* $\tilde{Q}$ of a quasi-copula Q is the function from $\mathbf{I}^2$ to $\mathbf{I}$ given by $\tilde{Q}(u,v) = u + v - Q(u,v)$]:

Theorem 6.3.5. *Suppose that ψ is a binary operation on distribution functions that is both induced pointwise by a two-place function Ψ from $\mathbf{I}^2$ into $\mathbf{I}$ and derivable from a function Z on random variables defined on a common probability space. Then precisely one of the following holds*:

1. $Z(x,y) = \max(x,y)$ *and Ψ is a quasi-copula*;

2. $Z(x,y) = \min(x,y)$ *and Ψ is the dual of a quasi-copula*;

3. Z *and Ψ are trivial in the sense that, for all x,y in $\mathbf{R}$ and all u,v in* $\mathbf{I}$, *either $Z(x,y) = x$ and $\Psi(u,v) = u$ or $Z(x,y) = y$ and $\Psi(u,v) = v$.*

Taken together, the results in Theorems 6.3.2, 6.3.3, and 6.3.5 demonstrate that "the distinction between working directly with distributions functions ... and working with them indirectly, via random variables, is intrinsic and not just a matter of taste" (Schweizer and Sklar 1983) and that "the classical model for probability theory—which is based on random variables defined on a common probability space—has its limitations" (Alsina et al. 1993).

We conclude this section by noting that many of the results in this section generalize to n dimensions [see (Nelsen et al. 1996) for details].

6.4 Markov Processes

As we have noted before, it is the copula of a pair (X,Y) of continuous random variables that captures the "nonparametric" or "scale-invariant" nature of the dependence between X and Y. Chapter 5 was devoted to a study of dependence between continuous random variables with a given copula. In this section, we will investigate the role played by copulas in another dependence structure—Markov processes.

Before reviewing the essential notions of Markov processes, we present a "product" operation for copulas first studied in (Darsow et al. 1992). This product is defined in terms of the two first-order partial derivatives of a copula $C = C(u,v)$, which we will now denote

$$D_1 C = \partial C / \partial u \text{ and } D_2 C = \partial C / \partial v . \qquad (6.4.1)$$

Definition 6.4.1. Let C_1 and C_2 be copulas. The *product* of C_1 and C_2 is the function $C_1 * C_2$ from $\mathbf{I}^2$ to $\mathbf{I}$ given by

$$(C_1 * C_2)(u,v) = \int_0^1 D_2 C_1(u,t) \cdot D_1 C_2(t,v) \, dt . \qquad (6.4.2)$$

Theorem 6.4.2. $C_1 * C_2$ *is a copula.*
Proof. For the boundary conditions (2.2.2a) and (2.2.2b), we have

$$(C_1 * C_2)(0,v) = \int_0^1 D_2 C_1(0,t) \cdot D_1 C_2(t,v) \, dt = \int_0^1 0 \, dt = 0 ,$$

and

$$(C_1 * C_2)(1,v) = \int_0^1 D_2 C_1(1,t) \cdot D_1 C_2(t,v) \, dt$$

$$= \int_0^1 D_1 C_2(t,v) \, dt = C_2(1,v) = v.$$

Similarly $(C_1 * C_2)(u,0) = 0$ and $(C_1 * C_2)(u,1) = u$. To show that $C_1 * C_2$ is 2-increasing, we compute the $(C_1 * C_2)$-volume of the rectangle $[u_1,u_2] \times [v_1,v_2]$:

$$V_{C_1 * C_2}\big([u_1,u_2] \times [v_1,v_2]\big)$$

$$= \int_0^1 [D_2 C_1(u_2,t) D_1 C_2(t,v_2) - D_2 C_1(u_1,t) D_1 C_2(t,v_2)$$

$$- D_2 C_1(u_2,t) D_1 C_2(t,v_1) + D_2 C_1(u_1,t) D_1 C_2(t,v_1)] dt$$

$$= \int_0^1 D_2 [C_1(u_2,t) - C_1(u_1,t)] D_1 [C_2(t,v_2) - C_2(t,v_1)] dt .$$

But as a consequence of Lemma 2.1.3, both $D_2[C_1(u_2,t)-C_1(u_1,t)]$ and $D_1[C_2(t,v_2)-C_2(t,v_1)]$ are nonnegative, and thus $V_{C_1*C_2}([u_1,u_2]\times[v_1,v_2]) \geq 0$. Hence $C_1 * C_2$ is a copula. $\square$

The *-product of copulas is a continuous analog of matrix multiplication and shares some of the properties of that operation, as the following example illustrates.

Example 6.7. Let C be a copula. Then

$$(\Pi * C)(u,v) = \int_0^1 D_2\Pi(u,t)D_1C(t,v)dt = \int_0^1 uD_1C(t,v)dt = uv;$$

similarly $(C * \Pi)(u,v) = uv$, and hence

$$\Pi * C = C * \Pi = \Pi.$$

Because $D_2M(u,t)$ is 1 for $t < u$ and 0 for $t > u$, we have

$$(M * C)(u,v) = \int_0^1 D_2M(u,t)D_1C(t,v)dt = \int_0^u D_1C(t,v)dt = C(u,v).$$

Similarly $(C * M)(u,v) = C(u,v)$, and hence

$$M * C = C * M = C.$$

So, if we view $*$ as a binary operation on the set of copulas, then Π is the null element, and M is the identity. Furthermore, we have (see Exercise 6.9)

$$(W * C)(u,v) = v - C(1-u,v) \quad \text{and} \quad (C * W)(u,v) = u - C(u,1-v)$$

(also see Exercises 2.6, 2.30, and Theorem 2.4.4), hence $*$ is not commutative. However, $*$ is associative—see (Darsow et al. 1992; Li et al. 1997) for proofs. Finally,

$$W * W = M \quad \text{and} \quad W * C * W = \hat{C}. \qquad \blacksquare$$

Before discussing Markov processes, we need to review some terminology and notation. A *stochastic process* is a collection or sequence $\{X_t | t \in T\}$ of random variables, where T denotes a subset of $\mathbf{R}$. It is convenient to think of the index t as *time*, and the random variable X_t as the *state* of the process at time t. For each s, t in T, we let F_s and F_t denote the distribution functions of X_s and X_t, respectively, and H_{st} the joint distribution function of X_s and X_t. The process is continuous if each F_t is continuous. Finally, for each s, t in T, we let C_{st} denote the subcopula of X_s and X_t, that is, for all x, y in $\mathbf{R}$, $H_{st}(x,y) =$

$C_{st}(F_s(x), F_t(y))$. When F_s and F_t are continuous, C_{st} is a copula, otherwise we extend C_{st} to a copula via the "bilinear interpolation" construction used in the proof of Lemma 2.3.5. Thus in this section, we will refer to *the* copula of X_s and X_t, even when the random variables fail to be continuous.

The process $\{X_t | t \in T\}$ is a *Markov process* if for every finite subset $\{t_1, t_2, \cdots, t_n\}$ of T and any t in T such that $t_1 < t_2 < \cdots < t_n < t$,

$$P\big[X_t \leq x | X_{t_1} = x_1, X_{t_2} = x_2, \cdots, X_{t_n} = x_n\big] = P\big[X_t \leq x | X_{t_n} = x_n\big], \quad (6.4.3)$$

i.e., in a Markov process, the conditional distribution functions only depend on the most recent time and state of the process. We will adopt the following notation for these conditional distribution functions:

$$P(x, s; y, t) = P\big[X_t \leq y | X_s = x\big]. \tag{6.4.4}$$

As a consequence of the Markov property (6.4.3), the conditional probabilities $P(x,s;y,t)$ in (6.4.4) satisfy the *Chapman-Kolmogorov equations,* which relate the state of the process at time t with that at an earlier time s through an intermediate time u:

$$P(x, s; y, t) = \int_{-\infty}^{\infty} P(z, u; y, t) \cdot \frac{\partial P(x, s; z, u)}{\partial z} \, dz, \quad (s < u < t). \tag{6.4.5}$$

If the conditional densities $p(x, s; y, t) = P\big[X_t = y | X_s = x\big]$ exist, then (6.4.5) takes the form

$$p(x, s; y, t) = \int_{-\infty}^{\infty} p(x, s; z, u) p(z, u; y, t) dz, \quad (s < u < t).$$

In this form we see that the Chapman-Kolmogorov equations can be viewed as a continuous version of the law of total probability, modified by the Markov property: if A_s, B_u, and E_t denote events at the times s, u, t, respectively ($s < u < t$), then

$$P\big[E_t | A_s\big] = \sum_u P\big[E_t | B_u\big] \cdot P\big[B_u | A_s\big],$$

the summation being over all possible events B_u at time u. The fact that the conditional probability $P\big[E_t | B_u\big]$ depends only on t and u ($u < t$), and not on events at times earlier than u, is the Markov property.

The main result of this section is the following theorem (Darsow et al. 1992), which relates the Chapman-Kolmogorov equations for a Markov process to the copulas of the random variables in the process.

Theorem 6.4.3. *Let* $\{X_t | t \in T\}$ *be a stochastic process, and for each s, t in T, let* C_{st} *denote the copula of the random variables* X_s *and* X_t. *Then the following are equivalent:*
  1. *The conditional distribution functions P(x,s;y,t) satisfy the Chapman-Kolmogorov equations (6.4.5) for all s < u < t in T and almost all x,y in* **R***;*
  2. *For all s < u < t in T,*

$$C_{st} = C_{su} * C_{ut}. \qquad (6.4.6)$$

For a proof of this remarkable result, see (Darsow et al. 1992). Theorem 6.4.3 is important not only because it provides a condition equivalent to the Chapman-Kolmogorov equations for Markov processes, but also because it is a new approach to the theory and yields a new technique for constructing such processes. As Darsow et al. (1992) observe,

> In the conventional approach, one specifies a Markov process by giving the initial distribution F_{t_0} and a family of transition probabilities $P(x,s;y,t)$ satisfying the Chapman-Kolmogorov equations. In our approach, one specifies a Markov process by giving all of the marginal distributions and a family of 2-copulas satisfying (6.4.6). Ours is accordingly an alternative approach to the study of Markov processes which is different in principle from the conventional one. Holding the transition probabilities of a Markov process fixed and varying the initial distribution necessarily varies all of the marginal distributions, but holding the copulas of the process fixed and varying the initial distribution does not affect any other marginal distribution.

The next two examples, from (Darsow et al. 1992), illustrate the use of Theorem 6.4.3 in the construction of Markov processes.

Example 6.8. Let T be the set of nonnegative integers, let C be any copula, and set $C_{mn} = C * C * \cdots * C$, the $(m-n)$-fold $*$-product of C with itself. Then the set $\{C_{mn}\}$ of copulas satisfies (6.4.6), so that a Markov process is specified by supplying a sequence $\{F_k\}$ of continuous marginal distributions. Processes constructed in this manner are similar to Markov chains. ∎

Example 6.9. In this example, we illustrate how one can calculate the copulas for a known Markov process, then create a new process via the same family of copulas but with new marginals. The transition probabilities for standard Brownian motion are given by

$$P(x,s;y,t) = \Phi\left(\frac{y-x}{\sqrt{t-s}}\right), \quad (s < u < t),$$

where Φ denotes the standard normal distribution function. From (5.4.1) we have

$$P(x,s;y,t)= \frac{\partial}{\partial x}H_{st}(x,y)= D_1 C_{st}\big(F_s(x),F_t(y)\big),$$

and thus it follows that

$$C_{st}\big(F_s(x),F_t(y)\big)= \int_{-\infty}^{x}\Phi\left(\frac{y-z}{\sqrt{t-s}}\right)dF_s(z) \qquad (6.4.7)$$

for $0 < s < t$. If we assume that $X_0 = 0$ with probability 1, then $F_t(x) = \Phi(x/\sqrt{t})$ for $t > 0$. Substitution into (6.4.7) yields

$$C_{st}(u,v)= \int_0^u \Phi\left(\frac{\sqrt{t}\Phi^{-1}(v)-\sqrt{s}\Phi^{-1}(w)}{\sqrt{t-s}}\right)dw. \qquad (6.4.8)$$

This is a family of copulas that satisfies (6.4.6). When u and v are replaced by non-normal distribution functions in (6.4.8), we obtain joint distribution functions with non-normal marginals for random variables in a Brownian motion process. ∎

We close this section by recalling that the Chapman-Kolmogorov equations (and hence (6.4.6)) are a necessary but not sufficient condition for a stochastic process to be Markov. Using an extension of (6.4.6) to higher dimensions, it is possible to use copulas to give a condition that is necessary and sufficient. See (Darsow et al. 1992) for details.

Exercises

6.9 Let C be a copula. Verify the claims in Example 6.5:
(a) $(W * C)(u,v) = v - C(1-u,v)$ and $(C * W)(u,v) = u - C(u,1-v)$;
(b) $W * W = M$ and $W * C * W = \hat{C}$.

6.10 Let C_1 and C_2 be copulas.
(a) Show that if $C = C_1 * C_2$, then $\hat{C} = \hat{C}_1 * \hat{C}_2$.
(b) Show that $W * C_1 = \hat{C}_1 * W$.
(c) Show that $C_1 = \hat{C}_1$ if and only if $W * C_1 = C_1 * W$.

6.11 Show that both the Fréchet and Mardia families (see Exercise 2.4) are closed under the *-product operation; i.e.,
(a) if C_{α_1,β_1} and C_{α_2,β_2} are Fréchet copulas, then so is $C_{\alpha_1,\beta_1} * C_{\alpha_2,\beta_2}$, and

$$C_{\alpha_1,\beta_1} * C_{\alpha_2,\beta_2} = C_{\alpha_1\alpha_2+\beta_1\beta_2,\alpha_1\beta_2+\alpha_2\beta_1} ;$$

(b) if C_α and C_β are Mardia copulas, then so is $C_\alpha * C_\beta$, and

$$C_\alpha * C_\beta = C_{\alpha\beta} .$$

6.12 Show that the Farlie-Gumbel-Morgenstern family (3.2.10) is closed under the *-product operation; i.e., if C_α and C_β are FGM copulas, then so is $C_\alpha * C_\beta$, and

$$C_\alpha * C_\beta = C_{\alpha\beta/3}.$$

6.13 Generalize Example 6.8 as follows (Darsow et al. 1992): Let T be the set of integers, and to each k in T assign any copula C_k. Then for $m \le n$ in T, set

$$C_{mn} = \begin{cases} M, & \text{if } m = n, \\ C_m * C_{m+1} * \cdots * C_{n-1}, & \text{if } m < n. \end{cases}$$

Show that upon assigning a continuous distribution function to each element of T, this procedure yields a Markov process.

References

Aczél J (1966) Lectures on Functional Equations and their Applications. Academic Press, New York

Ali MM, Mikhail NN, Haq MS (1978) A class of bivariate distributions including the bivariate logistic. J Multivariate Anal 8:405-412

Alsina C (1981) Some functional equations in the space of uniform distribution functions. Aequationes Math 22:153-164

Alsina C, Schweizer B (1988) Mixtures are not derivable. Found Phys Lett 1:171-174

Alsina C, Frank MJ, Schweizer B (2005) Associative Functions on Intervals: A Primer of Triangular Norms. World Scientific, Hackensack, in press

Alsina C, Nelsen RB, Schweizer B (1993) On the characterization of a class of binary operations on distribution functions. Statist Probab Lett 17:85-89

Armstrong M (2003) Copula Catalogue, Part 1: Bivariate Archimedean Copulas. URL: http://www.cerna.ensmp.fr/Documents/MA-CopulaCatalogue.pdf

Barlow, RE, Proschan F (1981) Statistical Theory of Reliability and Life Testing. To Begin With, Silver Spring, MD

Barnett V (1980) Some bivariate uniform distributions. Comm Statist A—Theory Methods 9:453-461

Beneš V, Štěpán J eds (1997) Distributions with Given Marginals and Moment Problems. Kluwer, Dordrecht

Bertino S (1968) Su di una sottoclasse della classe di Fréchet. Statistica 25:511-542

Bertino S (1977) Sulla dissomiglianza tra mutabili cicliche. Metron 35:53-88

Block HW, Savits T, Shaked M (1982) Some concepts of negative dependence. Ann Probab 10:765-772

Block HW, Ting M-L (1981) Some concepts of multivariate dependence. Comm Statist A—Theory Methods 10:749-762

Block HW, Costigan T, Sampson AR (1997) A review of orthant probability bounds. In: Johnson NL, Balakrishnan N (eds) Advances in the Theory and Practice of Statistics: A Volume in Honor of Samuel Kotz. Wiley, New York, pp 535-550

Blomqvist N (1950) On a measure of dependence between two random variables. Ann Math Statist 21:593-600

Blum JR, Kiefer J, Rosenblatt M (1961) Distribution-free tests of independence based on the sample distribution function. Ann Math Statist 32:485-498

Brindley EC, Thompson Jr WA (1972), Dependence and aging aspects of multivariate survival. J Amer Statist Assoc 67:822-830

Cambanis S (1977) Some properties and generalizations of multivariate Eyraud-Gumbel-Morgenstern distributions. J Multivariate Anal 7:551-559

Cambanis S, Simons G, Stout W (1976) Inequalities for $Ek(X,Y)$ when the marginals are fixed. Z Wahrscheinlichkeitstheorie und verw Gebiete 36:285-294

Capéraà P, Fougères AL, Genest C (1997) A non-parametric estimation procedure for bivariate extreme value copulas. Biometrika 84:567-577

Capéraà P, Genest C (1993) Spearman's ρ is larger than Kendall's τ for positively dependent random variables. J Nonparametr Statist 2:183-194

Carriere JF (1994) A large sample test for one-parameter families of copulas. Comm Statist Theory Methods 23:1311-1317

Chakak A, Koehler KJ (1995) A strategy for constructing multivariate distributions. Comm Statist Simulation Comput 24:537-550

Clayton DG (1978) A model for association in bivariate life tables and its application in epidemiological studies of familial tendency in chronic disease incidence. Biometrika 65:141-151

Clemen R, Jouini M (1996) Copula models for aggregating expert opinions. Operations Research 44:444-457

Conti PL (1993) On some descriptive aspects of measures of monotone dependence. Metron 51:43-60

Conway DA (1979) Multivariate Distributions with Specified Marginals. (Technical Report 145, Stanford University)

Conway DA (1983) Farlie-Gumbel-Morgenstern distributions. In: Kotz S, Johnson NL (eds) Encyclopedia of Statistical Sciences, Vol 3. Wiley, New York, pp 28-31

Conway DA (1986) Plackett family of distributions. In: Kotz S, Johnson NL (eds) Encyclopedia of Statistical Sciences, Vol 7. Wiley, New York, pp 1-5

Cook RD, Johnson ME (1981) A family of distributions for modeling non-elliptically symmetric multivariate data. J Roy Statist Soc Ser B 43:210-218

Cook RD, Johnson ME (1986) Generalized Burr-Pareto-logistic distributions with applications to a uranium exploration data set. Technometrics 28:123-131

Cox DR, Oakes D (1984) Analysis of Survival Data. Chapman and Hall, London

Cuadras CM (1992) Probability distributions with given multivariate marginals and given dependence structure. J Multivariate Anal 42:51-66

Cuadras CM, Augé J (1981) A continuous general multivariate distribution and its properties. Comm Statist A—Theory Methods 10:339-353

Cuadras CM, Fortiana J, Rodríguez-Lallena JA eds (2002) Distributions with Given Marginals and Statistical Modelling. Kluwer, Dordrecht

Cuculescu I, Theodorescu R (2001) Copulas: diagonals, tracks. Rev Roumaine Math Pures Appl 46:731-742

Dall'Aglio G (1956) Sugli estremi dei momenti delle funzioni de ripartizione doppia. Ann Scuola Norm Sup Pisa Cl Sci (3) 10:35-74

Dall'Aglio G (1960) Les fonctions extrêmes de la classe de Fréchet à 3 dimensions. Publ Inst Statist Univ Paris 9:175-188

Dall'Aglio G (1972) Fréchet classes and compatibility of distribution functions. Sympos Math 9:131-150

Dall'Aglio G (1991) Fréchet classes: the beginnings. In: Dall'Aglio G, Kotz S, Salinetti G (eds) Advances in Probability Distributions with Given Marginals. Kluwer, Dordrecht, pp 1-12

Dall'Aglio G (1997a) Joint distribution of two uniform random variables when the sum and the difference are independent. In: Beneš V, Štěpán J (eds) Distributions with Given Marginals and Moment Problems. Kluwer, Dordrecht, pp 117-120

Dall'Aglio G (1997b) The distribution of a copula when the sum and the difference of the univariate random variables are independent. In: Johnson NL, Balakrishnan N (eds) Advances in the Theory and Practice of Statistics: A Volume in Honor of Samuel Kotz. Wiley, New York, pp 517-533

Dall'Aglio G, Kotz S, Salinetti G eds (1991) Advances in Probability Distributions with Given Marginals. Kluwer, Dordrecht

Daniels HE (1950) Rank correlation and population models. J Roy Statist Soc Ser B 12:171-181

Darsow WF, Nguyen B, Olsen ET (1992) Copulas and Markov processes. Illinois J Math 36:600-642

Deheuvels P (1978) Caractérisation complète des lois extrêmes multivariées et de la convergence des types extremes. Publ Inst Statist Univ Paris 23:1-37

Deheuvels P (1979) La fonction de dépendence empirique et ses propriétés. Un test non paramétrique d'indépendence. Acad Roy Belg Bul Cl Sci (5) 65:274-292

Deheuvels P (1981a) A Kolmogorov-Smirnov type test for independence and multivariate samples. Rev Roumaine Math Pures Appl 26:213-226

Deheuvels P (1981b) A non parametric test for independence. Publ Inst Statist Univ Paris 26:29-50

Deheuvels P (1981c) Multivariate tests of independence. In: Analytical Methods in Probability (Oberwolfach, 1980) Lecture Notes in Mathematics 861. Springer, Berlin, pp 42-50

Devroye L (1986) Non-Uniform Random Variate Generation. Springer, New York

Drouet Mari D, Kotz S (2001) Correlation and Dependence. Imperial College Press, London

Durante F (2005) A new class of symmetric bivariate copulas. J Nonparametr Statist. In press

Durante F, Sempi C (2003) Copulæ and Schur-concavity. Int Math J 3:893-905

Durante F, Sempi C (2005) Copula and semi-copula transforms. Int J Math Math Sci. In press

Durbin J, Stuart A (1951) Inversions and rank correlations. J Roy Statist Soc Ser B 13:303-309

Edwardes MD deB (1993) Kendall's τ is equal to the correlation coefficient for the BVE distribution. Statist Probab Lett 17:415-419

Embrechts P, Lindskog F, McNeil A (2003) Modelling dependence with copulas and applications to risk management. In: Rachev S (ed) Handbook of Heavy Tailed Distributions in Finance. Elsevier, New York, pp. 329-384

Esary JD, Proschan F (1972) Relationships among some concepts of bivariate dependence. Ann Math Statist 43:651-655

Eyraud H (1938) Les principes de la mesure des correlations. Ann Univ Lyon Series A 1:30-47

Fang K-T, Fang H-B, von Rosen D (2000) A family of bivariate distributions with non-elliptical contours. Comm Statist Theory Methods 29:1885-1898

Farlie DJG (1960) The performance of some correlation coefficients for a general bivariate distribution. Biometrika 47:307-323

Feller W (1971) An Introduction to Probability Theory and Its Applications, Vol II 2nd edn. Wiley, New York

Ferguson TS (1995) A class of symmetric bivariate uniform distributions. Statist Papers 36:31-40

Féron R (1956) Sur les tableaux de corrélation dont les marges sont données, cas de l'espace à trois dimensions. Publ Inst Statist Univ Paris 5:3-12

Fisher NI (1997) Copulas. In: Kotz S, Read CB, Banks DL (eds) Encyclopedia of Statistical Sciences, Update Vol 1. Wiley, New York, pp 159-163

Fisher NI, Sen PK eds (1994) The Collected Works of Wassily Hoeffding. Springer, New York

Frank MJ (1979) On the simultaneous associativity of $F(x,y)$ and $x + y - F(x,y)$. Aequationes Math 19:194-226

Frank MJ (1981) The solution of a problem of Alsina, and its generalization. Aequationes Math 21:37-38

Frank MJ, Nelsen RB, Schweizer B (1987) Best-possible bounds for the distribution of a sum—a problem of Kolmogorov. Probab Theory Related Fields 74:199-211

Fréchet M (1951) Sur les tableaux de corrélation dont les marges sont données. Ann Univ Lyon Sect A 9:53-77

Fréchet M (1958) Remarques au sujet de la note précédente. C R Acad Sci Paris Sér I Math 246:2719-2720.

Fredricks GA, Nelsen RB (1997a) Copulas constructed from diagonal sections. In: Beneš V, Štěpán J (eds) Distributions with Given Marginals and Moment Problems. Kluwer, Dordrecht, pp 129-136

Fredricks GA, Nelsen RB (1997b) Diagonal copulas. In: Beneš V, Štěpán J (eds) Distributions with Given Marginals and Moment Problems. Kluwer, Dordrecht, pp 121-128

Fredricks GA, Nelsen RB (2002) The Bertino family of copulas. In: Cuadras CM, Fortiana J, Rodríguez Lallena JA (eds) Distributions with Given Marginals and Statistical Modelling, Kluwer, Dordrecht, pp 81-92

Frees EW, Valdez EA (1998) Understanding relationships using copulas. N Am Actuar J 2:1-25

Galambos J (1978) The Asymptotic Theory of Extreme Order Statistics. Wiley, New York

Genest C (1987) Frank's family of bivariate distributions. Biometrika 74:549-555

Genest C, Ghoudi K (1994) Une famille de lois bidimensionnelles insolite. C R Acad Sci Paris Sér I Math 318:351-354

Genest C, MacKay J (1986a) Copules archimédiennes et familles de lois bi-dimensionnelles dont les marges sont données. Canad J Statist 14:145-159

Genest C, MacKay J (1986b) The joy of copulas: Bivariate distributions with uniform marginals. Amer Statist 40:280-285

Genest C, Rivest L-P (1989) A characterization of Gumbel's family of extreme value distributions. Statist Probab Lett 8:207-211

Genest C, Rivest L-P (1993) Statistical inference procedures for bivariate Archimedean copulas. J Amer Statist Assoc 88:1034-1043

Genest C, Rivest L-P (2001) On the multivariate probability integral transform. Statist Probab Lett 53:391-399

Genest C, Quesada Molina JJ, Rodríguez Lallena JA (1995) De l'impossibilité de construire des lois à marges multidimensionnelles données à partir de copules. C R Acad Sci Paris Sér I Math 320:723-726

Genest C, Quesada Molina JJ, Rodríguez Lallena JA, Sempi C (1999) A characterization of quasi-copulas. J Multivariate Anal 69:193-205

Gideon RA, Hollister RA (1987) A rank correlation coefficient resistant to outliers. J Amer Statist Assoc 82:656-666

Gnedenko BV (1962) The Theory of Probability. Chelsea, New York

Gumbel EJ (1960a) Bivariate exponential distributions. J Amer Statist Assoc 55:698-707

Gumbel EJ (1960b) Distributions des valeurs extrêmes en plusiers dimensions. Publ Inst Statist Univ Paris 9:171-173

Gumbel EJ (1961) Bivariate logistic distributions. J Amer Statist Assoc 56:335-349

Hoeffding W (1940) Masstabinvariante Korrelationstheorie. Schriften des Matematischen Instituts und des Instituts für Angewandte Mathematik der Universität Berlin 5 Heft 3:179-233 [Reprinted as Scale-invariant correlation theory. In: Fisher NI, Sen PK (eds) The Collected Works of Wassily Hoeffding. Springer, New York pp 57-107]

Hoeffding W (1941) Masstabinvariante Korrelationsmasse für diskontinuierliche Verteilungen. Arkiv für matematischen Wirtschaften und Sozialforschung 7:49-70 [Reprinted as Scale-invariant correlation measures for discontinuous distributions. In: Fisher NI, Sen PK (eds) The Collected Works of Wassily Hoeffding. Springer, New York pp 109-133]

Hollander M, Wolfe DA (1973) Nonparametric Statistical Methods. Wiley, New York

Hougaard P (1986) A class of multivariate failure time distributions. Biometrika 73:671-678

Hürlimann W (2003) Hutchinson-Lai's conjecture for bivariate extreme value copulas. Statist Probab Lett 61:191-198

Hutchinson TP, Lai CD (1990) Continuous Bivariate Distributions, Emphasising Applications. Rumsby Scientific Publishing, Adelaide

Joe H (1990) Multivariate concordance. J Multivariate Anal 35:12-30

Joe H (1993) Parametric families of multivariate distributions with given margins. J Multivariate Anal 46:262-282

Joe H (1996) Families of m-variate distributions with given margins and $m(m-1)/2$ bivariate dependence parameters. In: Rüschendorf L, Schweizer B, Taylor, MD (eds) Distributions with Fixed Marginals and Related Topics. Institute of Mathematical Statistics, Hayward, CA, pp 120-141

Joe H (1997) Multivariate Models and Dependence Concepts. Chapman & Hall, London

Jogdeo K (1982) Concepts of dependence. In: Kotz S, Johnson NL (eds) Encyclopedia of Statistical Sciences, Vol 1. Wiley, New York, pp 324-334

Johnson ME (1987) Multivariate Statistical Simulation. Wiley, New York

Johnson NL, Kotz S (1972) Distributions in Statistics: Continuous Multivariate Distributions. Wiley, New York

Kamiński A, Sherwood H, Taylor, MD (1987-88) Doubly stochastic measures with mass on the graphs of two functions. Real Anal Exchange 13:253-257

Karlin S (1968) Total Positivity. Stanford University Press, Stanford

Kellerer HG (1964) Verteilungsfunktionen mit gegebenen Marginalverteilungen. Z Wahrscheinlichkeitstheorie und verw Gebiete 3:247-270

Kimberling CH (1974) A probabilistic interpretation of complete monotonicity. Aequationes Math 10:152-164

Kimeldorf G, Sampson A (1975a) One-parameter families of bivariate distributions with fixed marginals. Comm Statist A—Theory Methods 4:293-301

Kimeldorf G, Sampson A (1975b) Uniform representations of bivariate distributions. Comm Statist A—Theory Methods 4:617-627

Kimeldorf G, Sampson A (1978) Monotone dependence. Ann Statist 6:895-903

Kimeldorf G, Sampson A (1987) Positive dependence orderings. Ann Inst Statist Math 39:113-128

Kimeldorf G, Sampson A (1989) A framework for positive dependence. Ann Inst Statist Math 41:31-45

Klement EP, Mesiar R, Pap, E (2004) Transformations of copulas and quasi-copulas. In: López-Díaz M, Gil MÁ, Grzegorzewski P, Hryniewicz O, Lawry J (eds) Soft Methodology and Random Information Systems. Springer, Berlin, pp. 181-188

Kotz S, Johnson NL (1977) Propriétés de dépendance des distributions itérées, généralisées à deux variables Farlie-Gumbel-Morgenstern. C R Acad Sci Paris Sér I Math 285:277-280

Kowalski CJ (1973) Non-normal bivariate distributions with normal marginals. Amer Statist 27:103-106

Kruskal WH (1958) Ordinal measures of association. J Amer Statist Assoc 53:814-861

Lai TL, Robbins H (1976) Maximally dependent random variables. Proc Nat Acad Sci USA 73:286-288

Lancaster HO (1963) Correlation and complete dependence of random variables. Ann Math Statist 34:1315-1321

Lancaster HO (1982) Measures and indices of dependence. In: Kotz S, Johnson NL (eds) Encyclopedia of Statistical Sciences, Vol 2. Wiley, New York, pp 334-339

Lawless JF (1982) Statistical Models and Methods for Lifetime Data. Wiley, New York

Lee M-LT (1996) Properties and applications of the Sarmanov family of bivariate distributions. Comm Statist Theory Methods 25:1207-1222

Lehmann EL (1966) Some concepts of dependence. Ann Math Statist 37:1137-1153.

Lehmann EL (1975) Nonparametrics: Statistical Methods Based on Ranks. Holden-Day, San Francisco

Li H, Scarsini M, Shaked M (1996a) Bounds for the distribution of a multivariate sum. In: Rüschendorf L, Schweizer B, Taylor MD (eds) Distributions with Fixed Marginals and Related Topics. Institute of Mathematical Statistics, Hayward, CA, pp 198-212

Li H, Scarsini M, Shaked M (1996b) Linkages: a tool for the construction of multivariate distributions with given nonoverlapping multivariate marginals. J Multivariate Anal 56:20-41

Li X, Mikusiński P, Sherwood H, Taylor MD (1997) On approximations of copulas. In: Beneš V, Štěpán J (eds) Distributions with Given Marginals and Moment Problems. Kluwer, Dordrecht, pp 107-116

Li X, Mikusiński P, Sherwood H, Taylor MD (2002) Some integration-by-parts formulas involving 2-copulas. In: Cuadras CM, Fortiana J, Rodríguez Lallena JA (eds) Distributions with Given Marginals and Statistical Modelling, Kluwer, Dordrecht, pp 153-159

Lin GD (1987) Relationships between two extensions of Farlie-Gumbel-Morgenstern distributions. Ann Inst Statist Math 39:129-140

Ling C-H (1965) Representation of associative functions. Publ Math Debrecen 12:189-212

Long D, Krzysztofowicz R (1996) Geometry of a correlation coefficient under a copula. Comm Statist Theory Methods 25:1397-1404

López-Díaz M, Gil MÁ, Grzegorzewski P, Hryniewicz O, Lawry J eds (2004) Soft Methodology and Random Information Systems. Springer, Berlin

Lu J-C, Bhattacharyya GK (1990) Some new constructions of bivariate Weibull models. Ann Inst Statist Math 42:543-559

Makarov GD (1981) Estimates for the distribution function of a sum of two random variables when the marginal distributions are fixed. Theory Probab Appl 26:803-806

Marco JM, Ruiz-Rivas C (1992) On the construction of multivariate distributions with given nonoverlapping multivariate marginals. Statist Probab Lett 15:259-265

Mardia KV (1967) Families of Bivariate Distributions. Griffin, London

Marshall AW (1989) A bivariate uniform distribution. In: Gleser LJ, Perlman MD, Press SJ, Sampson AR (eds) Contributions to Probability and Statistics. Springer, New York, pp 99-106

Marshall AW (1996) Copulas, marginals, and joint distributions. In: Rüschendorf L, Schweizer B, Taylor MD (eds) Distributions with Fixed Marginals and Related Topics. Institute of Mathematical Statistics, Hayward, CA, pp 213-222

Marshall AW, Olkin I (1967a) A generalized bivariate exponential distribution. J Appl Probability 4:291-302

Marshall AW, Olkin I (1967b) A multivariate exponential distribution. J Amer Statist Assoc 62:30-44

Marshall AW, Olkin I (1988) Families of multivariate distributions. J Amer Statist Assoc 83:834-841

Melnick EL, Tennenbein A (1982) Misspecifications of the normal distribution. Amer Statist 36:372-373

Menger K (1942) Statistical metrics. Proc Natl Acad Sci USA 28:535-537

Menger K (1956) Random variables from the point of view of a general theory of variables. In: Proceedings of the Third Berkeley Symposium on Mathematical Statistics and Probability Vol II. University of California Press

Mikusiński P, Sherwood H, Taylor MD (1991) Probabilistic interpretations of copulas and their convex sums. In: Dall'Aglio G, Kotz S, Salinetti G (eds) Advances in Probability Distributions with Given Marginals. Kluwer, Dordrecht, pp 95-112

Mikusiński P, Sherwood H, Taylor MD (1991-92) The Fréchet bounds revisited. Real Anal Exchange 17:759-764

Mikusiński P, Sherwood H, Taylor MD (1992) Shuffles of Min. Stochastica 13:61-74

Moore DS, Spruill MC (1975) Unified large-sample theory of general chi-squared statistics for tests of fit. Ann Statist 3:599-616

Morgenstern D (1956) Einfache Beispiele Zweidimensionaler Verteilungen. Mitteilungsblatt für Mathematische Statistik 8:234-235

Muliere P, Scarsini M (1987) Characterization of a Marshall-Olkin type class of distributions. Ann Inst Statist Math 39 (2) A:429-441

Nelsen RB (1986) Properties of a one-parameter family of bivariate distributions with specified marginals. Comm Statist Theory Methods 15:3277-3285

Nelsen RB (1991) Copulas and association. In: Dall'Aglio G, Kotz S, Salinetti G (eds) Advances in Probability Distributions with Given Marginals. Kluwer, Dordrecht, pp 51-74

Nelsen RB (1992) On measures of association as measures of positive dependence. Statist Probab Lett 14:269-274

Nelsen RB (1993) Some concepts of bivariate symmetry. J Nonparametr Statist 3:95-101

Nelsen RB (1995) Copulas, characterization, correlation, and counterexamples. Math Mag 68:193-198

Nelsen RB (1996) Nonparametric measures of multivariate association. In: Rüschendorf L, Schweizer B, Taylor MD (eds) Distributions with Fixed Marginals and Related Topics. Institute of Mathematical Statistics, Hayward, CA, pp 223-232

Nelsen RB (1997) Dependence and order in families of Archimedean copulas. J Multivariate Anal 60:111-122

Nelsen RB (2002) Concordance and copulas: A survey. In: Cuadras CM. Fortiana J, Rodríguez Lallena JA (eds) Distributions with Given Marginals and Statistical Modelling, Kluwer, Dordrecht, pp. 169-177

Nelsen RB, Schweizer B (1991) Bounds for distribution functions of sums of squares and radial errors. Int J Math Math Sci 14:561-570

Nelsen RB, Úbeda Flores M (2004) A comparison of bounds on sets of joint distribution functions derived from various measures of association. Comm Statist Theory Methods 33:2299-2305

Nelsen RB, Úbeda Flores M (2005) The lattice-theoretic structure of sets of bivariate copulas and quasi-copulas. C R Acad Sci Paris Ser I. In press

Nelsen RB, Quesada Molina JJ, Rodríguez Lallena, JA (1997) Bivariate copulas with cubic sections. J Nonparametr Statist 7:205-220

Nelsen RB, Quesada Molina JJ, Rodríguez Lallena, JA, Úbeda Flores M (2001a) Distribution functions of copulas: A class of bivariate probability integral transforms. Statist Probab Lett 54:277-282

Nelsen RB, Quesada Molina JJ, Rodríguez Lallena, JA, Úbeda Flores M (2001b) Bounds on bivariate distribution functions with given margins and measures of association. Comm Statist Theory Methods 30:1155-1162

Nelsen RB, Quesada Molina JJ, Rodríguez Lallena, JA, Úbeda Flores M (2002a) Multivariate Archimedean quasi-copulas. In: Cuadras CM. Fortiana J, Rodríguez Lallena JA (eds) Distributions with Given Marginals and Statistical Modelling, Kluwer, Dordrecht, pp. 179-185

Nelsen RB, Quesada Molina JJ, Rodríguez Lallena, JA, Úbeda Flores M (2002b) Some new properties of quasi-copulas. In: Cuadras CM. Fortiana J, Rodríguez Lallena JA (eds) Distributions with Given Marginals and Statistical Modelling, Kluwer, Dordrecht, pp. 187-194

Nelsen RB, Quesada Molina JJ, Rodríguez Lallena, JA, Úbeda Flores M (2003) Kendall distribution functions. Statist Probab Lett 65:263-268

Nelsen RB, Quesada Molina JJ, Rodríguez Lallena, JA, Úbeda Flores M (2004) Best-possible bounds on sets of bivariate distribution functions. J Multivariate Anal 90:348-358

Nelsen RB, Quesada Molina JJ, Schweizer B, Sempi C (1996) Derivability of some operations on distribution functions. In: Rüschendorf L, Schweizer B,

Taylor MD (eds) Distributions with Fixed Marginals and Related Topics. Institute of Mathematical Statistics, Hayward, CA, pp 233-243

Oakes D (1982) A model for association in bivariate survival data. J Roy Statist Soc Ser B 44:414-422

Oakes D (1986) Semiparametric inference in a model for association in bivariate survival data. Biometrika 73:353-361

Oakes D (1994) Multivariate survival distributions. J Nonparametr Statist 3:343-354

Olsen ET, Darsow WF, Nguyen B (1996) Copulas and Markov operators. In: Rüschendorf L, Schweizer B, Taylor MD (eds) Distributions with Fixed Marginals and Related Topics. Institute of Mathematical Statistics, Hayward, CA, pp 244-259.

Plackett RL (1965) A class of bivariate distributions. J Amer Statist Assoc 60:516-522

Quesada Molina JJ, Rodríguez Lallena JA (1994) Some advances in the study of the compatibility of three bivariate copulas. J Ital Statist Soc 3:397-417

Quesada Molina JJ, Rodríguez Lallena JA (1995) Bivariate copulas with quadratic sections. J Nonparametr Statist 5:323-337

Raftery AE (1984) A continuous multivariate exponential distribution. Comm Statist A—Theory Methods 13:947-965

Raftery AE (1985) Some properties of a new continuous bivariate exponential distribution. Statist Decisions Supplement Issue No. 2, 53-58

Randles RH, Wolfe DA (1979) Introduction to the Theory of Nonparametric Statistics. Wiley, New York

Rényi A (1959) On measures of dependence. Acta Math Acad Sci Hungar 10:441-451

Roberts A, Varberg DE (1973) Convex Functions. Academic Press, New York

Rüschendorf L (1982) Random variables with maximum sums. Adv Appl Probab 14:623-632

Rüschendorf L (1985) Construction of multivariate distributions with given marginals. Ann Inst Statist Math 37 Part A:225-233.

Rüschendorf L, Schweizer B, Taylor MD eds (1996) Distributions with Fixed Marginals and Related Topics. Institute of Mathematical Statistics, Hayward, CA

Sarmanov IO (1974) New forms of correlation relationships between positive quantities applied in hydrology. In: Mathematical Models in Hydrology. International Association of Hydrological Sciences, Paris, pp 104-109

Scarsini M (1984) On measures of concordance. Stochastica 8:201-218

Schmitz V (2004) Revealing the dependence structure between $X_{(1)}$ and $X_{(n)}$. J Statist Plann Inference 123:41-47

Schweizer B (1991) Thirty years of copulas. In: Dall'Aglio G, Kotz S, Salinetti G (eds) Advances in Probability Distributions with Given Marginals. Kluwer, Dordrecht, pp 13-50

Schweizer B, Sklar A (1974) Operations on distribution functions not derivable from operations on random variables. Studia Math 52:43-52

Schweizer B, Sklar A (1983) Probabilistic Metric Spaces. North-Holland, New York

Schweizer B, Wolff EF (1981) On nonparametric measures of dependence for random variables. Ann Statist 9:879-885

Seeley RT (1961) Fubini implies Leibniz implies $F_{yx}=F_{xy}$. Am Math Monthly 68:56-57

Shaked M (1977) A family of concepts of dependence for bivariate distributions. J Amer Statist Assoc 72:642-650

Shea GA (1983) Hoeffding's lemma. In: Kotz S, Johnson NL (eds) Encyclopedia of Statistical Sciences, Vol 3. Wiley, New York, pp 648-649

Sherwood H, Taylor MD (1988) Doubly stochastic measures with hairpin support. Probab Theory Related Fields 78:617-626

Shih JH, Louis TA (1995) Inferences on the association parameter in copula models for bivariate survival data. Biometrics 51:1384-1399

Sklar A (1959) Fonctions de répartition à n dimensions et leurs marges. Publ Inst Statist Univ Paris 8:229-231

Sklar A (1973) Random variables, joint distributions, and copulas. Kybernetica 9:449-460.

Sklar A (1996) Random variables, distribution functions, and copulas—a personal look backward and forward. In: Rüschendorf L, Schweizer B, Taylor MD (eds) Distributions with Fixed Marginals and Related Topics. Institute of Mathematical Statistics, Hayward, CA, pp 1-14

Sklar A (1998) personal communication

Tong YL (1980) Probability Inequalities in Multivariate Distributions. Academic Press, New York

Vaswani SP (1947) A pitfall in correlation theory. Nature 160:405-406

Vitale RA (1978) Joint vs. individual normality. Math Mag 51:123

Vitale RA (1990) On stochastic dependence and a class of degenerate distributions. In: Block HW, Sampson AR, Savits TH (eds) Topics in Statistical Dependence, Institute of Mathematical Statistics, Hayward, CA, pp 459-469

Wald A (1947) Sequential Analysis. Wiley, New York

Whitt W (1976) Bivariate distributions with given marginals. Ann Statist 4:1280-1289

Widder DV (1941) The Laplace Transform. Princeton University Press, Princeton

Wolff EF (1977) Measures of Dependence Derived from Copulas. Ph.D. thesis, University of Massachusetts, Amherst

Wolff EF (1981) N-dimensional measures of dependence. Stochastica 4:175-188

List of Symbols

Symbol		Page
A_C	absolutely continuous component of C	27
$\beta, \beta_C, \beta_{X,Y}$	Blomqvist's coefficient	182
$C, C_{XY}, C_\theta, C_{\alpha,\beta}$	copulas	10, 25, 51
$\hat{C}$	survival copula	33
$\tilde{C}$	dual of a copula	34
C^*	co-copula	34
$\mathbf{C}$	the set of (2-dimensional) copulas	161
$\mathbf{C}^n$	a set of n-copulas	107
D_1, D_2	partial derivative operators	244
δ, δ_C	diagonal section of C	12
$\hat{\delta}, \tilde{\delta}, \delta^*$	diagonal sections of $\hat{C}, \tilde{C}, C^*$	35
Dom	domain	8
ε_a	unit step at a	17
F, G, F_X, F_k	distribution functions	17
$F^{(-1)}, G^{(-1)}$	quasi-inverses of F, G	21
$\overline{F}, \overline{H}, \overline{C}$	survival functions corresponding to F, H, C	32
$\gamma, \gamma_C, \gamma_{X,Y}$	Gini's coefficient	180
φ	generator of an Archimedean copula	112
$\varphi^{[-1]}$	pseudo-inverse of φ	110
Φ	standard normal distribution function	23
$\Phi^2, \Phi_C^2, \Phi_{X,Y}^2$	Hoeffding's dependence index	210
$G^\wedge, G^\vee$	upper, lower bound for G	228
H	joint distribution function	17
$\mathbf{H}$	a Fréchet-Hoeffding class	228
$\mathbf{I}$	unit interval [0,1]	8
$\mathbf{I}^2$	unit square [0,1]×[0,1]	8
$\mathbf{I}_n$	regular partition of $\mathbf{I}$	68
K_C	Kendall distribution function of C	163
λ_U, λ_L	upper, lower tail dependence parameters	214

M	Fréchet-Hoeffding upper bound copula	11
P	probability function	24
Π	product copula	11
$Q: \mathbf{I}^2 \to \mathbf{I}$	quasi-copula	236
$Q: \mathbf{C} \times \mathbf{C} \to [-1,1]$	concordance function	158
$\mathbf{Q}$	the set of quasi-copulas	238
$\mathbf{R}$	real line $(-\infty,\infty)$	7
$\overline{\mathbf{R}}$	extended real line $[-\infty,\infty]$	7
Ran	range	8
$\rho, \rho_C, \rho_{X,Y}$	Spearman's rho	167
S_C	singular component of C	27
$\sigma, \sigma_C, \sigma_{X,Y}$	Schweizer and Wolff's sigma	209
$\tau, \tau_C, \tau_{X,Y}$	Kendall's tau	158, 161
U_{ab}	uniform (a,b) distribution function	17
V_C, V_H	C-volume, H-volume (or measure) of a set	8
W	Fréchet-Hoeffding lower bound copula	11
X, Y, Z	random variables	24
$X_{(1)}, X_{(n)}$	order statistics	49
$(X,Y), \mathbf{X}$	random vectors	158, 222
$y = \tilde{y}(x)$	median regression curve of Y on X	217
Ω	set of generators for Archimedean copulas	115
$\oplus$	addition mod 1	61
$\otimes$	convolution	241
$\circ$	composition of functions	35, 136
$\prec$	point-wise or concordance ordering	39, 169
$\lfloor \cdot \rfloor$	integer part	61, 180
x^+	positive part of x, $x^+ = \max(x,0)$	70
$*$	product operation for copulas	244

Index

Absolutely continuous, 27
Absolutely monotonic, 152
Addition mod 1, 61
Ali-Mikhail-Haq copulas, 29, 40, 87,
    92, 100, 118, 131, 142, 148, 150,
    172, 183
Archimedean
    axiom, 122
    copula, 112
Association, measures of, 157, 207
Average
    quadrant dependence, 189
    likelihood ratio dependence, 203

Bertino copulas, 86
Bilinear interpolation, 19
Bivariate
    Cauchy distribution, 57
    exponential distributions, 23, 33,
        52, 58, 196, 200
    extreme value distributions, 28
    logistic distribution, 28, 93
    normal distributions, 57, 61, 75
    Pareto distributions, 33
Blomqvist's coefficient, 182

Chapman-Kolmogorov equations, 246
Circular uniform distribution, 55
Clayton copulas, 118, 128, 135, 138,
    142, 152, 154, 163, 218
Co-copula, 33
Comonotonic, 32, 95
Compatibility, 107
    direct, 105
Completely monotonic, 151
Component, 27

Comprehensive, 15
Concave copula, 102
Concordance, 157
    function, 158
    measures of, 168
    ordering, 39, 169, 188
Convex
    copula, 102
    sum, 72
Convolution, 241
Cook and Johnson copulas, 118
Copula, 10
    Archimedean, 112
        rational, 146
        strict, 112
    associative, 113
    compatible, 105
    concave, 102
    convex, 102
    convex sum, 72
    diagonal, 85, 166
    dual of, 33
    empirical, 219
    extreme value, 97, 143, 207, 216
    families of
        Ali-Mikhail-Haq, 29, 40, 87, 92,
            100, 118, 131, 142, 148, 150,
            172, 183
        Bertino, 86
        Clayton, 118, 128, 135, 138,
            142, 152, 154, 163, 218
        comprehensive, 15
        Cook and Johnson, 118
        Cuadras-Augé, 15, 39, 53, 102,
            215

Farlie-Gumbel-Morgenstern
(FGM), 77, 86, 100, 131,
133, 162, 168, 249
Frank, 119, 133, 153, 171, 185
Fréchet, 14, 104, 162, 168, 215,
248
Generalized Cuadras-Augé, 53
Gumbel-Barnett, 119, 133, 135
Gumbel-Hougaard, 28, 96, 98,
118, 137, 142, 153, 164
Iterated Farlie-Gumbel-Morgenstern, 82
Mardia, 14, 40, 104, 248
Marshall-Olkin, 53, 95, 98, 165,
168, 215
Pareto, 33, 118
Plackett, 91, 98, 171, 185, 197,
215, 218
Raftery, 58, 172, 215
harmonic, 101
homogeneous, 101
max-stable, 95, 97
multivariate, 42, 105
n-dimensional, 45
ordinal sum, 63
product, 11
product operation, 244
quasi-, 236
quasi-concave, 103
quasi-convex, 103
rational, 146
Schur-concave, 104, 134
Schur-convex, 104
shuffles of M, 67
survival, 32
symmetric, 38
with cubic sections, 80
with linear sections, 77
with quadratic sections, 77
Corner set monotonicity, 198
Countermonotonic, 32, 95
Cross product ratio, 87
Cuadras-Augé copulas, 15, 39, 53,
102, 215

Daniels' inequality, 175
Debye function, 171
Dependence
measures of, 207
properties, 186
Diagonal, 85
copula, 85
section, 12
secondary, 16
Dilogarithm, 172
Distribution
bivariate
Cauchy, 16
exponential, 23, 33, 52, 58, 196,
200
extreme value, 28
logistic, 28, 93
normal, 57, 61, 75
Pareto, 33
circular uniform, 55
elliptically contoured, 37
Gumbel's bivariate exponential, 23,
33
Gumbel's bivariate logistic, 28, 93
Marshall-Olkin bivariate exponential, 52, 196, 200
Plackett, 89
Raftery's bivariate exponential, 58,
172, 215
Vaswani's bivariate normal, 61
Distribution function, 17, 24
convolution of, 241
joint, 17
Kendall, 163
mixture of, 241
n-dimensional, 46
Domain of attraction, 97
Doubly stochastic measure, 26
Dual of a copula, 33
Durbin and Stuart's inequality, 176

Empirical
copula, 219
dependence function, 219

Exchangeable random variables, 38
Expected quadrant dependence, 190
Exterior power family, 142
Extreme value copula, 97, 143, 207, 216

Farlie-Gumbel-Morgenstern (FGM) copulas, 77, 86, 100, 131, 133, 162, 168, 249
n-copulas, 108
Fixed margins, 227
Frank copulas, 119, 133, 153, 171, 185
Fréchet copulas, 14, 104, 162, 168, 215, 248
Fréchet-Hoeffding
bounds, 11, 30
class, 228
Function
absolutely monotonic, 152
completely monotonic, 151
concordance, 158
Debye, 171
dilogarithm, 172
distribution, 17, 24
joint, 17
Kendall, 163
n-dimensional, 46
quasi-inverse of, 21
uniform, 17
unit step, 17
grounded, 9, 44
joint survival, 32
m-monotonic, 154
n-increasing, 43
pseudo-inverse of, 110
quasi-monotone, 8
subadditive, 135
survival, 32
2-increasing, 8

Generator, 112
exterior power family, 142
interior power family, 142

Gideon and Hollister's R_g, 213
Gini's coefficient, 180, 211
Grounded function, 9, 44
Gumbel's bivariate
exponential distribution, 23, 33, 113
logistic distribution, 28, 93
Gumbel-Barnett copulas, 119, 133, 135
Gumbel-Hougaard copulas, 28, 96, 98, 118, 137, 142, 153, 164

Harmonic copula, 101
Hoeffding's
dependence index, 210
lemma, 190
Homogeneous copula, 101
Horizontal section, 11

Interior power family, 142
Iterated Farlie-Gumbel-Morgenstern copulas, 82

Joint
distribution function, 17
survival function, 32
symmetry, 36

Kendall distribution function, 163
Kendall's tau, 158, 161
Kolmogorov's problem, 228

Laplace transform, 74, 154
Laplace's equation, 101
Left
corner set decreasing (LCSD), 198
tail decreasing (LTD), 191
Level curve, 124
Level sets of a copula, 12
Likelihood ratio dependence, 200
average, 203
Lipschitz condition, 11, 236
Lower tail dependence, 214

m-monotonic, 154
Mardia copulas, 14, 40, 104, 248
Marginal symmetry, 36
Margins
    of a 2-place function, 9
    of an n-place function, 44
Markov process, 246
Marshall-Olkin
    bivariate exponential distribution,
        52, 196, 200
    copulas, 53, 95, 98, 165, 168, 215
Max-stable copula, 95, 97
Measures
    of association, 157
    of concordance, 168
    of dependence, 207
Medial correlation coefficient, 182
Median regression, 217
Mixture, 241
Mutually completely dependent, 68,
    75

n-box, 43
n-copula, 45, 105
    Archimedean, 151
n-increasing function, 43
n-subcopula, 45
Negative
    lower orthant dependence (NLOD),
        222
    orthant dependence (NOD), 222
    quadrant dependence (NQD), 187
    upper orthant dependence (NUOD),
        222
Nondecreasing set, 30
Nonincreasing set, 30

Odds ratio, 87
Order
    concordance, 39, 169, 188
    Archimedean copulas, 135
    statistics, 29, 35, 49
Ordinal sum, 63
Orthant dependence, 222

Pareto copulas, 33, 118
Pearson's correlation coefficient, 23,
    165, 170
Plackett
    copulas, 91, 99, 171, 185, 197,
        215, 218
    distributions, 89
Positive
    likelihood ratio dependence (PLR),
        200, 203
    lower orthant dependence (PLOD),
        222
    part, 70
    orthant dependence (POD), 222
    quadrant dependence (PQD), 187
    regression dependence, 196
    upper orthant dependence (PUOD),
        222
Product copula, 11
Product of copulas, 244
Pseudo-inverse of a function, 110

Quasi-copula, 236
    n-dimensional, 239
Quasi-inverse of a distribution func-
    tion, 21
Quasi-monotone function, 8
Quadrant dependence, 187
    average, 189
    expected, 190

Radial
    error, 233
    symmetry, 36
Raftery's bivariate exponential distri-
    bution, 58, 172, 215
Random variable, 24
    exchangeable, 38
    mutually completely dependent, 68,
        75
Rational Archimedean copulas, 146
Regression, median, 217
Right corner set increasing (RCSI),
    198

Right tail increasing (RTI), 191

Schur
    concave copula, 104, 134
    convex copula, 104
Schweizer and Wolff's σ, 209, 223

Section
    diagonal, 12, 85
    horizontal, 11
    secondary diagonal, 16
    vertical, 12
Serial iterate, 151
Shuffles of M, 67
Singular, 27
Sklar's theorem, 17, 24
    n-dimensions, 46
Spearman's
    footrule, 186
    rho, 167, 185
Starlike region, 193
Stochastic monotonicity, 195
Stochastically increasing (SI), 196
Strict Archimedean copula, 112
Strict generator, 112
Subadditive function, 135
Subcopula, 10
    n-dimensional, 45
Subharmonic copula, 101
Superharmonic copula, 101
Support, 27
Survival
    copula, 32
    function, 32

Symmetric copula, 38
Symmetry, 36
    joint, 36
    marginal, 36
    radial, 36

Tail dependence, 214
Tail monotonicity, 191
Total positivity of order two
    bivariate (TP_2), 199
    multivariate (MTP_2), 224
Track, 236
2-increasing function, 8

Uniform distribution function, 17
Unit step distribution function, 17
Upper tail dependence, 214

Vaswani's bivariate normal distribution, 61
Vertical section, 12
Vertices
    of an n-box, 43
    of a rectangle, 8
Volume
    of an n-box, 43
    of a rectangle, 8

Yule's coefficient
    of association, 218
    of colligation, 185

Zero set and curve, 123

Mathematics of Financial Markets
Second Edition

R.J. Elliott and E.P. Kopp

This book presents the mathematics that underpins pricing models for derivative securities, such as options, futures and swaps, in modern financial markets. The idealized continuous-time models built upon the famous Black-Scholes theory require sophisticated mathematical tools drawn from modern stochastic calculus. However, many of the underlying ideas can be explained more simply within a discrete-time framework.

2005. 352 p. (Springer Finance) Hardcover ISBN 0-387-21292-2

Binomial Models in Finance

J. van der Hoek and R.J. Elliott

This book deals with many topics in modern financial mathematics in a way that does not use advanced mathematical tools and shows how these models can be numerically implemented in a practical way. The book is aimed at undergraduate students, MBA students, and executives who wish to understand and apply financial models in the spreadsheet computing environment.

2005. 307 p. (Springer Finance) Hardcover ISBN 0-387-25898-1

Statistics and Finance

D. Ruppert

This textbook emphasizes the applications of statistics and probability to finance. The basics of probability and statistics are reviewed and more advanced topics in statistics, such as regression, ARMA and GARCH models, the bootstrap, and nonparametric regression using splines, are introduced as needed. The book covers the classical methods of finance such as portfolio theory, CAPM, and the Black-Scholes formula, and it introduces the somewhat newer area of behavioral finance. Applications and use of MATLAB and SAS software are stressed.

2004. 473 p. (Springer Texts in Statistics) Hardcover ISBN 0-387-20270-6

Springer Series in Statistics   *(continued from p. ii)*

Huet/Bouvier/Poursat/Jolivet: Statistical Tools for Nonlinear Regression: A Practical Guide with S-PLUS and R Examples, 2nd edition.

Ibrahim/Chen/Sinha: Bayesian Survival Analysis.

Jolliffe: Principal Component Analysis, 2nd edition.

Knottnerus: Sample Survey Theory: Some Pythagorean Perspectives.

Kolen/Brennan: Test Equating: Methods and Practices.

Kotz/Johnson (Eds.): Breakthroughs in Statistics Volume I.

Kotz/Johnson (Eds.): Breakthroughs in Statistics Volume II.

Kotz/Johnson (Eds.): Breakthroughs in Statistics Volume III.

Küchler/Sørensen: Exponential Families of Stochastic Processes.

Kutoyants: Statistical Influence for Ergodic Diffusion Processes.

Lahiri: Resampling Methods for Dependent Data.

Le Cam: Asymptotic Methods in Statistical Decision Theory.

Le Cam/Yang: Asymptotics in Statistics: Some Basic Concepts, 2nd edition.

Liu: Monte Carlo Strategies in Scientific Computing.

Longford: Models for Uncertainty in Educational Testing.

Manski: Partial Identification of Probability Distributions.

Mielke/Berry: Permutation Methods: A Distance Function Approach.

Molenberghs/Verbeke: Models for Discrete Longitudinal Data.

Nelsen: An Introduction to Copulas, 2nd edition.

Pan/Fang: Growth Curve Models and Statistical Diagnostics.

Parzen/Tanabe/Kitagawa: Selected Papers of Hirotugu Akaike.

Politis/Romano/Wolf: Subsampling.

Ramsay/Silverman: Applied Functional Data Analysis: Methods and Case Studies.

Ramsay/Silverman: Functional Data Analysis, 2nd edition.

Rao/Toutenburg: Linear Models: Least Squares and Alternatives.

Reinsel: Elements of Multivariate Time Series Analysis. 2nd edition.

Rosenbaum: Observational Studies, 2nd edition.

Rosenblatt: Gaussian and Non-Gaussian Linear Time Series and Random Fields.

Särndal/Swensson/Wretman: Model Assisted Survey Sampling.

Santner/Williams/Notz: The Design and Analysis of Computer Experiments.

Schervish: Theory of Statistics.

Shao/Tu: The Jackknife and Bootstrap.

Simonoff: Smoothing Methods in Statistics.

Singpurwalla and Wilson: Statistical Methods in Software Engineering: Reliability and Risk.

Small: The Statistical Theory of Shape.

Sprott: Statistical Inference in Science.

Stein: Interpolation of Spatial Data: Some Theory for Kriging.

Taniguchi/Kakizawa: Asymptotic Theory of Statistical Inference for Time Series.

Tanner: Tools for Statistical Inference: Methods for the Exploration of Posterior Distributions and Likelihood Functions, 3rd edition.

van der Laan: Unified Methods for Censored Longitudinal Data and Causality.

van der Vaart/Wellner: Weak Convergence and Empirical Processes: With Applications to Statistics.

Verbeke/Molenberghs: Linear Mixed Models for Longitudinal Data.

Weerahandi: Exact Statistical Methods for Data Analysis.

West/Harrison: Bayesian Forecasting and Dynamic Models, 2nd edition.

Printed in Great Britain
by Amazon.co.uk, Ltd.,
Marston Gate.

448208 4R00160